Power from steam

A history of the stationary steam engine

Power from steam

A history of the stationary steam engine

Power from steam

A history of the stationary steam engine

RICHARD L. HILLS

CAMBRIDGE
UNIVERSITY PRESS

CAMBRIDGE UNIVERSITY PRESS
Cambridge, New York, Melbourne, Madrid, Cape Town, Singapore,
São Paulo, Delhi, Dubai, Tokyo, Mexico City

Cambridge University Press
The Edinburgh Building, Cambridge CB2 8RU, UK

Published in the United States of America by
Cambridge University Press, New York

www.cambridge.org
Information on this title: www.cambridge.org/9780521458344

First published 1989
First paperback edition 1993
Reprinted 1995, 1997

A catalogue record for this publication is available from the British Library

Library of Congress Cataloguing in Publication Data

Hills, Richard Leslie, 1936–
Power from steam: a history of the stationary steam engine
Bibliography: p.
Includes index.
1. Steam-engines – History. I. Title.
TJ461.H57 1989 621.1′6′09 89–7304

ISBN 978-0-521-34356-5 Hardback
ISBN 978-0-521-45834-4 Paperback

Contents

Preface

'The power of the steam-engine and its inconceivable importance as an agent of civilization, has always been a favorite theme with philosophers and historians as well as poets. As Religion has always been, and still is, the great *moral* agent in the civilizing world, and as Science is the great *intellectual* promoter of civilization, so the Steam-Engine is, in modern times, the most important *physical* agent in that great work.'* This was the position which R. H. Thurston ascribed to the steam engine in 1878. In its many forms, the steam engine became the major source of power in the nineteenth century, both in manufacturing industry and in transport whether on land or at sea. This book seeks to trace the technical development of the steam engine as applied in industry and centred on the cotton textile industry in the Lancashire region.

I have aimed to set the technical developments against a background of the scientific discoveries which led to a true understanding of the nature of heat and how best steam should be used in these engines. 'W. S. Jevons wrote in 1865, "The whole question of the steam engine is one of economy" – its development consisted in nothing but the quest for greater efficiency – and indeed, he thought, the material achievement of our civilization could be summed up as "the economy of power" '.† The story is traced through more efficient ways of developing power until the original reciprocating engines were replaced by steam turbines both in the textile mills themselves and in electricity generating stations. After a period of about two hundred years, the final reciprocating steam engines were replaced as a power source in mills by electric motors.

October 1988 — Richard L. Hills

* R. H. Thurston, *A History of the Growth of the Steam-Engine* (London 1878) p. 2.

† A. J. Pacey, 'Vis Viva and Power, A History of the Concept of Mechanical Energy, 1600–1800' (U.M.I.S.T., 1967) p. 48.

Acknowledgements

It has been more than twenty years since I was first introduced to a mill engine so that there will be many people who ought to be acknowledged but whom I will inadvertently omit. So I apologise to those many staff in the mills which I visited who gave me so much time and advice that I am unable to include mention of all of them here. Without actually sharing in their experiences of running these monsters and keeping them in meticulous order, I would not have been able to understand how the mill engines operated. There was also the involvement with Courtaulds Ltd of making a film about the last of these engines which taught me a great deal, not least from Mr Harrop, their fuel and power engineer.

Then the experience of removing some of these engines for preservation and being responsible for re-erecting them in the Museum of Science and Industry in Manchester gave me further insights into the construction and installation of these machines. I had as my Chief Engineers first Frank Wightman, a really experienced millwright of the old school, who needed all the tricks he had learnt over the years to effect the removal of these engines. They were erected by Sid Barnes who was always cheerful even under the most difficult circumstances. I wish to thank the other members of my engineering staff for their patience in working under an historian who had to supervise their craftsmanship. But once again the problems we encountered gave additional insights.

The librarians and archivists up and down the country gave their valuable time helping to find obscure volumes or documents. Some such places were the British Museum, the Science Museum, the Public Record Office in London; libraries in Birmingham, Bolton, Leeds, Manchester and, of course, my own staff at the Museum in Manchester, to mention but a few of the places from where I obtained assistance. Then there was the Smithsonian Institution in America which provided an invaluable source of archives including some with British connections.

Parts of this book have been given earlier as lectures to the Newcomen Society, Manchester Association of Engineers and elsewhere and I am grateful to the comments which members of these societies have made. I had many useful discussions with colleagues at the old Department of History of Science & Technology, U.M.I.S.T., which helped to develop some of the ideas put forward here. My thanks must go to Ron Fitzgerald and Brian Kershaw at the Industrial

Museum at Armley in Leeds who both read the original typescript and made valuable suggestions. But above all I must thank Professor Donald Cardwell for his support and encouragement over very many years. But for him, this book would never have appeared. To all these people, I wish to express my thanks for their help but I take responsibility for any short-comings. Finally the staff of the Cambridge University Press have turned my typescript into a presentable document and to them I owe a debt of gratitude as well.

October 1988 Richard L. Hills

List of figures

List of tables

1
The noblest machine

> England is the birth-place of the steam engine. Its invention has been a grand triumph over the material which nature has placed at our disposal. There is no limit to the sphere of its usefulness, nor can anyone measure the benefits which directly and indirectly accrue to society from its employment.

So wrote Michael Reynolds in his book *Stationary Engine Driving* in 1880,[1] yet, just over 100 years later in 1982, the last remaining rotative steam engine in the Lancashire cotton industry ceased driving the looms in Queen Street Mill at Harle Skye near Burnley. It was the end of the era of direct steam power in industry for, by that time, almost all the thousands of reciprocating steam engines that had powered the machines that made England the first industrial nation had been scrapped as obsolete and uneconomic. Yet, 100 years before Reynolds was writing, the rotative steam engine did not exist because it had not been developed out of the earlier forms of pumping engines.

I was fortunate when I went to Manchester in 1965 to study the history of technology and to establish an industrial museum for the city which had been the centre of the Industrial Revolution for I saw a variety of industries still powered by steam. For each, the basic steam engine had been adapted in the way most appropriate for the task it had to do. Some engines ran day and night, like those in waterworks or sewage pumping stations. Their task was to pump water up to the top of a watertower, or to an outfall in the sewage system, so the load was almost constant. Water cannot be hurried, so such engines did their duty in a slow stately manner, steadily, day in, day out. Once their charges had been started, it appeared that all the enginemen had to do was to oil the bearings regularly and keep a watchful eye and attentive ear to prevent anything going wrong.

Rolling mill engines were exactly the opposite and were the most dramatic to watch in the dark mills with the unearthly glow of the hot metal. The engine would be slowly idling round until the red-hot lump of iron was fed into the rolls. Then the engineman had to open the throttle fully as the rolls gripped the iron and squeezed it out. It looked as easy as squeezing toothpaste out of a tube, but a slight miscalculation on the part of the engineman and the rolls might jam through insufficient power, or the rod of red-hot iron could run amuck all over the rolling floor. Woe betide anyone who was in the way. Once the metal had passed through the rolls, power had to be shut off immediately to prevent the engine racing. One of the greatest rolling mill engines, the 12,000 h.p. engine from Vicker's River Don Works, has been preserved at the Kelham Island Industrial Museum in Sheffield. While it is regularly demonstrated under steam, no longer can there be seen the spectacular sight of it rolling red-hot 4 in thick steel armour plates. A hard skin of scale formed on the surface which had to be removed. As the plates passed through the rolls, branches of brushwood were thrown on top which exploded in sheets of flame as they were squashed against the metal and burst off the scale with showers of sparks cascading everywhere.

The skill of the engineman was equally necessary in colliery winding engines for he had to learn how to bring the cages up the shaft and stop them without jerking at precisely the correct level so that the tubs full of coal could be pushed on and off. At Astley Green Colliery near Leigh, the cage held three tubs, one above the other, each with three tons of coal, so that the total weight of each cage must have been over 30 tons. That mass had to be positioned to within half an inch, three times over in each wind, in little over half a minute, before that cage was sent down to bring the other up. There were different speeds for winding coal and men because no human frame could withstand the acceleration or deceleration applied to the coal. So the 3,000 h.p. of these engines had to be handled skilfully and gently or there would soon be trouble.

Yet, out of all the varied applications of steam power, the most interesting and possibly the most demanding for the engine itself was the textile mill engine. Absolute regularity and dependability were necessary to ensure a constant speed against a varying load for the whole of the working day, week in, week out, year after year. So many of the inventions in the development of the rotative steam engine seem to have originated with the textile mill engine. Possibly this was because it may be said with considerable justification that the cotton textile industry formed the major growth point of the Industrial Revolution and the history of the cotton textile industry, in England at any rate, coincided with the rise and fall of the rotative steam engine. The textile mills demanded amounts of power which were unprecedented in most other industries.

My first visit to a cotton spinning mill was almost symbolic. In the autumn of 1965, soon after I had arrived in the remote world of industrial Manchester having left behind the fertile fields of Kent, I found my way to Leigh, Lancashire, and Hall Lane and Brooklands Mills. I did not realise then how much of that scene through which I was passing, at that time new to me but in fact reaching back to the roots of the Industrial Revolution, would soon be swept away for ever.

The train I had taken to Leigh followed one of the many routes then open between Manchester and Liverpool. It traversed part of the original Liverpool and Manchester Railway, the first true 'inter-city' railway which opened with the first steam-hauled passenger service in 1830. While I travelled on one of the newly-introduced diesel railcars, the North West was the last bastion of steam traction on that very line which had launched the railway age. However, within three years, British Railways had ceased using steam locomotives altogether, and today Leigh has no railway.

From the train, I had been able to see those enormous palaces of the Industrial Revolution, the textile mills. The great bulk of their many storeys and their long length stood out in contrast with the rows of humble workers' cottages surrounding them. In that area, brick-built mills predominated. Those constructed after 1900 followed the fashion with Accrington brick and terracotta. Accrington brick was hard and bright red. Terracotta could be moulded easily into blocks to form decorative trimmings round the windows or the copings and came in various colours such as red, yellow or white. Obviously the millowners took a great pride in their mills and, as terracotta was glazed, both this and Accrington brick were kept clean by rain which washed off the grime caused by burning coal in the mill boilers. It was not too difficult to find a particular mill because it was the custom to use coloured bricks to spell out the name, or its symbol, on the tall chimney. Today, the chimneys, with their distinctive tops to help create a draught, have been pulled down as the textile industry has contracted and steam power been abandoned. Now there is very little trade with textiles left in Leigh and even these seemingly substantial permanent relics are fast disappearing into oblivion as no alternative uses have been found for these enormous mill buildings.

To reach the mills where I was going, I had to cross over the Leeds and Liverpool Canal. This stretch was opened in 1821, but it linked up with the Bridgewater Canal which had been opened for navigation in 1761 only a few miles away.[2] I paused to watch a barge full of coal pass underneath the bridge. The chug-chug of its diesel engine had replaced the clip-clop of horses hooves, but otherwise I could have been looking at a scene almost 200 years old. Some barges were driven by steam engines but now only pleasure boats pass through for the coal mines which provided the bulk of the freight have all been worked out and closed. The Lancashire textile industry was situated on the coal fields and its decline has coincided with the exhaustion of the coal. At the beginning of 1993, the last mine in the county was closed by the Government but was later placed on the list for possible reprieve.

The mills were situated by the side of the canal so that coal could be delivered directly for raising steam in their boilers to heat the spinning rooms and to power their steam engines. At one time, barges had brought the bales of raw cotton as well, and taken away the spun yarn too, but all that traffic had long ceased going by water. The canals performed another essential service to many mills, for from them was drawn water to fill their boilers and also to condense the steam after it had passed through the engines.

I spent all day being shown the various spinning processes. A sharp blow of even quite a blunt axe burst the metal bands surrounding the bales of cotton because the bolls of fibres were packed so tightly in them. Profit in spinning was made through knowledge of how best to blend different types of cotton to produce the quality or character needed in a particular yarn. The finest and strongest yarns were made from cotton with long fibres or 'staple' but this was the most expensive. The art was to know how much cheaper, short staple yarn could be added. So men took bales of different types of cotton, broke them open and filled large storage bins in horizontal layers. One side of the bin was taken down and the cotton pulled out by a rake, working vertically to achieve an initial mix.[3]

In the opening and scutching machines, the lumps of cotton began to be pulled apart and pieces of seed pod or dirt removed. Three cleaning stages might be necessary, depending upon the country of origin and the type of cotton. The cotton came off these machines in a broad roll or lap. When a predetermined length had been wound onto a rod, it was taken off and weighed. This was the point of quality control for the thickness of the final yarn. Cotton yarn was sold by 'counts' and the count number was determined by the number of hanks, each hank 840 yards long, which weighed 1 lb. Therefore, the higher the count number, the finer the yarn. The gearing on all the subsequent machines was related to each other to achieve the correct count from this lap.

The lap was taken to the carding engines where the fibres were finally disentangled from each other and were cleaned for the last time. The cotton was drawn off in the form of a thick loose sort of rope called a 'sliver'. The individual fibres were bent round the points of the card clothing and needed straightening so the sliver was passed through drawing rollers. To balance out any unevenness in the thickness of the sliver, six or eight were passed through the rollers at the same time and merged into one. This also helped to blend them if different types of fibres, such as man-made ones, were being added.

Drawn sliver had a more silky feel but was too thick for spinning directly on the machines available in the 1950s, so it had to be passed through three similar frames, the slubber, the intermediate and the rover. Here the fibres were lightly twisted to help them cohere together as they were drawn out and the 'roving' became thinner and thinner. The rovings were wound onto bobbins and the final bobbins were taken to the spinning machines. The mills I saw in Leigh were equipped with modern ring spinning machines, but, later, I was to see the traditional Lancashire mules. The yarn might be sent out from the mill direct from the spinning frames on tube or cop, or it might be checked for quality as it was wound onto cones or made up into a 'beam' for the weavers.

All these machines, humming and whirring away, needed power to drive them. It was nearing knocking-off time at five o'clock that winter's evening when my guide told me that Brooklands Mill was still driven by its original steam engine. We dashed round to the enginehouse and were just in time to see the engine tenter and his assistant preparing to close down for that night. The soft light from a few low-hung electric light bulbs lit up the brightly polished iron of the connecting rods and cranks as they flashed round. Flashed may give the wrong impression as this engine, built in 1893, rotated at a stately 52 r.p.m.

There were two cylinders, one bigger than the other, with shining covers. I had yet to learn about cross compounds and Corliss valve gears, but I was impressed with the vast size of the flywheel, 30 ft diam., disappearing into the gloom at the top of the enginehouse. The flywheel had the greatest diameter of any that I was to see. I was amazed at the feeling of power in those thrusting piston rods and yet at the quietness, for most of what little noise there was came from the clicking of the Corliss valve gear. Here was a machine that developed 1,200 h.p. and yet it was possible to hold a conversation in a normal tone of voice. How different from internal combustion engines.

The enginemen took a pride in their job and kept their charges scrupulously clean. Special cotton 'wipes' were placed under any part where oil might drip. Sometimes, heavy white cotton cloth was put on the floor and woe-betide anyone who stepped off it. Later, I was to receive many a black look as I tried to photograph these engines and had to stand in inaccessible places to fit as much as possible of the engine into the viewfinder. Brooklands Mill enginehouse was no exception. Long years of polishing wrought and cast iron give it a depth of silvery sheen. The blue of the planished steel lagging also has a mysterious glowing quality when well kept. The brasswork was, of course, always gleaming against the black background of the main castings of the cylinders and enginebeds. Part of the floor was paved with colourful mosaic while the lower parts of the walls had decorated glazed tiling. Here was the vital powerhouse for the

Figure 1 A view of the cross compound engine built by J. & E. Wood in 1893 for Brooklands Mill, Leigh. (Author.)

whole mill. If anything happened to go wrong with that engine, the whole mill stopped and everybody was thrown out of work. Therefore it is not surprising that, on such a vital piece of equipment, so much care was lavished.

The engineman sounded the warning bells in the mill to tell the operatives to stop their machines and disengage them from the main drives. Oil was pumped into the bearings and cylinders, and, a few moments later, the main steam valve was shut. The cranks and the great flywheel slowed down as the mighty machine came to a halt for that day. If the men had failed to stop her in the right position, she would be turned round so all was ready for starting in the morning. We went out from the gentle warmth of the engine room into the darkness and the cold night air. At that time, I did not realise that I was walking out of an almost by-gone era, the end of the first Industrial Revolution. Within two years, Brooklands Mill had closed. Its heart, the engine, was cold and still and soon went for scrap. Others in spinning mills followed in quick succession. A few survived longer in weaving sheds and the last at Queen Street Mill struggled on until 1982 where it has been preserved *in situ* with some of the looms.

Economy always has been one of the important factors deciding the choice of engineering equipment. The world of the industrialist always has been, and still is, intensely competitive. He has to reduce the selling price of his commodity below that of his competitors. If we take the example of our mill steam engine, we would expect to find that the millowner would select the type that was most economical, and probably therefore had the latest technical features. Yet this choice might conflict with some of his other requirements such as reliability. The latest designs often tended to be more expensive through patent royalties or tooling and production costs. Although mill engines were a source of pride to their owners and were lavishly equipped, we do not know what financial restraints were imposed on their design. The main objective was to power that mill at a steady rate so that the mill would run smoothly to ensure high-quality production, while, at the same time, the engine had to perform economically to keep production costs competitive.

There seems to be no connection between the warmth of that enginehouse and the spindles turning on their spinning frames throughout the mill. The first law of thermodynamics tells us,

> Heat and mechanical energy are mutually convertible; and heat requires for its production, and produces by its disappearance, mechanical energy in proportion of 772 foot-pounds for each British unit of heat.[4]

This is not at all obvious, for the heat in the steam driving an engine appears to be lost when the steam is condensed into water. The shafting and belting linking that engine with the spindles remain cold and do not convey heat from one place to another. The earliest textile mills were powered by waterwheels where there were no signs of heat. It was only when the hydrological cycle began to be understood that people began to realise that the heat of the sun caused the rain which later flowed down rivers to turn the waterwheels. The woman sitting at her spinning wheel or the horse patiently turning a horsewheel do not at first sight appear to be using heat until we pause to reflect that animals (humans included) are more efficient at converting their fuel, in this case corn, into energy than any steam engine.[5]

Out of this first law of thermodynamics also comes the principle of the conservation of energy. It is a fundamental principle in nature that, energy, though it may be made to assume different forms, cannot be destroyed but the sum total remains the same. Hence the heat which is carried into the engine in the steam is either transformed into useful work, such as turning the spindles, or it passes to waste in various forms, such as heating the enginehouse. The sum of the heat usefully employed by being converted into work such as spinning plus the heat which is wasted always equals exactly the heat which was applied.[6]

Osborne Reynolds, in his lectures to the Society of Arts in 1883, compared two types of energy, army and mob.[7] When we wind up the spring of a clock or watch, we are giving it energy. This energy, however, can be used only in one particular way, driving the clock, and Reynolds compared this to a disciplined force or army. All the energy put in is released again in useful work. Energy derived directly from heat is difficult, if not impossible, to control in the same way. We are all too well aware of the destructive power of a mob of football hooligans dashing around and creating havoc. In a similar way, the energy of heat tries to burst out everywhere before it gets to the place where it can be usefully employed. It has to be guided, directed and restrained.

This concept is going to lead us into the second law of thermodynamics which has been stated in various ways. In the transfer or the transformation of heat-energy, the total effect produced is directly proportional to the total quantity of heat present and acting. Equal quantities of heat cause equal effects.[8] Heat can pass only from the hotter body to the colder one unless there is some form of compensation. In other words, if we make a cold body even colder by taking heat out of it, we have to pass that heat on to a still colder body unless we put in more energy as in the case of the domestic refrigerator. With this is linked the concept that work or energy is created by a change in temperature. Work can be done only when there is a fall in temperature between the hot source and the cold sink which, in the case of a steam engine in a mill, are the fire in the boiler and the condensing water. This concept vitally affects how steam is generated and used.

In a waterwheel, we can use either a little water falling through a great height, or a large amount of water falling through less height, to power our machines. Provided the ratios are determined correctly, the power output will be the same and, in a correctly designed system, the efficiency will be the same. Likewise, in a steam engine, we can use either a little high-pressure steam or a larger volume of low-pressure steam to drive the same machinery. However, the efficiency of the low-pressure engine will be less than the high-pressure one because, in the final analysis, the power is derived from the change in temperature and high-pressure is the hotter. This, of course, assumes that both engines are designed to use the steam in the most efficient way.

Therefore, when considering our steam engine, if we want to run it in the most economical way, we must start with the burning of the coal itself. The chemical combination of carbon and oxygen account for most of the heat produced from coal. This will yield a fixed amount of heat at a certain temperature. Some of that heat will be wasted, as far as giving useful work is concerned, because the exhaust gases have to be carried away up the chimney, and that requires energy. So far it has never proved possible to burn coal actually inside the working cylinder of any heat engine (or at least not as a commercial proposition) as has been possible with gas or liquid fuels, so the heat from coal must be transferred to another medium, and the best has proved to be steam via water.

The fire is placed where the water inside a boiler can absorb the heat. Now the shell of the boiler has to be made from materials which conduct heat, such as copper, iron or steel, but they become weaker the hotter they get, so there is a limit to the temperature which they, and thus the steam inside, can reach. There are further limitations due to the relationships between steam and water, pressure and volume. At the normal pressure of the atmosphere, water will boil at 100 °C or 212 °F. That temperature will remain the same until the heat from the fire has turned all the water into saturated steam. The space occupied by this steam is much greater than that occupied by the water. One cubic inch of water becomes roughly one cubic foot of steam, or just under 2,000 times greater. If heat continues to be applied, the temperature of the steam will rise and the steam becomes superheated when it behaves like a perfect gas.

If the water is confined within the shell of a boiler and there is no outlet as steam is generated, pressure will start to rise. Now there is a definite relationship between the pressure and the temperature at which water can be turned into steam. In mountaineering huts high in the Swiss Alps, it takes much longer to boil an egg. Because atmospheric pressure is lower, the boiling point is lower. Therefore the water is cooler and the egg has to be boiled longer. To generate steam at the same temperature as the coal burning in the fire, namely roughly 1,500 °C or 2,800 °F, would be impossible because any metals used for the boiler would have melted. At only 200 °C or 400 °F, saturated steam reaches 250 lb absolute pressure and the pressure will rise in proportion to the temperature. Because the maximum pressure known to have been used in the boiler of a textile mill was 260 p.s.i., we are losing potential energy in our steam engine. The steam pressure at Brooklands Mill was 130 p.s.i. where it was raised in a range of four Lancashire boilers.

Before the engine could be started, a fire had to be lit in the grate and the whole boiler warmed up slowly. Soon the water boiled and gradually the temperature and pressure rose. Normally the fires were banked up overnight, but the stoker had to arrive early in the morning to see that the steam could be raised to full pressure in time to start up the mill. The engineer also had to arrive early to prepare his engine. On each boiler, there was a steam valve to isolate that particular one from the steam main so that it could be taken out of service for cleaning or maintenance without stopping the whole mill. Even though they were covered with insulation or lagging, the boilers radiated some heat which was lost. As the steam passed through the steam main, more heat was lost there, even though this too was lagged. At the engine, the engineer opened the by-pass valves and the cylinder drain cocks to allow a small amount of steam to pass through the cylinders and pipes to warm them up. The steam condensed on the cold metal and the resulting water had to be drawn off otherwise, when the engine was started, this water might fill the end of the cylinder. Even though relief valves were fitted, frightful damage was done when pistons hit incompressible water in the cylinders. Only when the cylinders had been warmed up to the temperature of the incoming steam could the engine be started. Here were other sources of lost heat.

Inside the engine room, was the main steam valve controlling the supply to the engine. Before starting, the engine was turned round, either by crowbar or, on the bigger engines, by a small subsidiary steam barring engine, until the piston had passed just beyond one dead centre. In this position, the inlet valve

Figure 2 The boilerhouse of Trencherfield Mill, Wigan, showing a range of four of the Lancashire boilers. The stoker had to fill the hoppers on the fronts of the boilers from where mechanical stokers fed the fuel into the furnaces. (Author.)

was open and the piston still had the greater part of the stroke to complete. The main steam valve was opened as quickly as possible, the piston began to move and the engine to turn. After a couple or so revolutions, the main valve would be eased back because there was, as yet, no load on the engine, and it would be taken up to full speed gently. This gave a couple of minutes for all the parts to be thoroughly warmed through by the steam. The starting bells were rung in the mill and the steam valve carefully opened as the load increased through the spinning machines being started. The drain cocks were shut and everything checked to see all was well for running that day.

But what was happening to the steam? When the inlet valve to the cylinder was opened, steam filled the clearance spaces between the valve and the piston. In spite of the lagging and the thickness of the cylinder cover, all this mass radiated heat which was one reason for keeping the cylinder covers polished. So the incoming steam had to replace this lost heat and some was condensed. More heat was lost as the engine began to turn and heat energy was converted into mechanical energy. The hot piston rod too took heat out of the cylinder as it passed in and out of the packings in the cylinder cover. More steam entered to continue to fill the cylinder as the piston moved along and also to replace that which had been condensed. Of course some of this turned to water as well when it touched the cold cylinder walls but some performed more work.

Meanwhile the piston had been accelerating and, with it, the rest of the machinery. The point came where the inlet valve closed, stopping the entry of the steam but the piston continued to move. The steam expanded and the pressure fell, and of course the temperature of the steam fell too. Parts of the cylinder still being uncovered by the movement of the piston might be colder than the ambient steam so these were heated. Having been warmed by the initial steam, those parts uncovered earlier were then hotter. So, in some places, the steam was being condensed while in others it was being heated. Further, that steam, which had been condensed first into water at a higher pressure, changed back into steam as the pressure was reduced. This complex mixture of processes carried on until the piston reached the end of the stroke and the exhaust valve opened.

The steam in the cylinder rushed out through the exhaust valve and, because it went into a space with a lower pressure, any water droplets might turn to steam again. Also the hot cylinder walls continued to heat the steam and so any water left inside the cylinder was converted into steam as well. By this time, the cylinder walls had become considerably colder than the original hot incoming steam and had to be reheated before any effective work could be done on the next stroke. In stationary engines at textile mills, there was a separate condenser where a jet of cold water reduced the exhaust steam to water. There were two advantages in this. First, this created a partial vacuum which extended the pressure range and, of course, the temperature range, at which the steam could be used. Secondly, the steam heated the cold injection water and some of this was pumped back to the boilers with a saving in fuel.

To discover the true theoretical principles behind the steam engine, and then to determine the best way to apply them, took many years of research, experiment and testing. Luckily, it was not necessary to know the correct theory to make a steam engine run, but, to obtain the highest efficiency and most economical employment of the heat and energy, application of the correct basic principles became more and more essential. The history of the rotative engine traces the application of theory to technology and shows how improvements were introduced through more correct knowledge. At the same time, we are dealing with people and the comprehension and application of that knowledge, so their feelings, and often personalities, colour the picture. Then, in the background, the steam engine has always held a particular fascination for people, possibly because of its warmth, possibly because of the friendly smell of the hot oil, possibly because it makes a gentle rumbling noise and possibly because its movements can be followed easily. These feelings were well expressed by M. A. Alderson who received a prize for his essay written in 1834.

> It is the property which the steam-engine possesses of regulating itself, and providing for all its wants, that the great beauty of the invention consists. It has been said that nothing made by the hand of man approaches so near to animal life. Heat is the principle of its movement; there is in its tubes circulation, like that of the blood in the veins of animals, having valves which open and shut in proper periods; it feeds itself, evacuates such portions of its food as are useless, and draws from its own labours all that is necessary to its own subsistence. To this may be added, that they now regulate so as not to exceed the assigned speed, and thus do animals in a state of nature. That the supply-valves, like the pores of perspiration, open to permit the escape of superfluous heat in the form of steam. The steam-gauge, as a pulse to the boiler, indicates that heat and pressure of the steam within, and the motion of the piston represents the action and the power of which it is capable. The motion of the fluids in the boiler represents the expanding and collapsing of the heart; the fluid that goes to it by one channel is drawn off by another, in part to be returned when condensed by the cold, similar to the operation of veins and arteries. Animals require long and frequent periods of relaxation from fatigue, and any great accumulation of their power is not obtained without great expense and inconvenience. The wind is uncertain; and water, the constancy of which is in few places equal to the wants of the machinist, can seldom be obtained on the spot, where other circumstances require machines to be erected. To relieve us

from all these difficulties, the last century has given us the steam-engine for a resource, the power of which may be increased to infinitude; it requires but little room – it may be erected in all places, and its mighty services are always at our command, whether in winter or in summer, by day or by night – it knows no intermission but what our wishes dictate.

The steam-engine, then, we may justly look upon as the noblest machine ever invented by man – the pride of the machinist, the admiration of the philosopher.[9]

2
The impellent force of fire

For hundreds of years, people have played about with steam toys. Around AD 100, Hero of Alexandria described the 'Aeolipyle', a primitive form of reaction turbine. Then the Greeks arranged that, in their temples, doors should open suddenly to reveal the god after the worshipper had lit a fire on the altar. The fire heated water in a closed 'boiler' so that, when it boiled, it was forced through a pipe into a bucket, rather like the present day 'Teasmade'. The extra weight in the bucket caused it to drop and so open the doors.

The first attempts

After 1600, there was renewed interest in the properties of steam but, at first, this was mainly an intellectual pursuit. We can trace two lines of development, one in which the pressure of the steam was applied against the surface of water to force that water up a pipe, and the other in which the steam pushed against a piston. Salomon de Caus came to England around 1609 and was involved in designing gardens at Hatfield House and Greenwich Palace where fountains were laid out. One idea he published for creating fountains was to fill a copper sphere with water and to heat it. A pipe with a control valve reached nearly to the bottom of the sphere so that, when steam pressure rose inside, water would be squirted out of the pipe. There do not appear to have been any safety valves, water level indicators or other safety features which we would consider necessary today.

This basic concept of a water forcing engine may have been developed by the Marquis of Worcester. In 1663, there was published *A Century of the Names and Scantlings of the Marquis of Worcester's Inventions* in which he described how:

> I have seen the water run like a constant Fountaine-streame forty foot high; one Vessel of Water rarified by fire driveth up forty of cold water. And a man that tends the work is but to turn two Cocks, that one Vessel of Water being consumed, another begins to force and refill with cold water, and so successively, the fire being tended and kept constant, which the self-same Person may likewise abundantly perform in the interim between the necessity of turning the said Cocks.[1]

It has been assumed from this description that this machine consisted of a boiler connected to a couple of receivers from which water could be forced by the admission of steam through control cocks.

The Restoration of Charles II in 1660 saw the exploration of scientific knowledge about the universe through the establishment of the Royal Society in 1663. At that time, the main theory concerning the nature of the world was still based on the Aristotelian concept consisting of four elements, earth, air, fire and water. It must have been realised that when water was 'rarefied', or made into steam, the steam was different from the air in the atmosphere. Sir Samuel Morland had calculated in the 1680s that

> The vapours from water evaporated by the force of fire demand incontinently a much larger space (about two thousand times) than the water occupied previously.[2]

About the same time, Roger North had written in one of his notebooks, 'They argue from hence yt the air, pure, & air made by water evaporated are different'.[3] Knowledge of both these properties of steam was essential for the successful development of the steam engine.

In the diary of North is a sketch showing a steam engine with two cylinders connected to a common boiler. Protruding out of the cylinders are plugs, one of which is described as 'blown up' and the other as 'sinking'. He gave the following account of its action.

> The Rising of ye pluggs are ordered to turne a wheel by a toothed barr, wch when at the top, is struck loos by a catch or snack and then ye barr falls downe, & with its weight turnes ye wheelwork, wch shutts out ye steam from that pipe or socket, by a stop cock, & pari passu opens the other and then that riseth in like manner, and so they play alternately without help.[4]

This is the earliest known steam engine to have automatic or self-acting valve gear and the earliest to have steam from a separate boiler acting on a moving piston but it seems to have been only a model. It is unfortunate that we cannot definitely attribute the design of this engine to anyone but it has been ascribed to Morland who is known to have been interested in steam engines around 1680.

The next stages in the development of the steam engine were to involve condensing the steam to form a vacuum. The experiments of Galileo Galilei and Evangelista Torricelli in 1643 had shown that it was impossible to draw water up a pump with a piston for more than about 28 ft. Torricelli proved that this was caused by the pressure of the atmosphere failing to sustain the column of water at that point. They further calculated that the pressure of the atmosphere at sea-level was about 15 p.s.i. so that a column of water weighing that amount for each square inch was all that could be raised below a piston drawn up a cylinder. If the piston were raised any higher, it would leave a vacuum beneath it.

Other people started to explore the properties of a vacuum. In Germany around 1672, Otto von Guericke had shown in his famous experiment that sixteen horses could not separate two copper hemispheres 20 in in diameter from which the air had been exhausted. Of more significance for the development of the steam engine, was his next experiment in which 50 men could not hold up a piston when the air underneath it inside the cylinder had been exhausted.[5] One of the people who brought news of these ideas and experiments to England was Robert Boyle. With the assistance of Robert Hooke, in 1658 he evolved an improved type of air pump with which he created his vacua and which still is preserved by the Royal Society for it is the prototype of those used today. Through experiments with his pump, Boyle formulated the law of the pressure of gas known by his name.

Another person who worked with Boyle was Denis Papin, a French Huguenot who fled to England in 1675. Papin had been experimenting with an engine in which a small charge of gunpowder was exploded in a cylinder to expel the air inside through valves. On cooling, the water would be drawn into the vacuum thus formed.[6] It was soon obvious that this was impractical, so Papin turned to the alternative way of creating a vacuum, by the condensation of steam. He constructed a vertical cylinder with a little water placed in the bottom below a piston. If a fire were placed underneath and the water were boiled, the pressure of the steam would force the piston up the cylinder to the top where it could be held by a catch. When the fire was removed and the cylinder cooled, a vacuum was created inside the cylinder. By releasing the

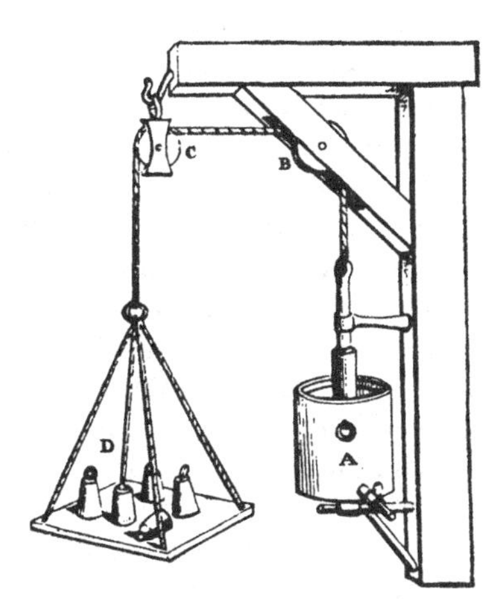

Figure 3 In this experiment, von Guericke exhausted the cylinder and showed that the pressure of the atmosphere would force a piston into it and lift heavy weights. (Bourne, *Steam Engine.*)

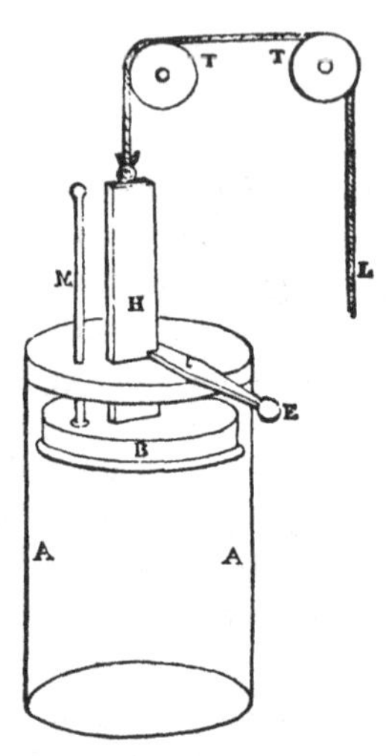

Figure 4 By generating steam in the bottom of this cylinder, Papin forced the piston up and held it with a catch. After cooling, the catch was released and the piston would draw up weights. (Bourne, *Steam Engine*.)

catch, atmospheric pressure would force the piston down the cylinder so that it could raise a considerable weight by means of a cord running over a pulley.

The Savery engine

The first steam engine which approached commercial success had no moving parts other than the valve gear. Thomas Savery alternately created a vacuum through condensing steam in a receiver which drew up water from below the engine and then employed steam pressure to force that water further up a stand-pipe to the required height. He took out a patent in 1698 for 'Raising water by the impellent force of fire'.[7] Savery demonstrated a model of his engine to King William III at Hampton Court[8] and it was probably the same engine which he showed at a meeting of the Royal Society on 14 June 1699.[9] Experiments in the late 1960s on model engines made in the Department of Mechanical Engineering at the University of Manchester Institute of Science and Technology showed that a small scale Savery engine could be made to work most convincingly. No doubt, when Savery demonstrated his models, he stirred up great enthusiasm for his project and may well have been able to raise money on the strength of this to enable him to erect his engines at Campden House, Kensington, and York Buildings in London. His treatise, *The Miners Friend*, is an early and excellent example of the salesman's glossy brochure. However, the larger Savery's engine became, the more the difficulties increased while the reverse was true of the next type of engine, the Newcomen.[10]

Savery made considerable developments to his engine. A drawing in the possession of the Royal Society[11] shows a boiler encased in brickwork supplying

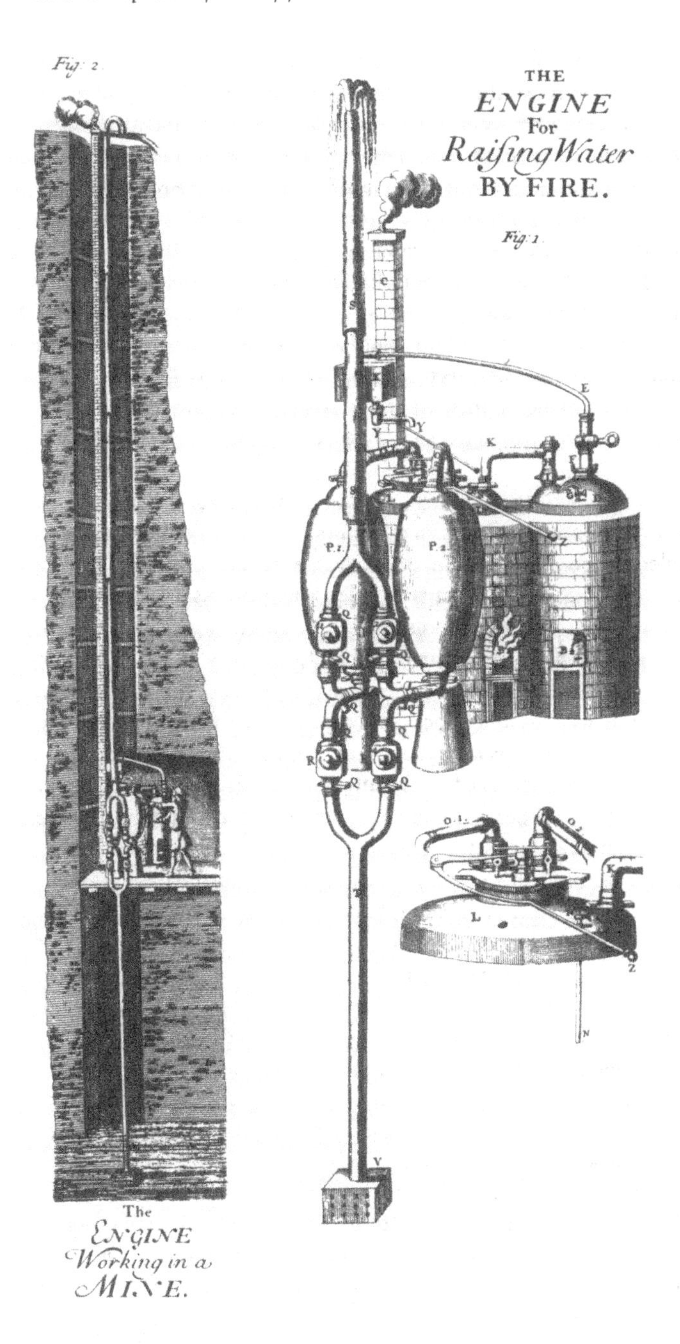

Figure 5 Savery's illustration of his engine in his publication *The Miners Friend*. (Savery, *Miners Friend*.)

two cylindrical receivers. The water feed pipes arched over and entered the top of the receivers so they would have been difficult to clear of air. The steam valves supplying each receiver were simple turn-cocks operated independently. There appears to be no provision for pouring cooling water over the outside of the receivers to condense the steam inside. The illustration in *The Miners Friend* shows a much more advanced design. Here the receivers were lozenge-shaped, presumably to withstand the higher pressures better. By this date, 1702, the pipes were arranged more neatly with both joining together at the bottom. This illustration shows the steam valve gear on top of the boiler which had a plate valve that the engine attendant could operate to let steam into either receiver alternately. Cold water could be taken from a cistern filled by the outlet pipe and poured over the outside of either receiver. Water from the pipe could be used also to fill a secondary boiler which in its turn could replenish the main one.

Just imagine the work of the engineman. He had to operate the steam cock and the cold water valve at least four times a minute, to check on the water level in the main boiler by operating the level cock (apparently the only safety feature on the whole engine), to refill the main boiler when necessary and also to operate the second boiler too. All this was in addition to stoking the fires of both boilers. It must have been an exceptional boy of 13 or 14 who could learn to operate a Savery engine in half an hour, as Savery claimed.[12] He would, of course, have been dressed in long frock coat and broad-rimmed hat. It is no wonder that people sought ways to make this engine function automatically.

Let us now consider the action of the engine. Steam was raised in the boiler by lighting a fire. The instructions Savery gave seem to be for starting the engine when empty yet he made no mention of having to wait for the pressure to rise. The first action was to blow the air out of each receiver in turn. However, once the air had been cleared from the system, Savery did not advocate blowing it through again from time to time as must have been necessary, for he said, 'It is much better to let none of the *Steam* go off, (for that is but loosing so much strength)'.[13] The mystery will remain over how he prevented his engine becoming 'wind-logged', or partially filled with air, and so the suspicion must remain that he never operated his engines for any great length of time.

Savery must have been using the steam in his engines under the most uneconomic conditions possible. There could be no lagging of the receivers to conserve the heat because cold water had to be poured over the outside. Then the steam pressure had to be sufficient to raise the column of water to the required height. The higher the water had to be raised, the higher the pressure and therefore the higher the temperature. Now, in the Savery engine, there could be no expansive working in which the pressure, and so temperature, could fall for the full working pressure had to be admitted throughout the stroke. Therefore the whole of the receiver had to be heated up to the working pressure every stroke and cooled down again to create a vacuum. The greater the head against which the engine was working, the greater must have been the temperature difference and therefore the greater the losses of heat with the steam. Also the greater the lift on the suction side, the greater was the need for more cooling water. A great deal of steam must have been wasted as it was condensed by the cold walls of the receivers. Indeed, Savery commented that it was possible to judge how far down the receivers the water had been forced because the receiver walls dried to that point and remained wet lower down.[14] It is difficult to estimate how much steam was condensed by the cold water in the receivers because that would depend upon so many unknown factors such as turbulence of both steam and water. What is clear is that the transfer, and so the loss of heat, must have been greater the more the pressure was raised.

Savery himself would have remained blissfully unaware of many of these problems through ignorance of modern thermodynamics, but he had other problems too. The engine at Campden House had a receiver which held 13 gallons and operated at about four strokes per minute. The water was raised 16 ft by suction and 42 ft by pressure. So the power developed was about 1 h.p. and the pressure around 19 p.s.i. For the engine in his own workshop at Salisbury Court in London, it has been estimated that the lift by suction was about 18–22 ft and the water was raised by pressure some 34 ft. The receivers were 3–4 ft high.[15] Steam pressure must have been 20–25 p.s.i. with a temperature of 126–130 °C (259–266 °F). Unfortunately no figures have been discovered for the engine at York Buildings nor for another engine at Sion House.[16]

In spite of Savery's claim that he had 'spared neither *Time*, *Pains* nor *Money* till I had absolutely *conquer'd* . . . the *oddest* and almost *insuperable* Difficulties',[17] he never did succeed in making his engine work water mills, drain fens, or even achieve its primary purpose of pumping water out of mines. We learn some of the reasons for this failure through John Theophilus Desaguliers who must have seen the one at York Buildings actually working for he commented,

> I have seen Captain *Savery* at *York*-Buildings make steam eight or ten times stronger than Common Air; and then its heat was so great, that it would melt common soft Solder; and its Strength was so great as to blow open several of the Joints of his Machine: so that he was forced to be at great Pains and Charge to have all his Joints solder'd with Spelter or hard Solder.[18]

Tinman's solder has a melting point depending upon its composition varying from 180 to 220 °C (356 to 428 °F) and a good plumber's solder may reach 270 °C (518 °F) before melting, but steam at 150 p.s.i. reaches 186 °C (366 °F). Savery found that he had to limit the pressure to 3 atmos which would give a steam temperature of 135 °C (274 °F), so at first sight, the pressure need not have been so low and the use of hard solder ought to have been unnecessary but solder becomes pasty and loses strength long before it actually melts. Later

Desaguliers used spherical boilers to withstand the pressure and it seems reasonable to assume that he was copying Savery.[19] These boilers were placed directly over the fire with no flues around them. As the water level fell, the upper portions of the plates would have become uncovered internally and, being thus exposed to the full heat of the fire, could easily have overheated, causing the solder first to go pasty and then melt. Overheating of the plates was a problem not unknown on the later wagon boilers when the plates next to the side flues became exposed through a drop in the water level.

Therefore Savery's problems may not have been caused only by the high pressure of the steam, but the high pressure points to something else. In 1716, Desaguliers and s'Gravesande carried out some experiments on a Savery engine which are described in Harris's *Lexicon Technicum*.

> We thought there was a great Waste of Steam by continually acting upon the Receivers without Intermission, it becoming useless till it had heated the Surface of the Water in the Receivers, and also to a certain Depth: and that if it were so contriv'd, that, after the Steam had press'd upon one Receiver full of Water, instead of being thrown upon another, it should be confin'd in the Boiler till the Receiver was refilled by the Atmosphere, and then turn'd upon the Water: That by this means its Confinement might give it so much Force, that it would push hard against the Surface of the Water, and have discharg'd a great deal of it, even before it had heated the Surface; . . . This Model soon shew'd us that one Receiver could be emptied three times, whilst two succeeding ones could be emptied but once a piece. So that by this means an Engine would be so simple, as to be more easily work'd, cost almost half less, and raise a third more Water.[20]

Savery's engine at Sion House had only one receiver and it would appear that, after Desaguliers's experiments, most attempts at building Savery engines concentrated on the single receiver type until well after 1800 and the much later development of C. H. Hall's pulsometer in 1876.[21] What Desaguliers had discovered was that, in order to transfer heat from the steam to raise the temperature of all the surfaces up to that necessary for the steam to be effective within a reasonable space of time, it was best to have the working temperature of the steam considerably higher. Also a good pressure of steam would have helped to impart greater velocity to the ascending water.

Thomas Newcomen's engine

The first type of steam engine that became a practical proposition avoided the problems associated with high-pressure steam. The inventor, Thomas Newcomen, was an 'ironmonger' at Dartmouth in Devon and in 1712 erected what must be considered as the world's first successful steam engine near Dudley Castle then in Staffordshire.[22] When the Manchester Museum of

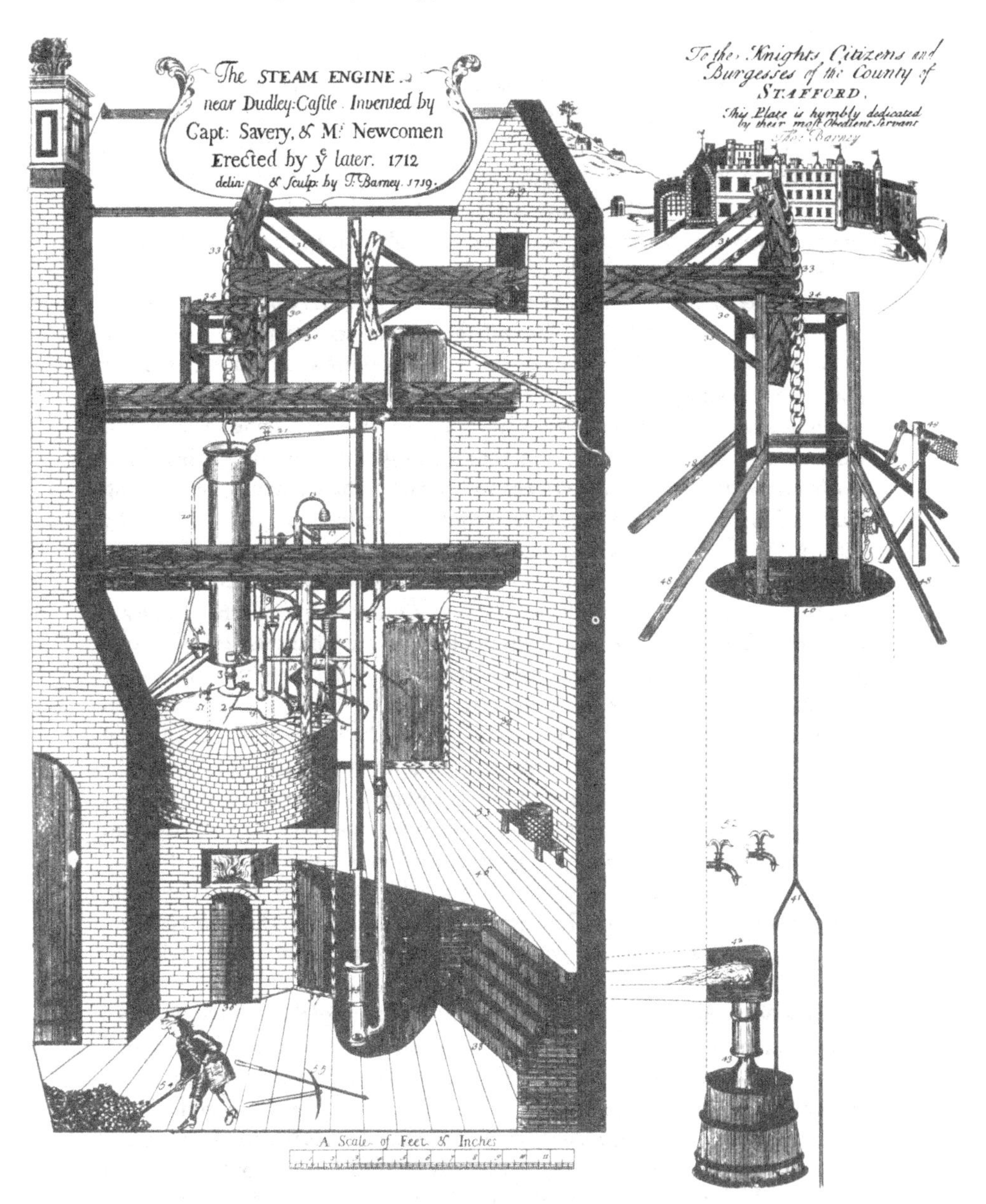

Figure 6 Barney's engraving of Thomas Newcomen's first engine built near Dudley Castle, Staffordshire, in 1712. (Science Museum.)

Science and Technology was founded in 1968, the Mechanical Engineering Department of the University of Manchester Institute of Science and Technology agreed to build a one-third scale model of this engine based on Thomas Barney's drawing of 1719. Our trials and tribulations to make our model run smoothly show what Newcomen himself must have experienced all those years ago, but we, at least, had the benefit of hindsight and knew that the engine could be made to work eventually, even though it took months of experimenting to make it operate successfully.[23]

Newcomen is known to have visited mines in Devon and Cornwall, supplying all sorts of ironwork and tools. Through these contracts, he must have been well aware of the struggles the miners were having to pump water out of the deeper shafts of the tin mines since those areas had few streams large enough for waterwheels and animals were not powerful enough. The steam engine he developed would only pump water and it is reasonable to assume that this was his objective from the start. He must have been well aware of the different types of pump then used in mining. The book of illustrations of machines published in Italy by Vittorio Zonca in 1607 shows various devices for raising water.[24] One is a variation of the ancient 'shadouf' in which a man pulls down one end of a beam pivoted in the middle and so lifts a scoop at the other end. In another, there are two men, one at either end of a beam similarly pivoted, but this device has two plunger pumps, one either side of the central support. In yet another, the pumps are bucket-type lift pumps. It was this last form that Newcomen employed on his first engine near Dudley Castle where the water had to be raised out of a coal mine 51 yd deep at the rate of 12 strokes a minute. On these pumps, during the up-stroke, the water was drawn up into the pump cylinder as the bucket was being raised. Valves in the bottom of the bucket opened on the down-stroke to let the water pass through to the upper side so that it was lifted on the next up-stroke. During the down-stroke, a foot valve in the bottom of the pump cylinder stopped the water flowing back into the mine. The buckets were suspended from a horizontal wooden beam, pivoted in the middle with the other end inside the enginehouse.

Surely Newcomen must have been well aware of one of the limitations of this type of pump; that it would not suck or draw the water up after it for more than about 28 ft because a vacuum would be formed above the water. Now, as today any boy knows who has played about with a bicycle pump, the piston will try to force itself back into the vacuum through the pressure of the atmosphere. Did Newcomen realise that he could use this force if somehow he could make, in a separate cylinder, a vacuum which would draw down a piston connected to the other end of the beam from which his water pumps were hanging? This is what he did when he placed his steam cylinder inside an enginehouse. The weight of the pump rods and buckets down the mine were heavier than the rod and piston in the steam cylinder at the other end of the beam, so that the engine rested with the beam normally 'out of the house', or with the pumps at their lowest position.

We have no actual knowledge about Newcomen's understanding of the scientific principles of his day, but he must have had a deep grasp of them judging by the way he made his engine work. Marten Triewald,[25] a Swedish engineer who worked on some of the earliest atmospheric engines, stated in his book that Newcomen had begun experimenting around 1700 and described a model.[26] It is unlikely that the owner of any mine, and almost certainly Newcomen himself, would have financed a project on the scale of that near Dudley Castle unless he had been sure of success. Again, it is doubtful if Newcomen would have developed an engine so far from Dartmouth as Staffordshire unless he had built a similar one nearer home. Joseph Carne, in his paper on the history of copper-mining in Cornwall,[27] states that one was erected at Huel Vor tin mine in Breage where it worked from 1710 to 1714. Maybe the vein did not prove viable and so the engine was stopped, or the experiments to make it run properly consumed so much coal that the inventor had to stop his trials. If this engine did exist, it must have worked sufficiently well to encourage Newcomen to embark on the one near Dudley Castle.

Triewald's description of a model with a steam boiler below a cylinder, inside the cylinder a piston connected to a beam, at the other end of the beam a weight to simulate the pumps, is the sort of thing that would be built by a person experimenting. However, model Newcomen engines are even more temperamental than their larger counterparts and how this model was developed into the world's first practical steam engine shows Newcomen's genius. Between the boiler and the piston, Newcomen must have fitted some form of valve to control the steam and then he must have installed a pipe, the eduction pipe, and another valve to let the condensate out of the bottom of the cylinder. On the Dudley Castle engine, this second valve is situated at the bottom of a long pipe in a hot well below the engine where there is no danger that, if the valve leaks, the condensate will be forced by atmospheric pressure back up the eduction pipe and into the cylinder and boiler. With the reduced scale of the Museum model, this did sometimes happen on the condensing stroke and illustrated one of the many difficulties Newcomen overcame so effectively.

To start his engine, Newcomen would have opened both the steam and condensate valves so that the steam would blow the air out of the inside of the cylinder. At some stage, the condensate valve was made into a self-acting flap valve so that it would open with the steam pressure against it. Then, after both valves were closed, the cold walls of the cylinder condensed the steam and the atmospheric pressure forced the piston down.[28] This is how the Museum engine could be started after it had stood and cooled. By working the steam valve manually, the engine would make three or four strokes and we can imagine how elated Newcomen felt to see his engine running. But then ours would stop and Newcomen must have found himself in the same predicament.

One reason for the engine stopping was air inside the cylinder which prevented the vacuum being formed. Desaguliers, writing in 1743,[29] said that quite early in their experiments, Newcomen and his assistant, John Calley, had found it necessary to seal the top of the piston with a layer of water, so that air would not be drawn into the cylinder past the sides of the piston. Even on the Museum engine, which has a drawn brass tube for its cylinder finished more accurately than anything Newcomen could have imagined in his wildest dreams and which has been fitted with an accurately turned and better packed piston, this water seal is essential. If the piston is allowed to run dry, air can be heard rushing into the cylinder and the engine will stop on that or the next stroke. The simplicity of this seal masks its vital importance and this must have been a very early discovery because, without it, the engine will not work. By using water, Newcomen made certain that, if any did get past the piston, it would help to condense the steam inside. Most other people would have resorted to more complicated ways of packing the piston instead of this simple idea which helped to improve the performance.

Air, however, will enter the cylinder from other sources. Some will be carried in by the steam from the boiler, while the various joints will leak a little. Desaguliers pointed out that there would be air in the water which was injected to cool the steam. Newcomen's solution to the problem of expelling this air was almost as simple as the last. Just outside the cylinder on the eduction pipe, he placed a cup with a valve in it which Desaguliers called a 'Snifting Clack, *because the Air makes a Noise every time it blows thro' it like a Man snifting with a Cold*'.[30] When the steam valve on the boiler is opened, the steam not only begins to push the piston up, but it also forces the condensate down the eduction pipe and blows the air out through the snifting valve. The weight of this valve is critical. Too light and it will let too much steam out and waste it, while one that is too heavy will prevent any steam or air being blown through. If the puff of steam and air is not blown out at each stroke, air will accumulate in the cylinder, gradually lessening the vacuum until the engine does not complete a stroke to reset the valves.

It is worth pausing to consider the design of the lower end of the cylinder. The steam pipe extends a long way through the bottom plate. This is partly to prevent the condensate running back into the boiler, but it also has the effect of passing the fresh incoming steam almost up to the bottom of the piston so it enters the cylinder above the condensate and air. In this way, the fresh steam pushes down on top of any air and spent steam and forces them down the eduction pipe.

Even after Newcomen had taken all these steps to clear the air out of the cylinder, the engine still would not have run properly because, after a few strokes, the cylinder would have become too hot to form a good working vacuum. The power of the engine is derived from the change in temperature of the steam at boiling point to the temperature of the condensate. The degree of vacuum that can be obtained will be related directly to the temperature, so that the power of the engine is determined by the amount it is heated and cooled.

When starting from cold, the engine will make a few good strokes until it becomes too hot but then it will not make any more until it has cooled down. Speeding up the cooling by pouring cold water over the outside was an obvious solution, as Savery had found, but it would have been slow. It is not known whether Newcomen and Calley tried this, but, with an open-topped cylinder and a boiler and fire underneath, they soon hit upon the idea of putting a jacket, made of lead, round the cylinder and pouring the cold water into that. This would not have been an economical solution because the incoming steam would have had to heat up this cold water a certain amount before it could fill the cylinder to commence the next stroke. While the engine in this form probably worked well enough to encourage Newcomen to continue experimenting, it is doubtful whether the performance would have been adequate to persuade anyone to build a large working version, but it did lead to the vital discovery which seems to have happened by accident.

Triewald described the event which changed the future prospects of the steam engine in this way:

> It happened at the last attempt to make the model work that a more than wished-for effect was suddenly caused by the following strange event. The cold water, which was allowed to flow into a lead-case embracing the cylinder, pierced through an imperfection which had been mended with tin-solder. The heat of the steam caused the tin-solder to melt and thus opened a way for the cold water, which rushed into the cylinder and immediately condensed the steam, creating such a vacuum that the weight attached to the little beam, which was supposed to represent the weight of water in the pumps, proved to be so insufficient that the air, which pressed with a tremendous power on the piston, caused its chain to break and the piston crush the bottom of the cylinder as well as the lid of the small boiler. The hot water which flowed everywhere thus convinced even the very senses of the onlookers that they had discovered an incomparably powerful force which had hitherto been entirely unknown in nature, – at least no-one had ever suspected that it could originate in this way.[31]

When the Museum engine is being controlled by hand and the pump is lightly loaded, the piston will rush down and, if it were not for springs fitted to the framework of the engine, would smash the steam inlet pipe and the condenser jet and might even break the bottom of the cylinder. Desaguliers confirms this, for he says that if the engineman

> does not open the Regulator soon enough, the Piston coming down with a prodigious Force will very probably strike against the Throat-Pipe and crush it to pieces.[32]

The discovery of cold water injection was the key to the success of the Newcomen engine and also its successors. There is nothing to suggest that it had ever been used before and this vital discovery probably occurred on the experimental model because it is so fundamental to the whole operation. Although Newcomen had to add a cold water tank at the top of the enginehouse and a pump to fill it, these were small disadvantages compared with the power his engine developed afterwards.

The siting of the cold water tank was vital on engines pumping water from mines because, as the desired vacuum was increased, not only did the quantity of injection water needed increase, but so did the proportion required to be injected at the start of condensation. Therefore, the cold water tank had to be placed high to obtain a good head and so give a good initial jet of water into the cylinder as soon as the valve opened.

After he had solved the problem of direct injection, Newcomen still had to make his engine work automatically before it could be accepted as a commercial proposition. Probably there were continual modifications to the valve gear and we will never know who invented what. The engine built by the Museum confirmed that the valve gear drawn by Barney in 1719 worked very well and had several features which made it excellent for its job. The valves were opened and shut through levers worked by adjustable pegs pushed into holes in a wooden 'plug rod' suspended vertically from the main beam.

The valve controlling the injection water was worked by the weighted 'F' lever, so called from its shape. One peg in the plug rod released a catch so the weight fell suddenly and quickly opened the valve, allowing the maximum amount of water to rush in at the start of the injection. Once started, less water was needed to maintain the vacuum, and the valve was slowly closed as the piston and beam descended, also pulling down the plug rod.[33] Another peg in the plug rod pushed down the tail of the 'F' lever until it closed the valve and was held by a catch ready to be tripped again.

The design of the injection valve was probably the weakest part of the engine, for, while one bearing was housed in the framework of the engine, the other was made in the valve itself which was shown supported only by the pipework. The packings on the Museum engine worked loose so we were glad to copy Newcomen exactly by putting a funnel and drain beneath it. Yet the valve itself functioned perfectly and was another amazing case of Newcomen arriving at the correct answer presumably empirically.

In Barney's drawing, the cold water pipe supplying the injection valve appeared merely to discharge over the top of the cold water tank, and some people have assumed that it must have been drawn inaccurately because the tank had no outlet from the bottom.[34] However, if the pipe were continued over the top of the tank and bent down inside nearly to the bottom, it worked perfectly as a syphon and converted the tank into a reservoir. The only disadvantage was that there was no cock to shut off the supply, so if any of the valves leaked when the engine was standing idle, all the water was syphoned out leaving none for starting.

Newcomen then had to find a way of controlling the steam valve by the engine itself. He arranged a lever in the shape of an upside-down 'Y' with a weight on the top. The arms of the 'Y' operated the valve so that the 'Y' lever itself could pivot a little before one or other of its arms came into contact with the valve lever. The weight on top was pushed over the dead centre by a peg on the plug rod so it overbalanced and suddenly knocked the valve open or shut.[35] In this way, the piston could move up or down the cylinder while the valve

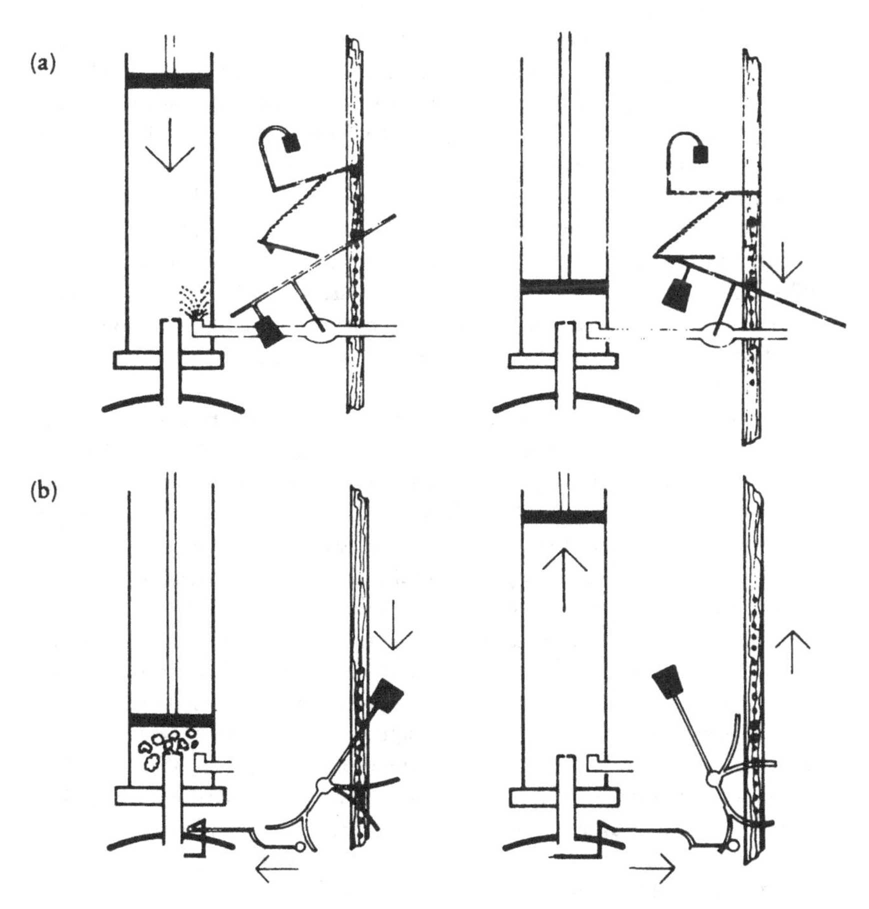

Figure 7 Diagrams showing the operation of the valve gear on Newcomen's first engine. (*a*) Water injection valve: (1) piston at top of cylinder, the catch is raised, letting the 'F' lever fall and open the valve; (2) as the piston descends the peg on the plug tree pushes down the 'F' lever till the catch holds and shuts the valve. (*b*) Steam valve: (1) at the bottom of the stroke, the peg on the plug tree pushes the 'Y' lever over until the weight overbalances and opens the steam valve; (2) at the top of the stroke, another peg pushes the 'Y' lever up until it overbalances the other way and shuts the steam valve. (Author.)

remained stationary until the end of the stroke when it suddenly changed, giving a snap action which many a later designer would have done well to copy. When the valve opened, the steam rushed in to destroy the vacuum and, most important, to drive the air out through the snifting valve. As the piston rose, the valve remained open to let in steam to fill the cylinder until the top of the stroke was reached when the valve tripped shut and remained so during the whole of the working stroke. This was yet another example of Newcomen's genius, for this type of valve mechanism was still being used by John Smeaton in the 1770s and probably even later.[36]

The round-topped boiler, on which of course the whole engine depended, showed equally interesting features. The Museum engine is powered by electric immersion heaters, which is a more efficient way of heating, but this is more than offset by the reduction in volume because the space inside is reduced by the cube of the difference in scale. Even allowing for this, Newcomen's boiler was probably too small and must have needed a large fire to keep the engine running. We started with twelve kilowatts of heating capacity which was insufficient so this was increased to sixteen which just ran the engine. Eighteen kilowatts made the engine work much better but all this heat produced about 0.145 h.p. when calculated in terms of water actually lifted.[37] This shows how inefficient Newcomen's engine must have been, but even this was better than the alternatives.

The reduction in size of the Museum engine meant that two boiler fittings would not work. The first was the stand-pipe through which the boiler was filled with warm water from the top of the cylinder. With the reduced scale, this did not have sufficient head. Then there was the buoy mechanism on Barney's drawing which suggests that the original engine ran short of steam. The buoy should prevent the 'F' lever opening until sufficient steam pressure had been raised. This again would not work through the reduced scale so we could not determine what it was meant to do on the original. Once the steam had reached the correct pressure just above that of the atmosphere, the 'F' lever would have been tripped at the wrong time. A safety valve was fitted but the least satisfactory feature of the boiler was the single pipe and cock to check the water level. This would show if the water level were above or below the end of the pipe, but not how far either way. Henry Beighton's drawing of the Griff engine (1717) had two cocks with different lengths of pipes, which, although an improvement, still would not show the precise level. We fitted a gauge glass because burnt-out immersion heaters are expensive to replace.

On later engines, to make them work smoothly, weights were added to the beam immediately behind the arch head and on the piston until the beam was in perfect equipoise. Then a weight equivalent to 1 p.s.i. of piston area was removed from the piston to give the pump end the necessary advantage. If the pump end were too heavy, the engine wasted power raising it and it would descend too quickly, a problem we found with the Museum engine. Before a stroke could begin, steam from the boiler had to raise the inside of the cylinder near to boiling point and drive out the air. This resulted in a considerable volume of condensed water which had to run out of the eduction pipe. As the piston rose up the cylinder, the cold cylinder walls also had to be heated, causing again more condensate. This was the reason why the steam valve had to remain open during practically the whole of the stroke. Indicator diagrams of atmospheric engines show that there was virtually no expansion of the steam inside the cylinder. Once the steam was cut off, condensation started very quickly to enable the power stroke to begin.

To commence the working stroke, injection water had to condense much of the steam and cool the cylinder walls until the pressure had dropped low enough to enable the weight or pressure of the atmosphere to overcome the weight of the pump rods and water at the other end of the beam. When large engines were running, it was observed that sometimes the cylinder seemed to be almost lifted off its supporting beams when the vacuum was being formed and before the pump rods started to move.[38] Newcomen achieved a high vacuum in his engine which worked with a loading up to 9.4 p.s.i. on the piston, but he must have used a large quantity of cooling water. Soon it was discovered to be more economical to load the engine less heavily at about 7 or 8 p.s.i. and make more strokes per minute. In 1765, Smeaton found that the optimum working conditions on his model engine were about 17 strokes per minute with 8 p.s.i. below atmospheric pressure.[39]

Once the working stroke had been completed, the vacuum had to be destroyed to enable the pump rods to draw the piston up again through their weight. This was achieved by the steam raising the temperature of the cylinder near to boiling point so it could help push the piston back up the cylinder against atmospheric pressure. The piston would not move until the vacuum had been broken. So it is clear that, while using the pressure of the atmosphere to provide power, the engine was in fact the earliest successful heat engine. No doubt Calley, Humphrey Potter and others helped Newcomen develop his engine into a successful working machine, yet to Newcomen must go the credit for pioneering this remarkable feat. The way he built it was so sound that it remained virtually unaltered until the vital inventions of James Watt in 1769 and 1784. It remained the basis of the Cornish pumping engines until the last of these was erected at the Dorothea Slate Mine at Nantlle in North Wales in 1906, roughly 200 years after Newcomen made his first experiments.

We can but surmise at the elation Newcomen must have felt when he saw his engine running, but a shock was in store for him. In 1699, an Act of Parliament had extended Savery's patent for an extra 21 years beyond the normal 14, that is to say until 1733. Savery died in 1713 but the group of Proprietors he had formed to exploit his patent carried on and realised that the terms of this patent for 'Raising water by the impellent force of fire' covered Newcomen's engine as well. Newcomen was forced into some form of partnership and had to build his engines under their auspices.[40] By the expiry of the patent over 100 engines had been built in England alone and before Newcomen himself died in 1729, his engines were at work in almost the length and breadth of Britain, in Austria, Belgium, France, Germany, Hungary, Sweden and possibly Spain but he himself received no honour in his lifetime.[41]

3
Common old smoaking engines

Looking today at the massive scale and crudity of early atmospheric engines, we find it hard to imagine that Newcomen was working at the limits of engineering technology available to him. The cylinders in particular were difficult to bore accurately, a problem which remained for many years. An account for the construction of an atmospheric engine in 1727 for Edmonstone Colliery in Midlothian shows that the cylinder alone cost £250 out of £1,007 for the engine which did not include the cost of building the enginehouse nor the labour charges of the engine erectors.[1]

The bucket pumps fitted to atmospheric engines were more suitable for raising a little water a great height rather than a large quantity of water a small distance. Then, the smaller engines did not work as well and were more expensive to run in proportion to larger ones which in any case consumed so much fuel that they were economic only either where fuel was cheap or where the product was very valuable. In the Cornish mines, the value of the ore had to pay for the expense of bringing from South Wales coal which happened to have a high calorific content. At this period, coal was transported in large lumps so that smaller pieces were regarded as waste and it was this waste which supplied the boilers of the engines pumping dry the coal mines. This is why so many of the atmospheric engines erected in other places like the London waterworks at York Buildings had short lives because they were too costly to run. So engineers looked to see how they could improve the performance of both the Savery and the atmospheric engines, how they could make engines with small power output and also how they could make steam engines drive other types of machinery in addition to pumping water.

Savery himself had envisaged,

> occasioning Motion to all sorts of Mill Work . . . and for the Working of all sorts of Mills where they have not the Benefit of Water nor constant Windes.[2]

He would have used the intermediary of a waterwheel and pumped water with his engine to operate a form of pump storage scheme. He recommended building

> your *Mill-House* thirty six foot high, in which you may make what Motions and what sort of *Mills* you please. By the side of which

> House without, may be placed your *Water-wheel* of thirty-two, thirty three, or thirty four foot *Diameter*... The *Engine* may be fixed in any corner of the Mill-House 20, 21, 22 feet or more from the Level of the Pond.[3]

As far as is known, Savery never constructed any engine on this principle.

Triewald had the same idea and wrote,

> The fire-machine can also be applied to all kinds of mills, e.g. grain- and saw-mills, in shipbuilding yards, and in fortresses, and generally in all places where there are no waterfalls. Item, in mining districts to supply blast-furnaces where there is plenty of wood and iron but no waterfall; equally to forges or tilt hammers.[4]

In 1726, Triewald tried to adapt the engine he erected at Dannemora in Sweden for pumping water out of the iron ore mines to work a hoist to raise ore as well. This is the first recorded application of a steam engine to produce direct rotary motion.[5]

Triewald made his engine drive a 'Stagenkunst', or series of reciprocating rods which oscillated two racks. The racks engaged in and turned lantern pinions. The Stagenkunst was moved in one direction by the engine and returned in the other by a counterweight. Because the winding drum needed to be driven in both directions to bring up the buckets of ore alternately, one rack turned it one way through a ratchet device and the other worked in reverse. The racks could be lifted out of gear during the non-working stroke of the engine.[6] This mechanism was too complicated to be worked by a man in the correct sequence with the strokes of the engine and in any case gave an intermittent rotary motion. Yet this sort of layout can be traced for another 60 years, through, for example, the 1740 patent of John Wise who had a 'double tumbling wheel, to which the above chains and dogs are fixed',[7] and to Patrick Miller and William Symington's steam boat in 1788. On the paddlewheel shafts of their boat were loose pulleys with pawl and ratchet devices which were turned by chains from a pair of atmospheric engine cylinders.[8] Although their boat moved at about five miles an hour, it was unsuccessful and such mechanisms were unsuitable for producing rotary motion.

One of the difficulties in producing rotary motion from an atmospheric engine lay in the way it actually worked. When writing about the adaptation of the crank to such an engine, John Farey stated,

> The difficulty of applying it to use arose from the want of regularity in the action of the old engine. An engine to work a crank, must at all times make exactly the same length of stroke; and to perform well, all these strokes must be performed in an equal period of time. The old engines had very little exactness in either of these particulars. From the nature of the detent which opened the injection-cock, and the

> great friction of turning it, the degree to which it was opened was not constantly the same in the succeeding strokes; and a very small difference of opening would materially influence the quantity of injection, and consequently the vacuum and velocity with which the piston would descend. The boilers also of the old engines were always made too small, so that the least alteration in the intensity of the fire made the engine vary its speed.[9]

Therefore the atmospheric engine gave about the worst possible combination of events, a stroke which altered in its length and duration, which varied in power and which gave a power pulse in one direction only. It is no wonder that many people thought that a waterwheel was preferable and concentrated on ways of improving steam engines to supply waterwheels.

Much earlier, in 1707, Papin had built an engine which probably ought to be considered more as a refinement of the type proposed by the Marquis of Worcester rather than that of Savery because it used only the pressure part of the cycle. Farey commented that this made it inferior to Savery's.[10] On it Papin fitted a weighted-lever-type of safety valve which he had invented in 1682 for his digester.[11] His intention was to turn a waterwheel by a stream of water issuing under pressure from a jet. Because he was using only one receiver and there would be a period when it had to be refilled, Papin fitted an air-vessel on the forcing side of the engine to give a constant stream of water. The pressure at which it was intended that this machine should work gave an hydraulic head of 64 ft. The water issued out of the jet at a considerable velocity and so the engine was, in effect, a high pressure one. To improve the efficiency and performance, he attempted to separate the steam from the water inside the receiver by a copper piston floating on top of the water. Inside this float, Papin proposed to place a red-hot weight of iron to keep the steam warmer but this idea was found to be impracticable, as indeed was the engine itself.

Jacob Leupold also built an engine to operate a waterwheel using the pressure part of the cycle only. He arranged two receivers to be below the level of the bottom of the waterwheel so that the water could fill them without the need for a vacuum. He published plans for a high-pressure engine with pistons in his *Theatrum Machinarum* in 1729[12] and his Savery-type engine had certain similarities, especially in the valve gear. The water was forced up from the receivers into a tank to supply an overshot waterwheel but there is no indication how the valve gear was driven.[13] Once again, it is doubtful if this engine was a commercial proposition.

One of the earliest people to investigate the performance of both Savery and atmospheric engines was Desaguliers who made the Savery engine a commercial proposition. On a Savery engine, he was quick to include Papin's safety valve in the form of a 'Steel-yard Weight'. Desaguliers was also quick to adopt Newcomen's crucial discovery of directly injecting cold water straight into the

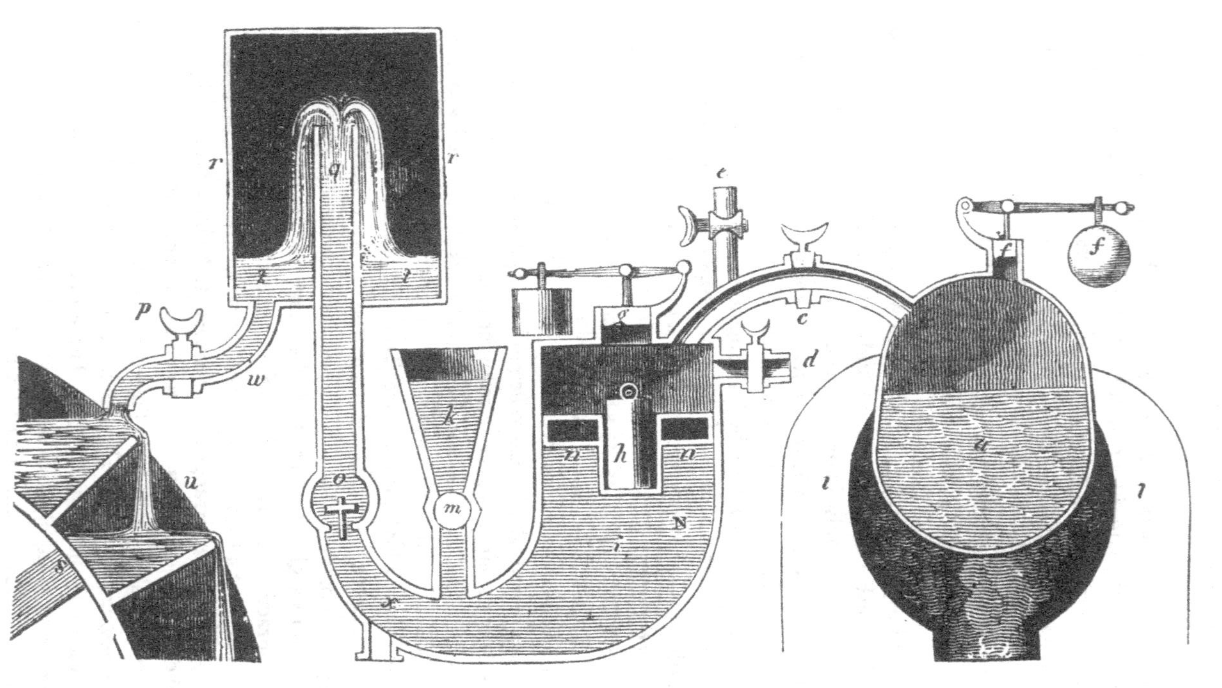

Figure 8 Papin's Savery-type engine which forced water against a waterwheel. The float inside the cylinder had a red-hot iron weight placed in it. (Stuart, *Steam Engine*.)

steam inside the receiver instead of pouring it over the outside. 'A Jet of cold Water will spout in thro' the spreading Plate I among the Steam, which it will immediately condense'.[14] Direct injection was included on all later engines and the performance of Savery engines became competitive with those of Newcomen regarding economy as well as being of course much simpler.

Desaguliers said that, in 1717 or 1718, he had erected seven or eight Savery engines, the first for Czar Peter I for his garden at St Petersburg. It raised the water 29 ft out of the well by suction and forced it up 11 ft by steam pressure because that was as much as he dared risk with soft solder.[15] We see here the beginning of the line of development that would be followed later, where the suction part of the cycle was becoming more important than the pressure.

In 1728, a direct comparison was made in Desaguliers's garden of his own Savery engine and a Newcomen-type erected by a Mr Gun-Jones. The Savery engine raised 10 tuns an hour about 38 ft high but the atmospheric engine did not perform so well.

> He made his Boiler of the exact size of mine, and his Cylinder was six Inches Bore, and about two feet in Length. When his Model of Leaver Engine was finish'd, it raised but four Tuns per hour into the same Cistern as mine. It cost him 300 l; and mine, having all Copper Pipes had cost me but 80 l.[16]

Unfortunately no figures are given for fuel consumption, but it will be evident from the difference in prime cost and the simplicity of the Savery engine why people were interested in it.

The subsequent development of the Savery engine was to follow four courses. One was to lessen the height to which the water was forced by steam pressure and finally to use only the suction part of the cycle. The second was to use the water pumped by the engine to turn a waterwheel and so to provide rotary power. This was later to become the most popular way of employing the Savery engine. The third range of improvements was to try and prevent the hot steam pressing directly upon the cold surface of the water in the receiver. The final line of development was to make the engine work automatically.

One of the first persons to try and make the Savery engine easier to operate was the Frenchman, Gensanne. By an ingenious system of levers and falling weights, he was able to operate the steam and injection valves simultaneously. By about 1744, several engines had been erected in France, one in Fresne near Condé, one at Sars near Charleroi where it drained a coal mine, and a third at the lead mines near Namur.[17] However, it would appear that a person was still necessary to work the levers so these engines were not self-acting.

In 1751, Smeaton reported to the Royal Society about the adaptations by de Moura of Portugal.[18] Inside the receiver, he fixed a light copper ball as a float on the end of a lever. As the water level rose and fell, the float actuated the steam and injection cocks. The joint where the pivot of the lever passed through the receiver wall was made air-tight by being immersed in a water trap. The float was counterpoised by a weight on the outside of the receiver. A series of chains and counterweights connected the lever to the steam and injection cocks so that the steam valve was opened with a snap action when the receiver was almost full of water.

Other people took up the idea of a float to separate the steam from the water, and Farey recommended that the best was

> a box of thin copper, hollow within, and very closely soldered: it should be circular like a grindstone, and made to fit the inside of the cylindrical receiver as nearly as possible, without actually touching its sides; The upper and the under surfaces should be covered with varnished wood, or the inside of the box may be filled with burnt cork, to prevent the heat penetrating through the float to the water.[19]

A floating separator made it impossible to use the variations in water level inside the receiver to actuate the valve gear. Also it would be interesting to know whether there was any tendency for the float to tilt and jam. The float became virtually redundant in the later development of the Savery engine which was to use the suction part of the cycle only.

In the meantime, the numbers of atmospheric engines increased steadily to around 150 in 1740. The early ones produced around 10 h.p. and might reach a duty of up to 6½ million pounds of water raised 1 ft through the burning of one bushel (84 lb in Newcastle upon Tyne but 88 lb in London) of coal. Most

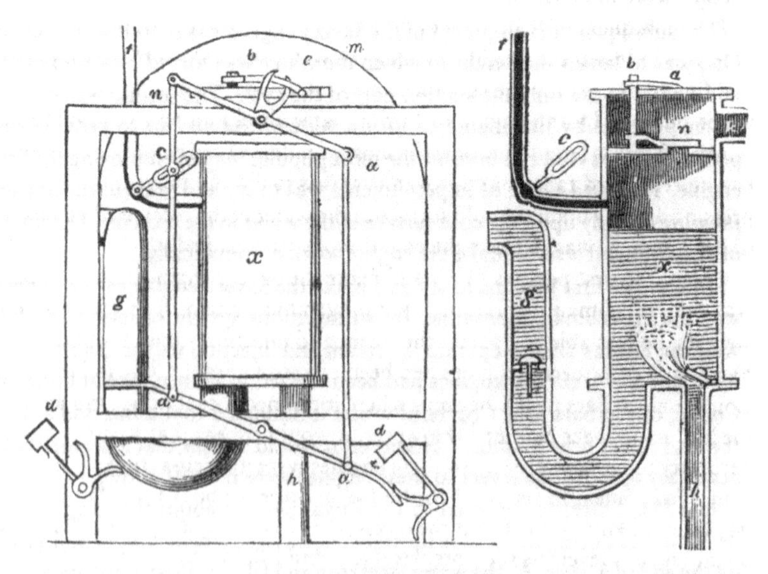

Figure 9 In Gensanne's engine, levers operated the steam and water injection valves simultaneously. (Stuart, *Steam Engine*.)

had a much lower duty than this, achieving only 4–5 million.[20] This was to become the standard way of measuring the performance of engines. Iron-masters relied on waterwheels to drive the bellows which produced the blast for their furnaces and, as more furnaces were built to cope with the growing demand for iron, this source of power became inadequate. During 1754, a drought at one of the furnaces in the Weald made the owner fear he would have to employ men to turn the waterwheel by walking on it.[21] Sometimes horse-driven pumps and then atmospheric engines were introduced to pump the water back up to the higher reservoirs to keep the wheels running. The first example of these 'water-returning engines' was installed at Coakbrookdale in 1742. In 1765, Smeaton improved Carron Ironworks on the Forth of Clyde with better blowing machinery which included an atmospheric returning engine.[22]

The increasing demand for power during the second half of the eighteenth century turned people's attention to improving all sorts of steam engines. Smeaton probably took the development of the atmospheric engine to the limits of practicability through the tests he carried out on his model engine in 1770. The experience he gained enabled him to raise the performance of the engine he erected at Long Benton colliery, Northumberland, to an unprecedented 9.1 million in 1774. In the following year, his engine for the Chacewater Mine in Cornwall achieved a duty of 10 million and developed 72 h.p.[23] In about 1784, in spite of Smeaton's doubts, a direct non-rotative blowing engine worked by an atmospheric engine was introduced to the Carron Ironworks.[24] This must be one of the earliest uses of direct steam power other than pumping water.

Interest in the Savery engine continued too. William Blakey tried to improve the performance by separating the steam and water cycles. Unfortunately, the description in his patent of 1766[25] is confused because he seems to be describing at least three different machines. The drawing shows that the main receiver was situated in the flue from the fire that generated the steam where it could be kept warm. Steam was let into this receiver to blow out the air through another tank or receiver at one side. To start the engine, a third receiver was filled with water and some of it was allowed to pass through a valve to condense the steam in the first. The resulting vacuum drew more water into the third receiver from the well. At this point, the description in the patent becomes confused because the cold water was drawn into the first receiver through another pipe out of the third, thereby negating the advantages of trying to separate the steam and water cycles. Steam pressure next forced the cold water into the second receiver and up the outlet pipe. It might have worked better had the second receiver been connected directly to the third.

Blakey had erected an engine in his garden in 1760[26] and had continued experimenting with it over the next few years. In 1765, he made a working model in London which was copied and applied at the silk mill in Derby though for what purpose we do not know. This would seem to be the first time that steam power was used in the textile industry. Blakey went on to develop his

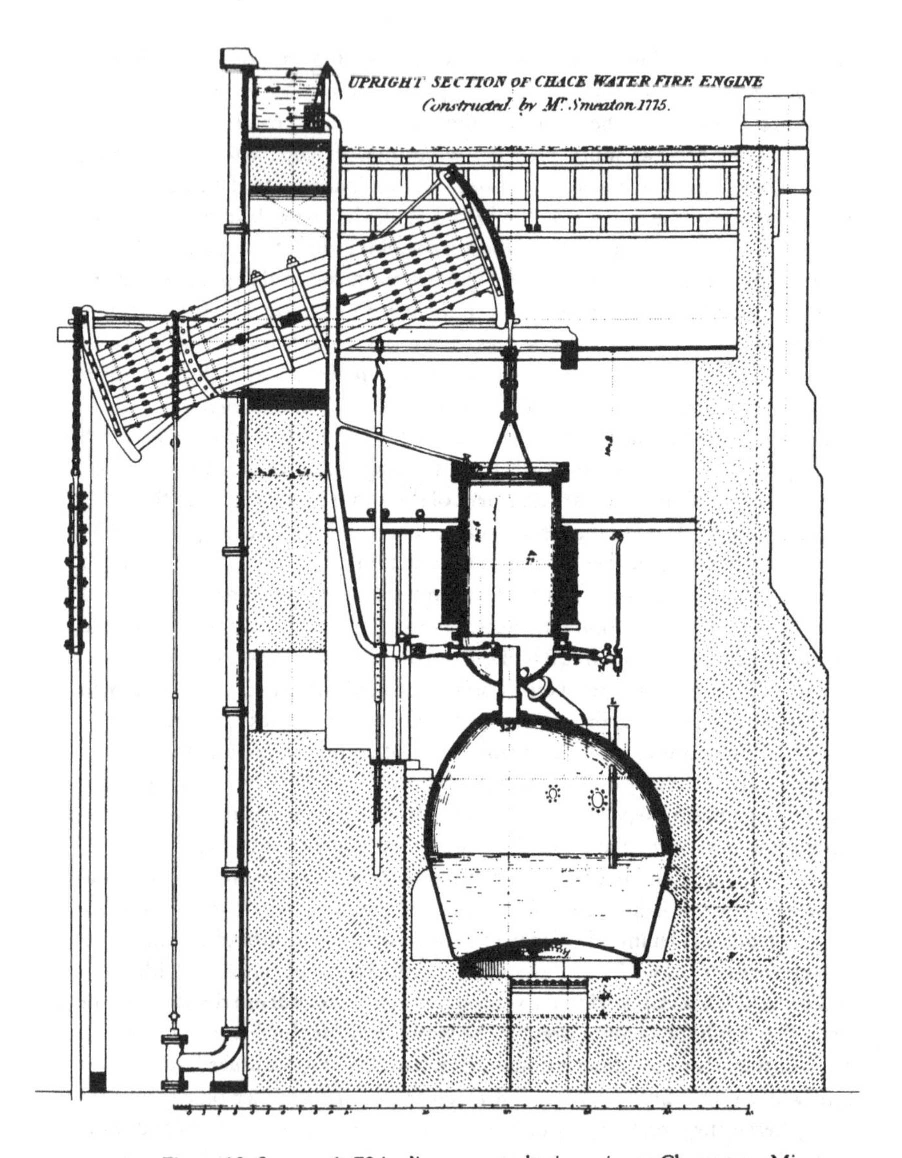

Figure 10 Smeaton's 72 in diam. atmospheric engine at Chacewater Mine, erected 1775. Smeaton experimented with laminated beams but they were not successful. (Farey, *Steam Engine*.)

engine further, for in August 1769, the *Gentleman's Magazine* described a drawing of an engine with an improved layout, in which he had a float controlling the air injection valve. He tried two receivers, one above the other, connected by a pipe, and had the idea of separating the steam from the water by a cushion of air. He also tried oil on top of the water to form an insulating disc. The lower receiver was the one to be filled with water. As the water level in it rose through the action of the vacuum, the float inside opened a valve which let in air to destroy the vacuum. The air filled the higher receiver too. As soon as the noise of the air entering stopped, the cold water injection valve was closed manually which allowed the steam pressure to build up again. 'As the steam increases, being much lighter than air, it keeps uppermost, and forces the air on the water which is in the receiver'.[27] Next the air was blown through by steam into the ascending pipe with the water when the injection cock could be opened once again. Under actual operating conditions, Farey commented,

> This engine was not found to answer any better than Savery's because the air will not make that complete separation between the steam and the cold water which the inventor expected; and there is a great loss of power to compress the air sufficiently to make it lift the column of water.[28]

In fact, when one of these engines was being demonstrated in Cornwall probably with a form of tubular boiler which Blakey had developed, it exploded. This has been claimed as the first application of a tubular boiler. Blakey also advocated using waste heat from furnaces for raising steam. On one of his early engines for grinding corn through a waterwheel, he claimed that it could be run

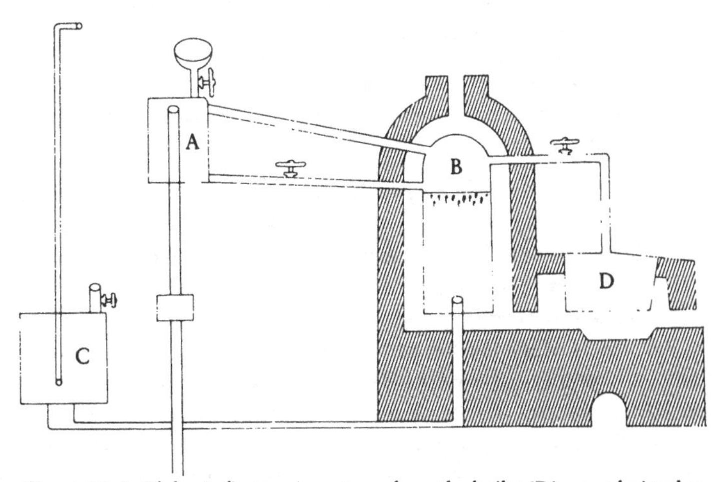

Figure 11 In Blakey's first engine, steam from the boiler 'D' was admitted to the main receiver 'B'. Water was drawn first into receiver 'A' and finally forced out of receiver 'C' up the standpipe. (Based on patent drawing.)

with little more fuel than was burnt in the baker's ovens from which he drew his waste heat. After building several engines in France and Holland, in 1780 he installed one at a rolling mill (possibly the Horse Hay Works) with two waterwheels, one of 15 and the other of 16 ft diam. Here again steam was raised by the waste heat from the furnaces,[29] possibly another first for a Savery engine.

In the light of subsequent history, one of the most important people to consider using a Savery engine was Matthew Boulton when he wished to supplement the water resources of the Hockley Brook which powered his new Soho Manufactory. He had purposely moved out of Birmingham in 1765 to avoid operating an expensive horse mill, but the growth of the Manufactory presented him with a similar dilemma once more. So his mind turned to employing a steam engine to lift water from the tail race of the waterwheel back to the mill dam.[30]

It appears that Boulton had a model of a Savery engine made towards the end of 1765 or early 1766 which he sent to Benjamin Franklin for his comments. In a covering letter (22 February 1766) Boulton wrote,

> My engagements since Christmas have not permitted me to make any further progress with my fire engine but as the thirsty season is approaching apace necessity will oblige me to set about it in good earnest. Query which of the steam valves do you like best? Is it better to introduce the jet of cold water at the bottom of the receiver or at the top? . . . I therefore beg, if any thought occurs to your fertile genius which you think may be useful or preserve me from error in the execution of this engine you'll be so kind as to communicate it to me.[31]

Franklin made little constructive comment, but Erasmus Darwin had heard about it and wrote in his usual enthusiastic vein.

> Your mode of a steam engine, I am told, has gained so much approbation in London that I cannot but congratulate you on the mechanical fame you have acquired by it.[32]

The model was received back in Soho and Boulton continued experiments with it until he heard about the engine on which Watt was working. After Watt had visited Soho in 1768, Boulton abandoned his trials and decided to wait until Watt's engine had been perfected.

The demand for rotary power was to increase by leaps and bounds through the inventions which mechanised the cotton industry. At the beginning of the eighteenth century, there were only three machines in the textile industry which were driven by power. They were fulling stocks for cleaning and compacting woollen cloth, raising machines which brought up the surface and silk throwing machines. It was not the traditional English wool industry which was mechanised first but the upstart cotton. Lancashire achieved the title, home of the cotton industry, in the early eighteenth century, principally through the import of raw cotton to Liverpool which was carded and spun by the women at their spinning wheels and woven by the men at their looms working in cottages on the edge of the Pennines.[33] The streams running down these hills provided soft clean water for the bleaching and finishing processes such as dyeing and printing. The poor hill farms barely supported those struggling to make their living there so no doubt the inhabitants were glad to earn a little extra through the textile industry. There was a ready outlet for the cloth they produced, not only at home in the rest of England but cotton goods sold well in Africa in exchange for slaves which were sent to the plantations in the West Indies and in the southern states of North America too.

Production of yarn and cloth was very slow because the only equipment available was the hand wheel that twisted a single yarn at a time while the looms were still very primitive. In 1733, John Kay of Bury developed his 'flying shuttle' which enabled one weaver to produce more quickly broad cloth that had required at least two men previously. The flying shuttle became more popular after 1760 when Kay's son, Robert, added the 'drop box' which stored shuttles with different coloured or types of wefts so that they could be selected when wanted. Because weavers now could weave so much weft, the other processes of textile production had to be mechanised to keep pace with demand. Over the decade commencing in 1760, several inventions were made in spinning and carding which, in association with continual improvements in other fields and particularly in the steam engine, promote the belief that the Industrial Revolution began at that time, an arbitrary though useful date.[34]

The necessity for improved spinning techniques giving greater productivity ultimately led to the first successful spinning machine, the 'spinning jenny' of James Hargreaves.[35] This originated in 1764. The first model could spin eight yarns at once, a great improvement on the old single thread wheels, while later ones were built to spin 80 or more. This never progressed beyond a hand-operated machine which remained almost entirely in cottages and so did not alter the structure of the industry. But the spinning machine invented by Richard Arkwright and patented in 1769 was driven by power.[36] The patent specification stated that the 'water frame', as it came to be called, should be driven by horsepower, and this is what Arkwright used at his first mill in Nottingham. Then in 1771, he moved to Cromford where he built a mill with a waterwheel that produced about 10 h.p.

The spectacular growth in power-driven textile machinery really started after 1775 when Arkwright had solved another problem, that of mechanised carding. He took out a second patent that year in which he specified machines for carding, drawing and roving the cotton, all essential stages in preparing the cotton for his water frame. It is significant that Arkwright and his partners built no more mills until after this second patent and, up to then, neither did they sell any licences to others for spinning. But once the system was proved, growth was rapid, particularly in and around the Derwent valley between Matlock and

Table 1. Imports of raw cotton 1771–90[39]

Year	lb cotton imported
1771–75	4,764,589 average per year
1776–80	6,766,613 average per year
1781	5,198,778
1782	11,828,039
1783	9,735,663
1784	11,482,083
1785	18,400,384
1786	19,475,020
1787	23,250,268
1788	20,467,436
1789	32,576,023
1790	31,447,605

Derby, with mills at Belper, Darley Abbey, Masson, Milford, Wirksworth, to name but a few, all of which were driven by water power.[37] Further afield, Arkwright had partnerships in mills at Ashbourne in Derbyshire, Birkacre in Lancashire, Bromsgrove in Worcestershire and New Lanark near Glasgow, as well as Shudehill in Manchester.

Other people cashed in on the boom, especially after Arkwright's 1775 patent had been declared null and void and his 1769 patent had expired in 1783. His water frame was particularly suited for spinning cotton yarn that could be used in the warp of coarse to medium quality fabrics where it was often combined with weft from the jenny. Another hand-operated machine, the 'mule', had been brought to a state of practical operation in 1779 by Samuel Crompton.[38] This could spin fine yarns and opened up a vast new area of trade. Therefore in the 1780s, we find two spinning systems side by side, the full power-driven system with the water frame, and the jenny or mule system which soon had the carding (using Arkwright's machines) done by power and the spinning by hand. At first, separate carding mills driven by horses or small waterwheels supplied the spinners working on jennies or mules in their own homes, but later all the machines were moved into mills for closer supervision.

The phenomenal increase in the cotton textile industry can be judged by the volume of cotton imported between 1771 and 1790, see Table 1. The cotton trade from England began to dominate world markets while, in England itself, it soon eclipsed the much older woollen industry and cotton goods became the largest export. Around 1790, the mule began to be adapted so that some parts of its spinning cycle could be driven by power and this further increased the demand for power. At first, the mill owners had to use either horses or water-wheels. Horses soon became tired and, in any case, mills outgrew the power that could be supplied by animals. The cotton industry had to spread far afield in search of suitable sites where water could be harnessed from rivers and this was why Arkwright's empire was so scattered.

So the cotton industry was transformed by a series of inventions from a domestic to a factory system. Trade depended upon good communications because the raw material had to be imported and the finished product exported. First the Mersey and Irwell rivers were made navigable to Manchester in 1720 which helped to cheapen rates of carriage between Liverpool and Manchester.[40] Then in 1761 the Duke of Bridgewater launched the canal era with his canal to bring coal from Worsley to Manchester.[41] Within the next 40 years, the North West was criss-crossed with canals. Local resources played a vital role, for, without rivers and water, there could have been no finishing of the fabrics, no power for waterwheels and, in the early part of the Industrial Revolution, no transport for heavy goods. But the rivers quickly proved to be inadequate to supply all the power that was needed. Then underneath everything was coal. This was to draw the cotton industry back into Lancashire where it will be noticed that all the major textile towns were situated on coal seams which stretched down into Cheshire at Stockport and Poynton.[42] Some millowners, like the Ashton's of Hyde, also owned coal mines literally under their mills. Coal was to provide the solution to the power crisis for it was the primitive steam engine to which Arkwright and his partners turned for their mill on Shudehill near the middle of Manchester.

In 1783, Shudehill near the highest part of Manchester of that day was a strange place to choose for a textile mill because there was no large stream or river which could drive it. It was even stranger when a waterwheel 30 ft diam. and 8 ft wide was erected. Two storage ponds were built and, instead of letting the water which had passed round the wheel flow away downstream, it was pumped back up to the higher reservoir to be recycled over and over again by a steam pumping engine. This engine had two pumps, each 31 in diam. with a stroke of 7 ft 9 in, and a steam cylinder 64 in diam. with the stroke presumably the same as the pumps. It worked at 11–12 strokes a minute and consumed about five tons of coal a day. In 1788, it was described as being of the 'old construction' which means it must have been an atmospheric type. Therefore the earliest steam-powered cotton spinning mill was driven by the earliest successful type of steam engine.[43]

Those millowners who wanted a simpler engine turned to the Savery type. In 1784, two were installed by Joshua Wrigley at the cotton spinning mill of Joseph Thackery at Chorlton-on-Medlock just to the south of Manchester because the river Medlock did not have sufficient flow to drive the waterwheel. John Aikin, writing about the Manchester area at that time, described these engines.

> The new-invented steam engines by a single cylinder closed above, pushing over water to an overshot-wheel, which returns to the

> reservoir, suppose a common pump-spring, were a great improvement, and employed to advantage as the application of machinery to several branches of business was extended. For by this means, there is less occasion for horses, and any power may be applied by enlarging the diameter of the cylinders, as one of twenty-four inches will force over more than sixty gallons at a stroke. This improvement, which is as simple as ingenious, was the invention of a common pump-maker, Wrigley by name, of this town, who never applied for a patent, but imparted freely what he invented to those who thought proper to employ him.[44]

These engines were examined by Smeaton and details of their sizes are given by Farey who recorded a duty of 5½ million pounds for the larger one which developed 5 h.p. and raised the water 19 ft. The smaller one lifted water about 14 ft and produced 2⅔ h.p.

> As neither of these engines were constructed with the best boilers such as have since been brought into use in modern steam engines, it is probable that a greater effort might be produced from the fuel.[45]

Wrigley formed a partnership with John Derbyshire for erecting such engines and also with a millwright, Joseph Young, who built one for the cotton manufacturers Marshall & Renolds in Manchester. Wrigley had quite an extensive business throughout Lancashire and Yorkshire, for, in 1791, James Watt Junior writing from Manchester mentioned that Wrigley had 'orders for 13 Engines for this town & neighbourhood, all of them for working Cotton Machinery of one kind or another by the medium of a Water wheel'.[46] This was a larger number of engines than Boulton & Watt had in hand.

Unfortunately we do not know how many Savery engines Wrigley built and he probably supplied some atmospheric ones too, nor do we know the details of his design. Farey merely stated,

> to render the engine complete, it must be made self-acting, that it may perform all the functions of opening and shutting its valves, and supplying itself with water, without any assistance from the attendant, who will then only have to keep up the fire.
>
> Several engines upon this principle, with various improvements taken from other steam-engines, were erected by Mr. Joshua Rigley, many years ago, at Manchester, and in other parts of Lancashire, to impel the machinery of some of the earliest manufactories and cotton-mills in that district. The engine usually raised the water about 16 to 20 feet high; and the water descending again, gave motion to an overshot water-wheel. This is Mr. Savery's original project for working mills by his engine: but Mr. Rigley contrived his engines to work without an attendant, the motion of the water-wheel being made to

> open and shut the regulator, and injection cock, at the proper intervals. They continued in use for some years, but were at length given up in favour of better engines.[47]

Watt did not hold a high opinion of Wrigley for he commented, 'Joshua Wrigley is erecting a rotative Engine at Manchester, but if he does it no better than he does his cotton mills he will be beshet the rev^{r}'.[48]

While the early cotton spinning mills were situated on some of the best water power sites in the country, even Arkwright found that his mills at Cromford did not have enough water resources. This was why millowners resorted to auxiliary steam engines to augment the water supply during a drought. The spread of power spinning into fibres other than cotton caused an ever deepening power shortage. The firm of Davison & Hawksley of Nottingham were the first to spin worsted by power. In 1789, Boulton & Watt sent them estimates for an 8 h.p. engine to cost between £490 and £500, but they decided to install a Savery engine instead, probably because it was cheaper. However this mill burnt down within three or four years and for their next one at Arnold they chose a double-acting atmospheric type patented by Francis Thompson.[49] John Marshall of Leeds formed a partnership with Benyon and in 1790 started a mill at Holbeck for spinning flax, again the first of its type. It was powered by a waterwheel, the water being supplied by a Savery engine.[50]

In the correspondence surviving in the Boulton & Watt Collection at Birmingham, Savery engines are mentioned quite often by engine erectors who had been asked to replace them by Watt's separate condenser model. Too often the owners had had an unfortunate experience like Messrs Salvin of Durham.

> They have an engine upon the Saverian plan (which has been finished about a month) which was to have been equal to the power of about 20 horses; but upon trial, it is found to fall short ½ of that power & to burn an immense quantity of coals; as will readily suppose.[51]

Their waterwheel was 12 ft in diameter. In the following year, 1792, Beverley Cross & Co. of Leeds complained,

> I am concerned with some friends at this Place in Building a Cotton Mill but unfortunately have been ill advised with regard to our Engine, one of the sort called Blow Engines being recommended to us which, now we are ready to work, we find insufficient.[52]

Both these examples of Savery engines were quite large, in the region of about 20 h.p., and both failed disastrously.[53] Smaller ones seem to have been more successful and the Manchester firm of Nightingale Harris & Co. was still using one to turn its machinery in May 1796.

One of the most successful Savery engines driving a waterwheel, which used only the vacuum part of the cycle, was built by Peter Keir in 1793–4 for his

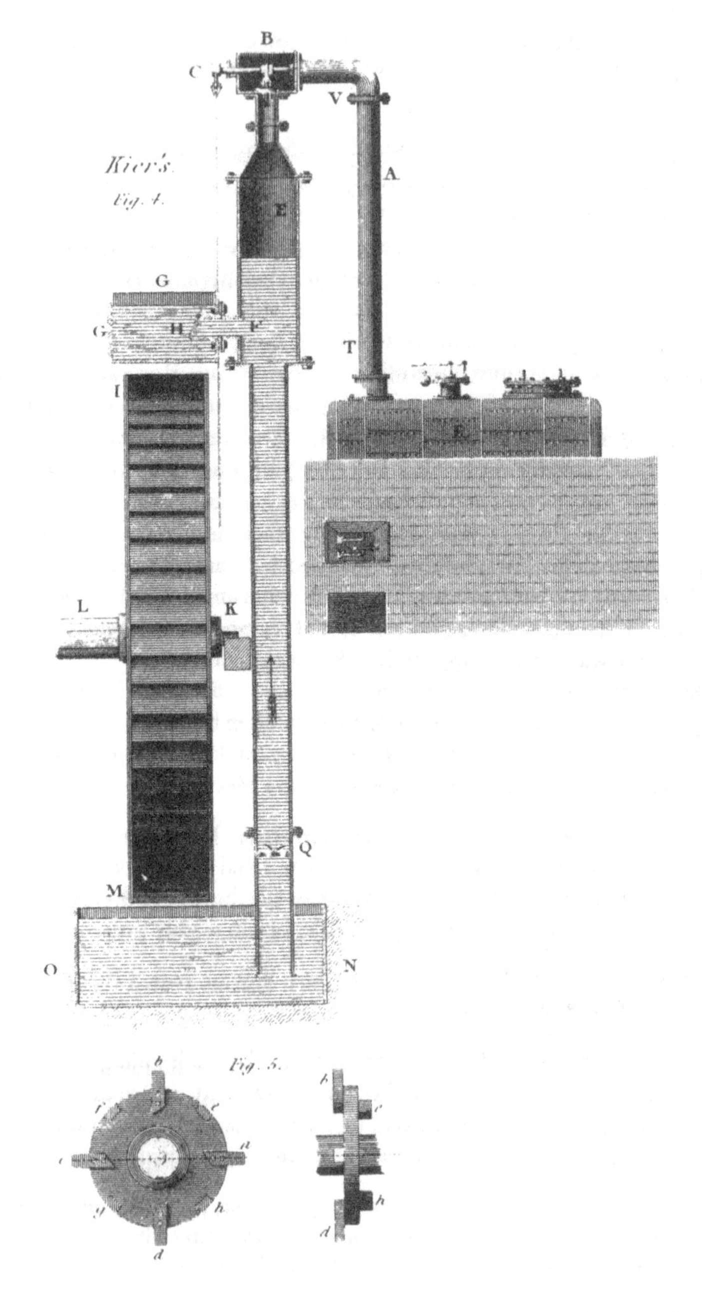

Figure 12 Keir's Savery engine, waterwheel and boiler. (Rees, *Cyclopaedia.*)

workshop at Camden Town in London,[54] possibly supplied by Wrigley. There was an overshot waterwheel 18 ft diam., 2 ft wide which made 10 r.p.m. Steam was supplied by a wagon-type boiler 7 ft long, 5 ft wide and 5 ft deep. Cams on the axle of the waterwheel operated the valve gear and drove a feedwater pump. A vacuum was created by one cock admitting a jet of water into the receiver which raised the water into the receiver above the level of the penstock. Then another valve opened to admit a little air before the main stream valve opened. Keir found that this worked better than admitting steam at a pressure of 2–3 lb. G. Birkbeck, when describing this engine, said,

> It still, however, remains a subject for investigation, whether the air so admitted, forms a stratum between the surface of the water and the

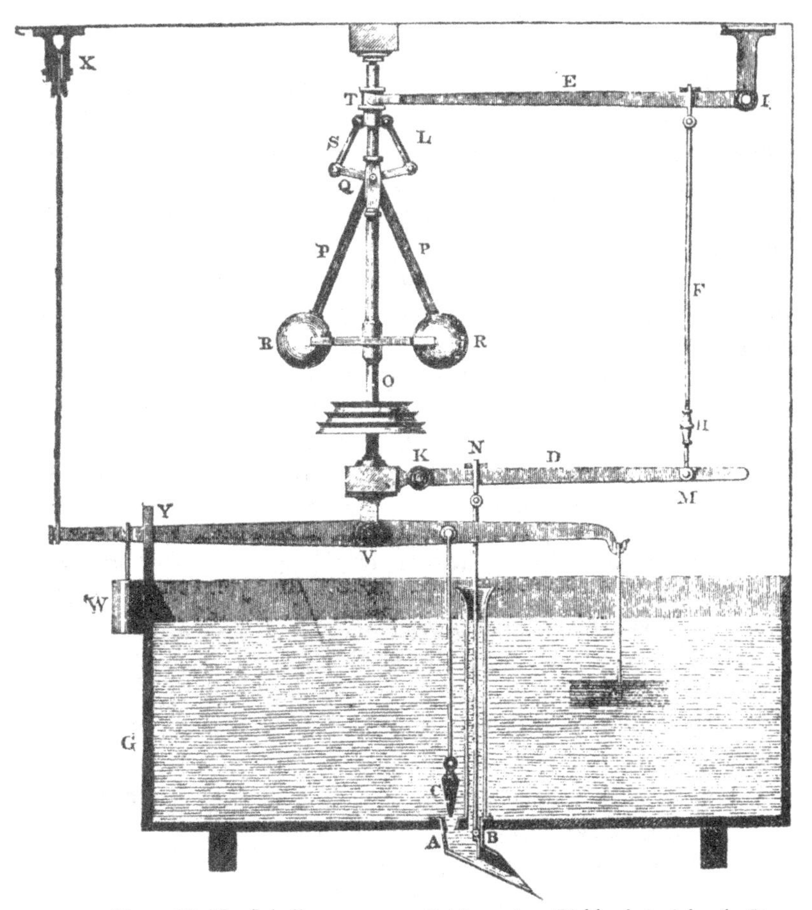

Figure 13 The flyball governor on Keir's engine. (Birkbeck & Adcock, *Steam Engine.*)

> steam, or whether, by admixture, it suddenly becomes expanded by the caloric of the steam and thus by an increase of elasticity expells the water more forcibly through the pipe.[55]

Presumably the air quickly destroyed any remaining vacuum without having to heat up the engine to the same extent as the steam. The steam following the air had time to warm up the parts again. The air was expelled together with the water through a flap valve situated under the normal level of the water in the pentrough.

Keir found that his engine worked well and burnt four bushels of good coal in 12 hours. The average performance at 10 lifts per minute, throwing 7 cu. ft of water into the pentrough 22 ft above the reservoir, equalled a duty of 17,325,000 lb. In allowing for the fact that at least one-third of the power must be deducted to allow for the losses within the waterwheel, an effective duty of over 11,550,000 lb was achieved which was equal to the best atmospheric engines. This economy was ascribed to the air valve.[56]

The speed of Keir's waterwheel was regulated by a flyball governor, which was probably its first application to waterpower. It had several interesting features for a turnbuckle lengthened or shortened one of the rods connecting the shuttle on the sluice mechanism so that the flow of water onto the wheel could be adjusted manually for the expected load. The governor then took over for minor variations and rotated at 60–70 r.p.m. There was a stone float linked to another sluice gate which could be lifted by a cord to stop the wheel. This also came into action if the water level fell too low and there was a danger of the outlet flap valve of the delivery pipe from the Savery engine becoming uncovered. Keir had thus produced an extremely ingenious engine which he used for over 28 years and he also built another for a Mr Lane near Coventry. In 1799, John Nuncarrow proposed adding a separate condenser and air pump to a Savery engine with a waterwheel[57] but by this time, this type of engine was outclassed by others of improved design.

References to Savery engines supplementing waterwheels in the textile industry are more frequent than those to atmospheric engines, possibly because in early days the mills were quite small. In 1792, James and William Carlisle built a mill at Bolton where they installed a beam engine for returning the water.[58] By this time, the rotative steam engine driving the machinery directly had proved its reliability. There had been considerable reluctance to trust rotative engines. In 1781, Smeaton had advised the Honourable Commissioners of His Majesty's Victualling Office,

> In compliance with your order of the 14th May last, desiring me to give my opinion, which of the two methods I prefer of constructing a mill to be worked by steam (one by the intervention of water the other by a crank), and the reason of that preference . . . I have fully and duly considered the business, and find that in point of quantity [of coal] to

> be consumed for raising a power sufficient to do this business ... there will be no material difference; but in point of convenience and good effect in the practical part, it appears to me that the difference will be very considerable, for, in the first place, I apprehend that no motion can ever act perfectly steady and equal in producing a circular motion, like the regular efflux of water turning in a water-wheel, and much of the good effect of a water-mill is well known to depend upon the motion communicated to the mill-stones being perfectly equal and smooth, as the least tremor or agitation takes off from the complete performance.
>
> Secondly, all the fire-engines that I have seen are liable to stoppages, and that so suddenly, that in making a single stroke the machine is capable of passing from almost full power and motion to a total cessation ... In the raising of water (a business for which a fire-engine seems peculiarly adapted), the stoppage of the engine for a few strokes is of no other ill consequence than the loss of so much time, but in the motion of mill-stones grinding corn, such stoppages would have a particular ill effect.[59]

For spinning cotton, it was particularly important that the turning power should be regular, for, otherwise, there would be thick and thin places in the yarn. This is a problem which we will meet constantly in following the development of the steam engine.

Smeaton was soon proved to be wrong about the consumption of coal in engines powering waterwheels. There was the inefficiency of the waterwheel which had to be added to the inefficiency of the engine, whether of Savery or atmospheric type. In a letter dated 1790, John Southern, the manager of Boulton & Watt's engine operations at Birmingham, wrote to Peter Ainsworth of Bolton,

> They wish you to observe that a 10 horse engine applied to pumping water for waterwheels, will not produce the effect of 10 horses in the machinery; it may take a 16 or a 20 horse engine to do that according to the construction etc. of your waterwheels. It will be much your interest to erect an engine to turn the machinery directly, if you can, in manner as Mr. Drinkwater's or Mess. Peel Answ[th]. Nr. Warrington. An engine to pump water comes more expensive, and does little more than half the effect when that is thrown on a wheel.[60]

The high consumption of coal was one of the reasons why the power system at Simpson's Shudehill Mill was soon modernised. In 1790, the original atmospheric engine was supplemented with a 6 h.p. Boulton & Watt rotative type, but, a year later, Barton (Simpson's manager) was inquiring from Peter Ewart, who was Boulton & Watt's representative in Manchester, about replacing the atmospheric engine and waterwheel with a large Watt rotative type. At that time, there were 4,000 spindles in the mill, and they hoped that one 40 h.p. engine could be placed in the same spot as their old one.

> Their old Boilers, Cylinder, Beam and Waterwheel will all be worn out in about 9 months and they wish to determine as soon as possible upon the plan to be adopted, in order that they may be prepared in time.[61]

To Ewart fell the task of building the new engine while retaining the old one at work as long as possible to minimise disruption to the spinning. This 40 h.p. engine was the largest which Boulton & Watt had built for a cotton mill and the foundations were started in January 1792. In April, Ewart wrote to Boulton & Watt,

> Mr. Simpsons are very impatient to have some of their materials upon the spot, for they have been quite ready for some time and the Old Engine is giving way every day.[62]

Most of the materials had arrived before that July so that the first steam engine to work in a cotton mill must have been replaced before the end of the summer of 1792 after working for about nine years.

4
The economy of power

The person who originated the vital inventions and discoveries for making the first successful rotative engine was James Watt, born at Greenock on the Firth of Clyde in 1736. His paternal grandfather had been a teacher of navigational mathematics while his father and uncle were practised surveyors. In addition, his father's business included marine engineering. On his mother's side, he was related to George Muirhead, a professor of Classics at Glasgow University. With such a background, it was only natural that Watt's interests would lie in similar areas. No one could have imagined that, when Professor James Anderson asked Watt to repair a model of an atmospheric engine belonging to the Natural Philosophy class of the College of Glasgow University during the 1763–4 session, it would be a turning point in the history of civilisation.[1] Watt found that the boiler, although correct in scale, could not supply enough steam to work the model for more than a few strokes. Through his experiments, he found that, at every stroke, the quantity of steam consumed was several times greater than the volume of the cylinder. He also discovered that a great deal more water was needed to condense the steam than he reckoned ought to be necessary if the laws of proportions for mixing liquids of different temperatures applied.

It had long been known that much steam was wasted in atmospheric engines through the heating and cooling of the cylinder at every stroke. It was also recognised that the type of material used for the casting of the cylinder, and also the thickness, affected the steam consumption. Desaguliers wrote in 1744,

> Some people make use of Cast Iron Cylinders for their Fire-engines; but I would advise nobody to have them; because, tho' there are Workmen that can bore them very smooth, yet none of them can be cast less than an Inch thick, and therefore they can neither bee heated nor cool'd so soon as others, which will make a Stroke or two in a Minute difference whereby an eighth or tenth less Water will be rais'd. A Brass Cylinder of the largest Size, has been cast under ⅓ of an Inch in Thickness; and, at the long run the Advantage of heating and cooling quick will recompense the Difference in first Expense; especially when we consider the intrinsick Value of the Brass.[2]

The model, on which Watt experimented, survives and has been preserved in the Hunterian Museum of Glasgow University. A recent investigation has shown that, if it represents an engine with a cylinder bore of 48 in and a stroke of 7 ft, then the cylinder walls are five times the scale thickness.[3] Therefore the model was ideal for drawing attention to the great defect of the atmospheric engine. It would seem that Watt thought, wrongly, that all of the heat was lost from the outside surface of the cylinder and that, therefore, the loss was dependent upon the surface area. Of course some heat is lost in this way, but much more is lost heating up the metal itself. Watt thought that the scale effect of the model increased the surface area in relation to the volume and it was this which caused greater condensation of steam. Watt was able to measure the amount of steam actually consumed, and so calculated its thermal capacity. The realisation of how great was this consumption of steam, and therefore the implied losses, began to make him consider how this could be reduced.

In the meantime, he had stumbled on another paradox; that the amount of water needed to condense the volume of the steam inside the cylinder was far greater than that which he had calculated if the laws of proportions were correct.[4] If an ounce of boiling water at 100 °C (or 212 °F) were mixed with an ounce of water at 11 °C (or 52°F), the result was two ounces of tepid water at 55 °C (or 132 °F). However, if the ounce of water at 11 °C (52 °F) were mixed with an ounce of steam at 100 °C (212 °F), the normal law of mixtures broke down. Watt continued to experiment and found that steam at ordinary atmospheric pressure could heat about six times its own weight of water from room temperature to boiling point. He sought an explanation from his friend Joseph Black, who at that time was Professor of Anatomy and Chemistry at the University of Glasgow. Following Herman Boerhaave's precept, academically inclined medical men studied chemistry and this included the study of heat. Among the medical men who had been trained in Scotland and made notable contributions to the study of heat were George Martine, William Cullen, Joseph Black, John Robison, William Irvine and Erasmus Darwin. Watt was therefore unique among the early power engineers in having had direct and indirect contact with men who made scientific studies of heat.

In 1762, Black had discovered that a change of state in any substance, for example the change from a solid to a liquid or a liquid to a gas, required a certain addition, or diminution, of heat, and that the quantity of heat was different for different substances and also was different according to the nature of the change.[5] When water is heated over a fire in say a kettle, it gradually becomes hotter until boiling point is reached, but then the temperature will cease rising. The water boils and steam bubbles off for as long as the kettle remains over the flames and until it becomes dry. Heat has been used to turn the water into steam but, because there is no change in temperature, Black called this 'latent heat'. He had been teaching this new doctrine to his student classes and, during the summer of 1764, explained it to Watt.

Watt's measurement of the latent heat of evaporation enabled him to compute exactly how much water should be injected to condense all the steam without cooling down the cylinder. But he found that, when the engine was operated at this theoretical maximum efficiency, there was a serious loss of power. The engine was now economical but no longer powerful. The thermometer confirmed that the temperature of the condensing water coming out of the cylinder was very high. Watt knew, from the experiments of Cullen, that tepid water would boil in a vacuum. He therefore carried out a further series of experiments to determine the temperature of the boiling point of water at various pressures.[6] From this, he realised that he could not have the walls of the cylinder boiling hot, which was necessary to reduce steam consumption, and also a vacuum with a pool of hot water at the bottom, because the water would boil, releasing steam.

The dilemma facing Watt was summed up quite simply. For maximum economy, the cylinder must be kept hot all the time, but for maximum power, it must be cooled down once every cycle. Watt almost literally stumbled on the solution in May 1765, as he recounted about 50 years later.

> It was *in the Green of Glasgow*. I had gone to take a walk on a fine Sabbath afternoon . . . I was thinking upon the engine at the time . . . when *the idea came into my mind, that as steam was an elastic body it would rish into a vacuum, and if a communication was made between the cylinder and an exhausted body, it would rush into it, and might be there condensed without cooling the cylinder*. I then saw that I must get quit of the condensed steam and injection water, if I used a jet as in Newcomen's engine. Two ways of doing this occurred to me. First the water might be run off by a descending pipe, if an outlet could be got at the depth of 35 or 36 feet, and any air might be extracted by a small pump; the second was to make the pump large enough to extract both water and air . . . *I had not walked further than the Golf-house when the whole thing was arranged in my mind.*[7]

Watt was remembering events which had happened many years before and it was to take over ten years of intermittent experiments before he developed a satisfactory engine.

There are, however, various points to notice in Watt's statement. He realised that he could condense the steam in a separate chamber or vessel from the main cylinder and so keep the cylinder hot and the condenser cold. He realised that he would have to remove the condensate to prevent flooding the engine and also any air to prevent windlogging. There would be a smaller volume of water to be removed if he used a surface condenser so that then he could pump both water and air out with quite a small pump. In the Boulton & Watt Collection, there is a drawing of a small boiler; a vertical cylinder with a piston inside, the piston rod passing out through the bottom cover with a weight hanging on the end; a surface condenser; and another smaller cylinder with a pump piston inside it. The drawing was prepared probably in 1795 or 1796 for patent litigation and it claimed to show Watt's first experimental separate condenser.[8]

Watt described this model:

> I took a large brass syringe 1¾" dia. and 10" long, made a cover and bottom to it of tin-plate, with a pipe to convey steam from the upper end to the condenser (for to save apparatus I inverted the cylinder). I drilled a hole longitudinally through the axis of the stem of the piston, and fixed a valve on its lower end, to permit the water which was produced by the condensed steam on first filling the cylinder, to issue. The condenser used on this occasion consisted of two pipes of thin tin-plate, ten or twelve inches long, and about one-sixth inch diameter, standing perpendicular and communicating at top with a short horizontal pipe of large diameter, having an aperture on its upper side which served for the air and water pump; and both the condensing pipes and the air pump were placed in a small cistern filled with cold water. The steam pipe was adjusted to a small boiler. When steam was produced it was admitted into the cylinder and soon issued through the perforation of the rod, and at the valve of the condenser. When it was judged that the air was expelled, the steam cock was shut, and the air pump piston rod was drawn up, which leaving the small pipes of the condenser in a state of vacuum, the steam entered them and was condensed. The piston of the cylinder immediately rose and lifted up a weight of about 18 lbs which was hung to the lower end of the piston rod.[9]

We made a replica of this apparatus at the North Western Museum of Science and Industry. To show how effective it was, we first tried to demonstrate it without steam. The cock to the boiler was shut, that between the cylinder and the condenser opened and the air pump worked vigorously. The weight did not move. But when the system was filled with steam and the boiler cock shut again, one gentle stroke of the air pump was sufficient to create a vacuum and the weight shot up. If his apparatus worked as well as ours, Watt must have felt elated.

Further experiments caused Watt to believe that his invention held great promise and could be developed into a full-sized engine with outstanding economy. Black lent money to Watt so he could carry on experimenting and, what proved to be more important still, introduced him to John Roebuck of Birmingham. Roebuck had established the Carron Iron Foundry in Scotland in 1759 and leased the coalfield at Borrowstones from the Duke of Hamilton. These mines were continually flooded and more powerful and economical pumping engines were needed urgently. In 1768, Roebuck agreed to take over Watt's debts and to bear the cost of a patent in return for a two-thirds share in it.

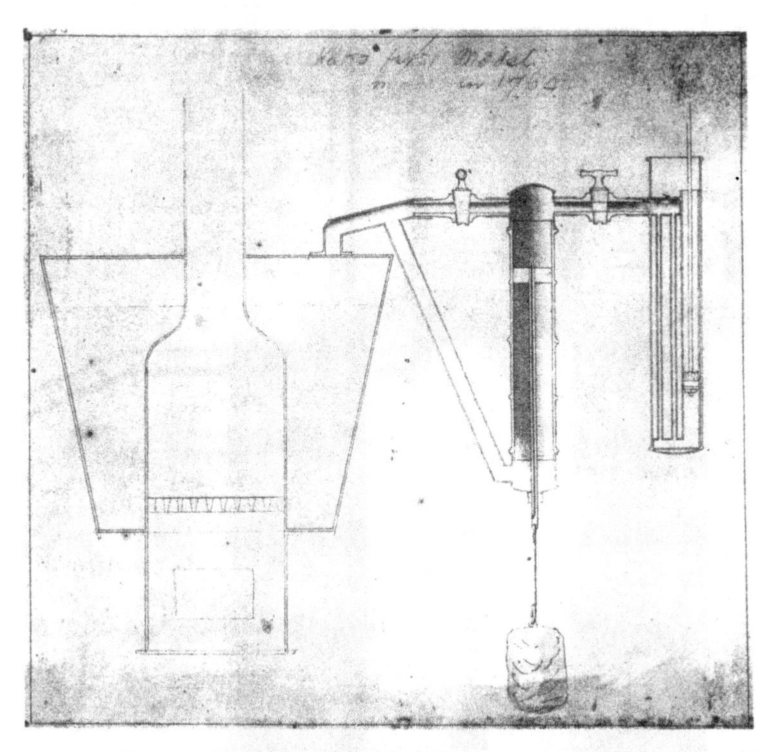

Figure 14 Watt's drawing of the boiler, syringe and condenser he used in his first experiments. (Boulton & Watt Collection, Birmingham.)

It was not until the beginning of 1769 that Watt's patent was granted 'for a method of lessening the consumption of steam and fuel in fire-engines'.[10] Watt had to visit London to apply for his patent but returned to Scotland and to Roebuck's house at Kinneil where he began to erect an experimental engine. In November 1768, he wrote to Roebuck,

> I think the best place will be to erect a small house in the glen behind Kinneil. The burn will afford us plenty of cold water, and we will be more free from speculation than we can be about Bo'ness.[11]

The ruins of the small house are still there. All this time, Watt had had to carry on with other work in order to earn his living and so could not concentrate on his engine full time. An illustration of an early engine, possibly at Kinneil, shows an inverted cylinder with the piston rod lifting the weight directly, similar to his first model. Another shows a cylinder resting on the ground with a chain going over a pulley wheel to pumps on the other side of a wooden frame. Both engines had a surface condenser and cylinder for an air pump. The air pump and steam inlet valve appear to have been worked manually. The engine erected in

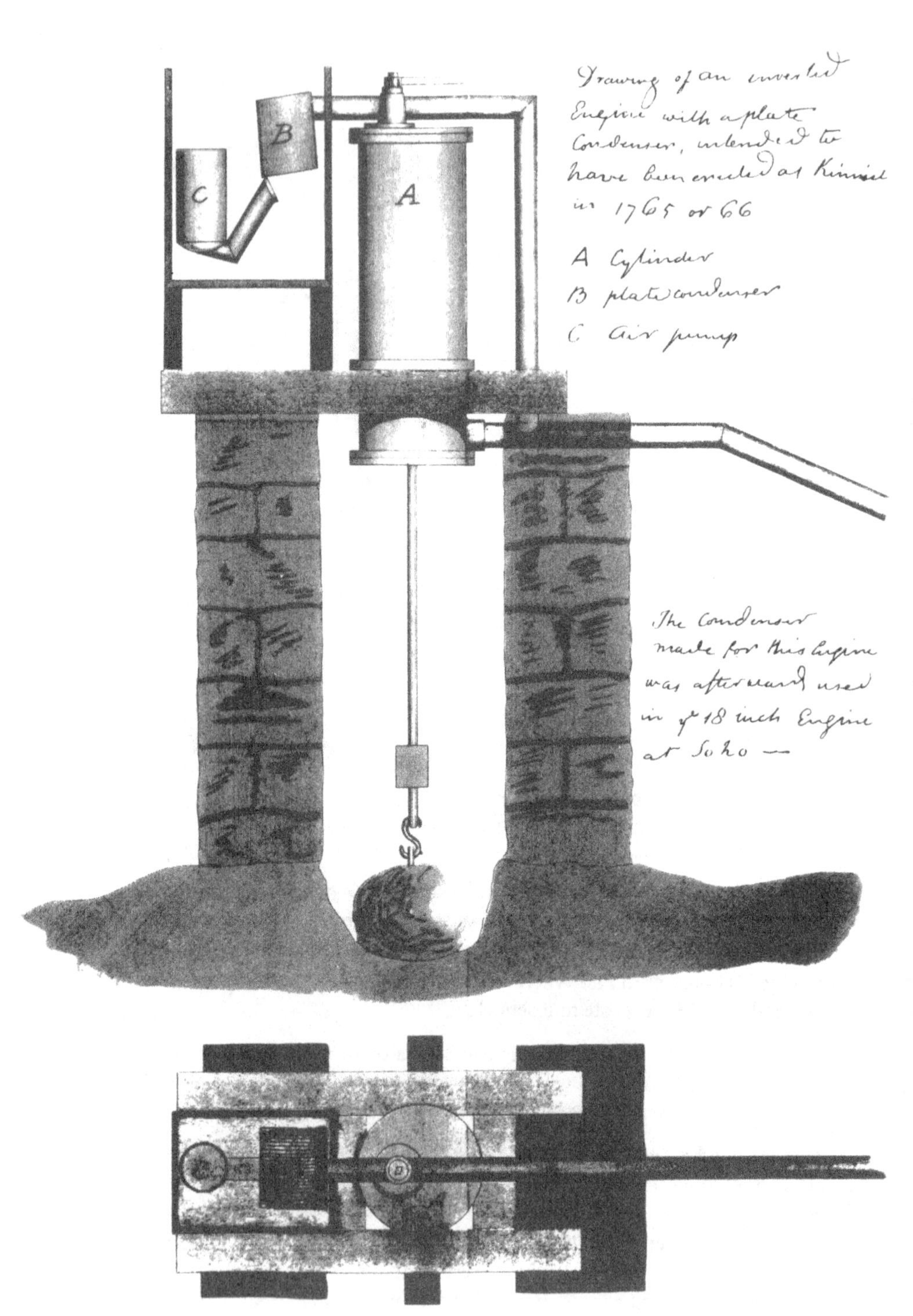

Figure 15 A drawing of one of Watt's early engines with an inverted cylinder lifting a weight directly. (Muirhead, *James Watt*.)

November 1768 was more conventional, for it had a beam, but this was off centre, for the stroke of the pump was 6 ft but that of the steam piston only 4 ft.[12] In April 1769, Watt received a new cylinder, 18 in diam. and 5 ft stroke but found it was not bored accurately. A drawing, possibly for the 1769 patent, shows a beam with linkage to the air pump but still no automatic valve gear.[13]

There is the possibility from comments in Watt's letters that the condenser air pump was at first worked manually. In February 1769, he proposed driving this pump by a waterwheel.[14] If indeed these first experimental engines were controlled manually, they must have been very difficult to operate for it is not easy to regulate the formation of the vacuum and work the valve gear. The purpose of the waterwheel may have been to create the vacuum first and then try to work the engine. While there were great advantages in using a surface condenser because the volume of water to be pumped out was less than with a jet type and also no air could be drawn in through a water jet, there were many practical difficulties in both constructing and operating a surface condenser. This feature was retained on the first few engines supplied commercially,[15] but was quickly replaced by the jet type. Surface condensers were not fitted on steam boats until 1831. Even then, they gave trouble with the different rates of expansion of the various parts so they leaked and it was found that they needed forced circulation of the cooling water.[16]

Watt's surface condensers were immersed in a tank of cold water which presumably he connected to the burn at Kinneil. When later he changed to jet condensing, he still surrounded the condenser with cold water. This idea was sound for it prevented air being drawn in through any leaks in joints but the water for the jet was drawn out of this cistern. The hot steam coming into the condenser also warmed the surrounding water so a lot of cold water had to be pumped through continually to keep it cool. Even so, the surrounding water sometimes became too hot to maintain the vacuum properly. Farey recognised this problem when he recommended that the injection valve should be placed low in the cistern where 'it will take the water in the coldest state'.[17] This difficulty was experienced with the Watt-type rotative engine in the science museum in Manchester and we realised a better design would have been to have a separate cold water tank to supply the injection directly.

Watt experienced all sorts of other difficulties with his Kinneil engines. One major problem was making the piston steam-tight without excessive friction. Even on his first model, Watt had used steam from the boiler instead of the atmosphere to push the piston along. This gave him two advantages. First it meant that no cold air entered the cylinder and so the efficiency was improved. He also encased the cylinder at a later period in a steam jacket to help keep it hot. Secondly, the use of steam would lead to more powerful engines which could be more easily controlled. The pressure of steam could be regulated and of course later was increased far above atmospheric pressure. On the other hand, Watt had now to seal his piston against a gas and not a liquid, e.g. the water which Newcomen had poured onto his piston. When we examine the numerous patents taken out over the next 150 years for improvements to pistons, piston rings and packings, it is not surprising that Watt had difficulties making his badly bored cylinders steam-tight. Cork, paper, paste-board, leather, hemp ropes, all were tried to no avail. Watt's engine, while still showing signs of fulfilling its early promise, was still not a commercial proposition and work on it came to a halt.

Meanwhile, Roebuck's financial affairs were becoming desperate. In March 1773, he was no longer able to meet his obligations and eventually was declared bankrupt. He had not honoured some of the payments he had agreed with Watt who then accepted the Kinneil engine instead. At this point, Matthew Boulton reentered the scene. As Watt was returning from taking the oath for his patent in London, he stopped in Birmingham during the autumn of 1768 where he met Boulton. By this time, Boulton's Soho Manufactory had become famous for the high quality of all sorts of small metal objects such as fancy buttons, steel watch chains, plated wares, etc., which were produced there. But the Hockley Brook provided insufficient power, particularly during the summer months. We have seen how Boulton had been considering some form of supplementary steam engine but, when he heard about Watt's experiments, he delayed until Watt had perfected his. Boulton was also interested in joining the partnership and made his famous offer to Watt.

> To remedy which and produce the most profit, my idea was to settle a manufactory near to my own by the side of our canal where I would erect all the conveniences necessary for the completion of the engines and from which manufactory we would serve all the world with engines of all sizes. By these means and your assistance, we could engage and instruct some excellent workmen (with more excellent tools than would be worth any man's while to procure for one single engine) and could execute the invention 20 per cent cheaper than it would be otherwise executed and with as great a difference of accuracy as there is between the blacksmith and the mathematical instrument maker.[18]

While this suggestion was not taken up at that time, Boulton remained interested and eventually was able to acquire Roebuck's two-thirds share of the partnership in settlement of his debt. Then Boulton proposed that Watt should move both himself and his engine down to Birmingham and the Soho Manufactory where Watt arrived with his family on 31 May 1774 and resumed his steam engine experiments.

The Kinneil engine was reerected at Soho but its tin cylinder soon collapsed.[19] A cast-iron one was ordered from John Wilkinson of Bersham who had just built a new boring machine which was more accurate than any other. Earlier, Smeaton had designed a boring machine for Roebuck's Carron Foundry which gave a truly circular hole, but, as the boring head was supported at one end only and rested on wheels, it could deviate from the correct alignment as it passed through the cylinder. Wilkinson's boring bar was supported at both ends and was advanced by a rack along the cylinder. This new cylinder from Bersham solved, for the time, the problem of steam leaking past the piston, and the engine became a success at last. Boulton insisted that the period of the 1769 patent should be extended and this was granted for another 25 years by Parliament on 22 May 1775. The famous partnership of Boulton & Watt was formed on 1 June.[20] Their first pumping engine was set to work at Bloomfield Colliery, Tipton, Staffordshire, on 8 March 1776.

It was probably soon after this in 1778 that the idea of putting a valve to cut off the steam from the boiler as it went into the top of the cylinder was tried.[21] The steam that had filled the cylinder on the working stroke was passed round to the underneath of the piston on the return up-stroke before being finally condensed on the next down-stroke. The next stage was to use the steam expansively. As early as 1769, Watt had written to William Small about his rotary steam wheel,

> I mentioned to you of still doubling the power of the effect of the steam . . . by using the power of steam rushing into the vacuum at present lost . . . for open one of the steam valves, and admit steam until one-fourth of the distance between it and the next valve is filled with steam, shut the valve and the steam will continue to expand and to press round the wheel with diminishing power, ending in one-fourth of its first exertion.[22]

He made actual experiments on his pumping engines in 1776 and included the idea in his patent of 1782 but steam pressures were too low for any great economies to be made.

What had Watt achieved? An atmospheric engine of the old style which was considered one of the best in the district was tested by Smeaton in 1772 at Long Benton Colliery near Newcastle-upon-Tyne. It was found to be capable of a duty of 5,044,158 lb. The duty of Smeaton's replacement engine was 9,636,660 lb. In 1778, Smeaton tested one of Watt's engines with a 20 in diam. cylinder and 5 ft 9 in stroke and found it had a duty of 18,902,136 lb.[23] Watt had roughly trebled the efficiency of the old atmospheric engine with his separate condenser.

So far, Watt had replaced only Newcomen's original cylinder with his own and added the separate condenser. The engine remained suitable only for pumping water. In 1769, Watt had included in his patent specifications ideas for a direct rotary engine or 'steam wheel' as well, but, although he spent a great deal of time on it, he never achieved success and neither did many others until the advent of the steam turbine over 100 years later. During the 1770s, the demand for rotative engines to drive mills directly increased and other people started to build them. In his 1740 patent, Wise had suggested 'a fly [to keep] the said shaft in its proper motion to make it circular'.[24] This idea seems to have been forgotten, and many people, including Watt, did not understand the potential of the flywheel, 'the regulating power of which I did not then fully appreciate'.[25]

We have discussed already how the piston on the Newcomen engine paused at the top of its stroke while the vacuum was being created and how irregular its action might be. The counterbalance weights might return it at a different speed to its downward movement, and then there were periods of no power at the beginning and end of its movement to be considered as well. The origins of a successful rotative engine start with Matthew Wasborough's patent in 1779. He applied

> pullies, wheels, or segments of wheels to which are fastened rochets and clicks or palls, so contrived that while the engine or machine moves in a perpendicular, horizontal or diagonal direction, that the rotative motive shall be effected . . . Likewise in some cases I add a fly or flys, in order to render the motion more regular and uniform.[26]

It was thought that this was the first time that a flywheel had been added to a steam engine and several were built; one in Wasborough's own works at Bristol, another at Southampton, and one for James Pickard at Birmingham. Farey says that these engines 'were subject to such irregularities as rendered them of little use'.[27]

It was Pickard who replaced the 'rick-racks' with a crank and produced quite a successful engine which he patented in 1780. He adapted a beam engine and turned the pump rod into a connecting rod. In his patent, he said that the connecting rod could be linked directly to the piston but just how is not described.

> When a regular uniform motion is required, a smaller wheel F, G, is applied, which, working with the large wheel C, D, and having just twice the number of coggs or teeth, consequently makes just twice the number of revolutions; about its axis or centre. A weight H, is also applied . . . such . . . that the said weight H, will always descend when the crank and the spear [connecting rod] are parallel.[28]

What Pickard was trying to do was to use a counterweight to carry the piston over the dead centres of the engine. He soon substituted Wasborough's flywheel which made the counterweight unnecessary. It was this combination of Wasborough's flywheel and Pickard's crank which led to the development of a successful type of single-acting rotative atmospheric engine which answered better than any tried before.[29]

On a pumping engine, the timing of the stroke was controlled by the valve gear, or the 'regulators' as the valves were called at least from the time of Desaguliers, for the next stroke would not start until the admission valve opened. This was not so on these new rotative engines, where the inertia of the flywheel kept the engine turning and the regulators were controlled by the rotation of the engine. Provided they did not open too soon, the engine would run once it had been started even if the valves were not set in their optimum positions. The slow movement of the piston at the beginning of the stroke through the action of the crank gave more flexibility in the timing and period when the vacuum could be formed. It was probably these points that Watt had not realised when he thought of incorporating a flywheel.

In 1795, it was claimed that most of the steam engines in and about Manchester were this simple, rotative atmospheric type, built by the principal engine builder of the area, Bateman & Sherratt. They were described at that time as being

> In general of a small size, very compact, stand in a small place, work smoothly and easy, and are scarce heard in the building where they are erected.[30]

Figure 16 Single-acting rotative atmospheric engine. (Farey, *Steam Engine.*)

These engines had become so numerous even by 1789 that Peter Drinkwater, when inquiring about an engine from Boulton & Watt, said he would

> have wrote to you much sooner on this subject had I not been incommoded on all sides by threats of a prosecution for erecting a nuisance – & indeed the prejudice is not yet much – if at all abated – the fact is we already have a great number of the common old smoaking Engines in & about the Town which I confess are far from being agreeable – & the public yet are not all inclined to believe otherwise than that a Steam Engine of *any sort* must be highly offensive.[31]

From the millowner's point of view, the greatest problem with these engines was that they had only one power stroke, when the piston was descending. The crankshaft had to be given enough speed in that stroke to raise the piston back up to the top of the cylinder again. If this did not happen, there was a danger that the engine might not get past top dead centre and might even turn backwards with disastrous consequences to the cotton spinning. In order to give some power when the piston was ascending, the connecting rod might be made into a counterweight equal to half the descending force of the piston, so that the force on the crankpin was equal ascending and descending. This, of course, worked properly only when the load was constant, and these engines, which rotated at quite high speed, ran fairly well. However, they were not so suitable for conditions where the load was changing continually.[32] To try and avoid infringing Watt's patent for the separate condenser in the years before 1800, condensing was done in the pipe just beyond the cylinder and a simple type of open-topped air pump fitted, which, being nearly as large as the cylinder and being placed between the pivot of the beam and the crank, helped to return the piston to the top of its stroke through the atmospheric pressure on the surface of the air pump bucket or piston.[33] These engines 'are now used in cotton mills, and for every purpose of the waterwheel, where a stream is not to be got'.[34]

Seeing that other engine builders were developing successful rotative engines, Boulton urged Watt to turn his inventive powers to producing one for their partnership. This was, after all, the reason why Boulton originally had been interested in Watt's patent. In 1781, he wrote,

> The people in London, Manchester and Birmingham are *steam mill mad*. I don't mean to hurry you but I think . . . we should determine to take out a patent for certain methods of producing rotative motion from . . . the fire engine.

Ever the business man, in another letter he wrote,

> There is no other Cornwall to be found and the most likely line for the consumption of our engines is the application of them to mills which is certainly an extensive field.[35]

Watt had been pondering upon various solutions and submitted them in a patent specification on 25 October 1781.[36] The title of the patent reads

> for certain new methods of applying the vibrating or reciprocating motion of steam or fire engines, to produce a continued rotative or circular motion round an axis or centre, and thereby to give motion to the wheels or mills or other machines.

The word 'continued' has been underlined because the five ideas drawn out by Watt were all concerned with smoothing out the uneven power cycle of the atmospheric or his own single-acting engine.

In his first method, the engine rotated a 'swashplate' by lifting a beam underneath it which fell back again through the heavy weight of the arch head and so raised the steam piston at the other end of another beam.[37] His eccentric device also was returned to its starting position by the descent of a weight fixed to it.[38] In his third method, he used what was, in effect, a heavy unbalanced crank but tried to conceal it in a flywheel.[39] His fourth method followed a similar idea but with two single-acting clyinders.[40] This should have given a smoother action because the steam cylinders each powered one-third of a rotation and a weight the remaining third. Even Watt's famous 'sun and planet' motion, his fifth method which he was to develop later in his first rotative engines to avoid trespassing on Pickard's patent, was conceived originally as a way of obtaining a smooth rotary motion where

> the gravity of the wheel D, or of the connecting rod AB, or of any other weight connected with them, causes the wheel D to descend on the other side of the wheel E, and thereby continues the motion it had impressed upon it, whereby the wheel D completes its revolution round E.[41]

In his fully developed rotative engine, Watt abandoned the weight for this would act best only against one load and it was superseded by the double-acting cylinder and flywheel. The sun and planet motion gave the additional advantage that the output shaft rotated at twice engine speed. This was very desirable in textile mills where the line-shafting had to be speeded up and also gave the flywheel better characteristics in ironing out variations in the movement of the piston.

Similarly, in his next patent in 1782, Watt did not use its most important features in quite the same way as he outlined them in the specification. He claimed novel ideas for 'new improvements upon steam or fire engines for raising water and other mechanical purposes'.[42] His first objective seems to have been to give the pumping engine a smoother and more economical cycle. One of the problems worrying people was that, once a sufficient vacuum had been created to overcome inertia and to start moving the mass of engine parts such as the beam, the pump rods and the water, atmospheric pressure, or in

Watt's case steam, would continue to accelerate this mass for the duration of most of the stroke and the piston might smash the bottom of the cylinder.[43] Most of the ideas suggested in this patent consisted of mechanical contrivances which would increase the resistance at the beginning of the stroke with a false load to slow it down but which would transfer this load to give assistance at the end of the stroke and even out the power.

These contrivances were to be employed in conjunction with using the steam expansively in the cylinder. In the Newcomen engine, the steam valve had to remain open for practically the entire ascent of the piston, really so that the incoming steam could reheat the cylinder walls. In the Watt engine, this was no longer necessary because the cylinder was hot already. This enabled Watt to consider using steam expansively which gave greater economy at the expense of loss of power and also at the expense of uneven power. The uneven power resulted from the pressure on the piston being greater at the beginning of the stroke than at the end so that Watt proposed

> applying combinations of levers, or other contrivances, to cause the unequal powers wherewith the steam acts upon the piston, to produce uniform effects in working the pumps or other machinery required to be wrought by the said engine: whereby certain large proportions of the steam hitherto found necessary to do the same work are saved.[44]

If the valve admitting steam to the cylinder were shut part way through the stroke, the steam would continue to expand but its pressure would fall. Watt followed Boyle's law to determine the pressure drop which may not have been so far off the mark as might be expected because his cylinders were steam-jacketed and his low-pressure steam was being reheated continually. He found that

> It appears that only one-fourth of the steam necessary to fill the whole cylinder is employed, and that the effect produced is equal to more than one-half of the effect which would have been produced by one whole cylinder full of steam, if it had been admitted to enter freely above the piston during the whole length of its descent.[45]

While he had halved the power, he had quartered the fuel consumption. By reducing the pressure at the end of the stroke near to that of the vacuum in the condenser, he had also reduced the temperature, a point that was partially masked by the jacketing, and so produced a thermodynamically more efficient engine. We will return to this point in our discussions later.

Some of the mechanical contrivances suggested by Watt for equalising the power stroke in his expansive engine appear again in one form or another for operating the valve gear in the Corliss and Porter engines 60 or more years later! In this 1782 patent, Watt also suggested using the power of the steam to push the piston both up and down, in other words a double-acting engine. Here he is

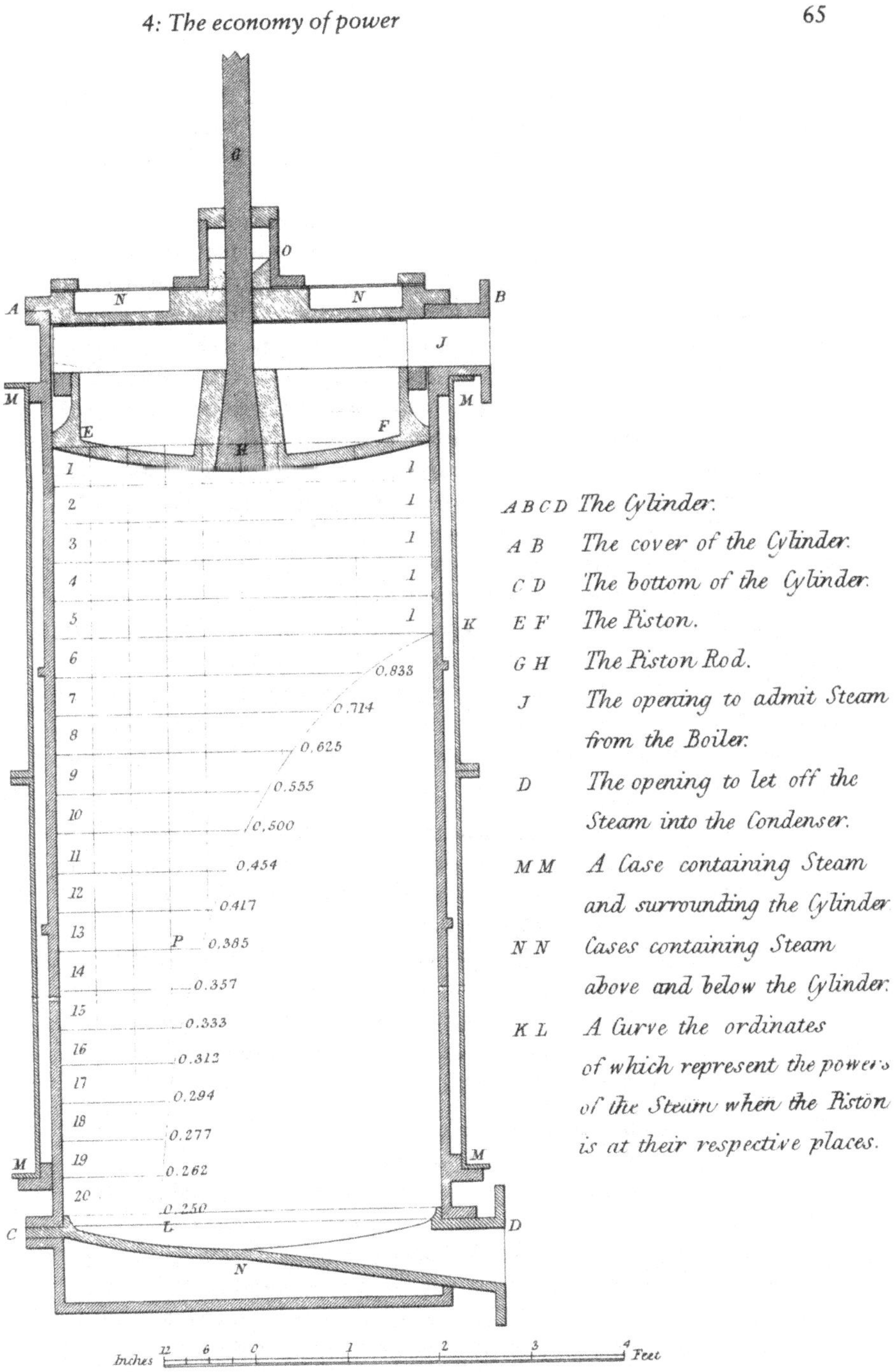

Figure 17 Watt's diagram showing how the pressure of steam falls in a cylinder after early cut-off. (Muirhead, *James Watt*.)

patenting a much earlier idea because there is a drawing of a double-acting engine amongst the papers submitted in 1775 to Parliament for the extension of his 1769 patent.[46] He realised that he could thus double the performance or power in the same time with a cylinder the same size.

It is possible that Watt set out to conceal what he was really trying to do because this double-acting engine shown in the patent specification was pumping water which needed only a single-acting one. He fitted a heavy weight in the arch head at the cylinder end to balance the pump rods, 'which weight ought to be equal, or nearly so, to the force or power of the steam when acting in the ascending direction'.[47] In this way, he really negated the whole purpose of a double-acting engine. The drawing shows a rotative engine with a sun and planet gear and a connecting rod to drive a flywheel as well as pumps operated through an arch head and chain. The chain at the other end of the beam was replaced by gear teeth, or a sector, on the arch head and a rack on the end of the piston rod. The rack and sector was another part of this vital patent.[48] Remove the water pump and we have the first type of rotative engine which Watt set up at the Soho Manufactory during the later part of 1782.[49] The first rotative engine erected elsewhere was a single-acting one put up for Wilkinson at Bradley to work a hammer which was running by the end of March 1783.

To obtain a more even turning force in atmospheric engines, other people had proposed using two cylinders, working alternately. In 1779, Falck tried to solve the problem of rotary motion by connecting the piston in each of two cylinders to a rack pushing round a gearwheel.

> The improvement which he suggests is to use two cylinders into which the steam is admitted alternately by a common regulator-valve, which always opens the communication of the steam to one cylinder, whilst it shuts up the opening to the other.
>
> The piston rods of these two cylinders are formed with teeth like racks, which work into teeth at opposite sides of a cog-wheel, situated between the two racks, in such a manner that the pistons which are attached to them shall ascend and descend alternately, but they will always move in opposite directions to each other.[50]

The common gearwheel between the racks oscillated and was attached to a beam. At one end of the beam was a connecting rod which operated a crank and a flywheel. Several were built by Bateman & Sherratt for the Manchester area.

This firm was later to produce another type of atmospheric engine which Farey considered to be the best design of double-acting engines of this arrangement because it was quite compact and the weights of the reciprocating parts balanced each other.[51] In June 1789, one was installed at the cotton mill of Joseph Thackery and John Whitehead at Garrat on the river Medlock to replace the earlier Savery engines and it continued in use for over 30 years.[52]

Bateman & Sherratt were involved with Thompson who patented another type of double-acting atmospheric engine in 1792.[53] His engine had two cylinders, one above the other, and the piston rod, which of course had two pistons, passed through a stuffing gland in the top of the higher cylinder. The open ends of the cylinders faced each other so that the atmosphere could force the lower piston down and the higher one up, with the result that a double-acting engine was achieved. The piston was attached to the arch head at the end

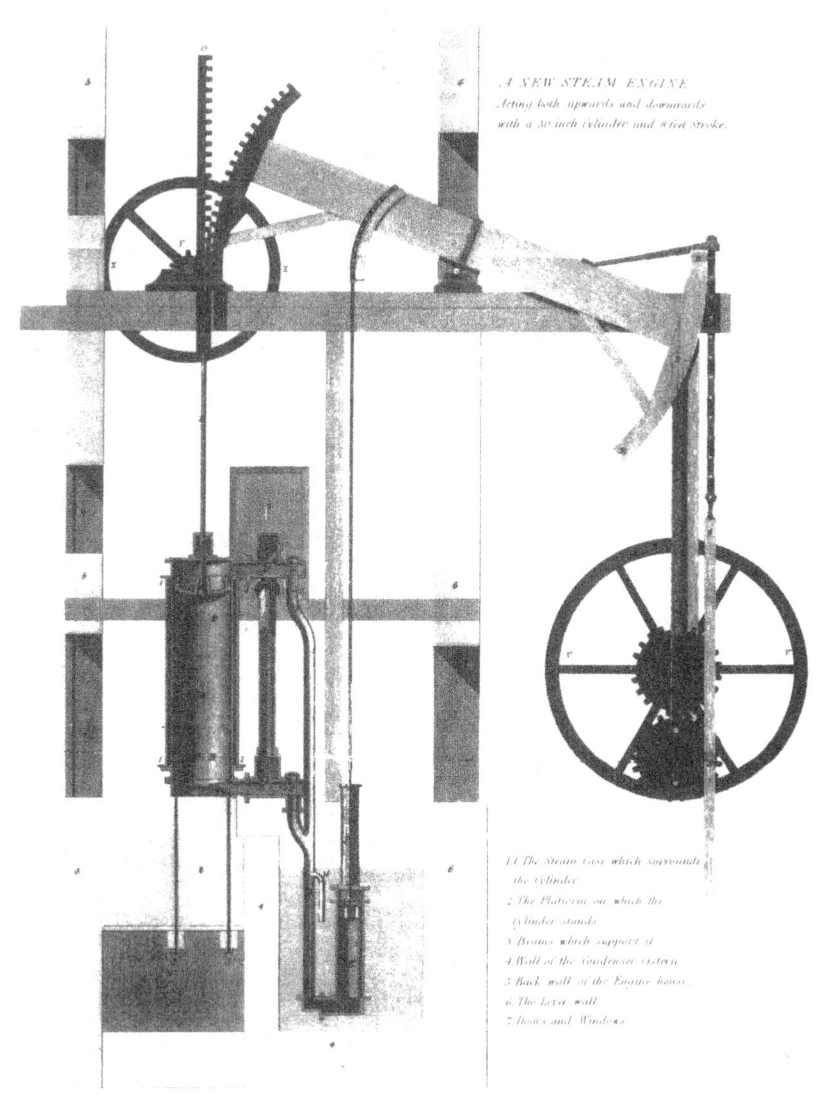

Figure 18 Watt's rack and sector connecting the piston to the beam and his sun and planet gear for giving rotary motion. (Muirhead, *James Watt.*)

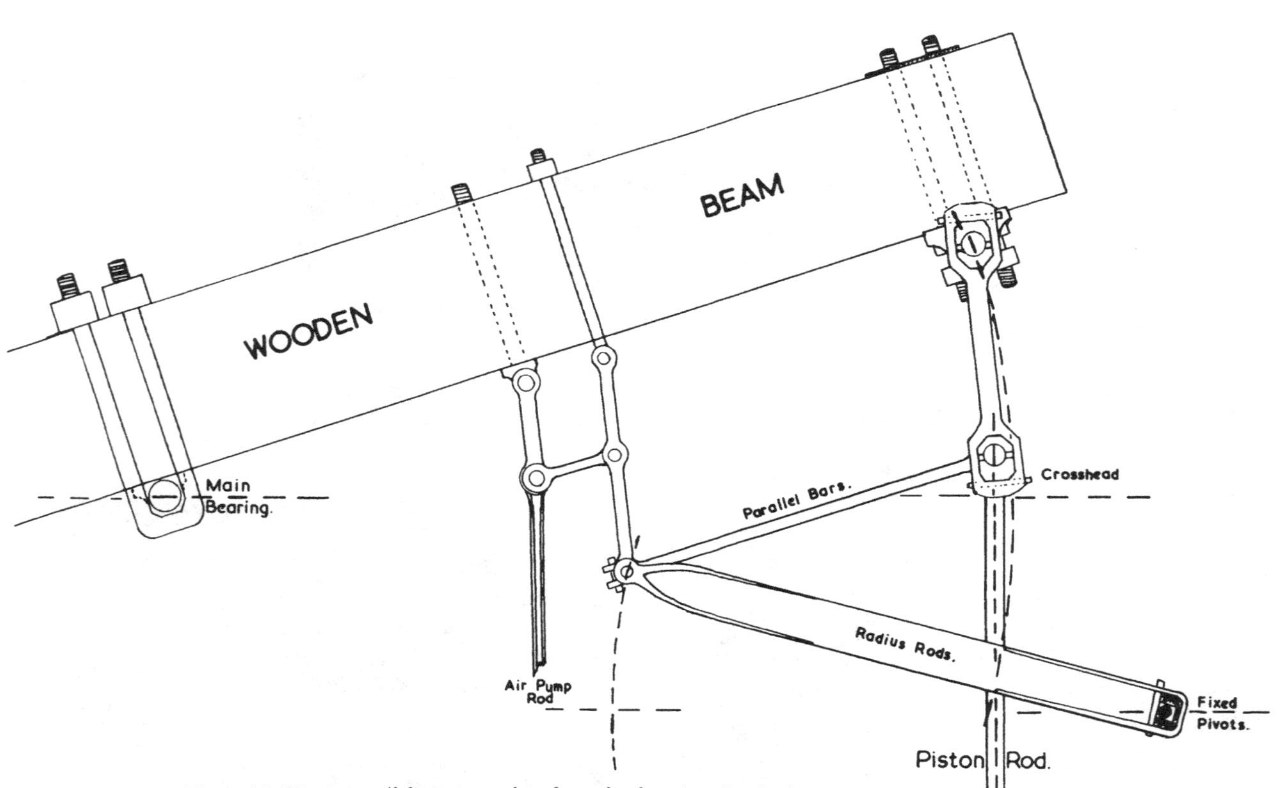

Figure 19 Watt's parallel motion, taken from the drawings for the first of Robinson's engines at Papplewick. (Boulton & Watt Collection.)

of the main beam by a double chain arrangement. John Kennedy had one of these Thompson engines at his Ancoats cotton mill and T. Houldsworth had another at the Leavers Street cotton mill, both in Manchester. Both were built around 1793, but the type never achieved much popularity. This was due partly to the slump in that year, but they were found to be inadequate for driving the larger mills because they did not run smoothly. Davison & Hawksley, of Arnold near Nottingham, had Watt replace the Thompson cylinders with one of his patent design after only three years use and found that they cut their fuel consumption by half.[54]

In the meantime, Watt had patented the final link in the evolution of the rotative engine. On 30 June 1784, he wrote,

> I have started a new hare! I have got a glimpse of a method of causing a piston rod to move up & down perpendicularly by only fixing to it a piece of iron upon the beam, without chains or perpendicular guides or untowardly frictions, arch heads or other pieces of clumsiness... I think it a very probable thing to succeed, and one of the most ingenious, simple pieces of mechanism I have contrived.[55]

Within a few months, Watt had developed his original three bar motion into the full parallel motion. The advantages were that the piston rod could be directly joined to the end of the beam moving in a straight line. The parallel motion was economical in space for it could be contained within the usual framework of the engine.[56] Additional links could be fitted to it to give as many straight line motions (of different lengths) as were needed. So Watt joined up both his main piston rod and the air pump rod while later Arthur Woolf could add another piston rod.

It seems as if once more Watt tried to conceal the importance of one of his inventions, for, when he patented the parallel motion in 1784, it was only a small part of another omnibus patent covering a steam road carriage, a rotary engine and at least three other principles for obtaining straight line motions.[57] The parallel motion was a simple and elegant solution for producing a rotative engine well within the engineering capacity of that period. It set the firm of Boulton & Watt firmly on the road to fame and fortune. No wonder Watt wrote in 1808,

> Though I am not over anxious after fame, yet I am more proud of the parallel motion than of any other mechanical invention I have ever made.[58]

5
The devil of rotations

> 'Look round the metropolis', exclaims Sir Humphrey Davy, 'our towns, even our villages, our dockyards, and our manufactories; examine the subterraneous cavities below the surface, and the works above; contemplate our rivers and our canals, and the seas which surround our shores, and everywhere will be found records of the eternal benefits conferred on us by this great man'.[1]

The Boulton & Watt rotative engine made a dramatic impact on the standard of civilisation and was one of the crucial machines which helped to launch the Industrial Revolution. This engine became the standard design for providing rotative power everywhere, even for a short period on board ships. It was a design which other manufacturers found they had to copy or else face failure. Some tried other types, but

> In many instances, the makers were obliged to give up the pursuit, after having made a few engines. Others who had better means of execution, and who took care to study Mr. Watt's models very closely, succeeded so far as to establish themselves in the business.[2]

Even the Watt engine needed many years of trials and modifications before it became the paragon and envy of everybody.

The total number of engines built by the Boulton & Watt partnership up to 1800 was 496, of which 38 per cent were pumping and 62 per cent rotative, mostly for the textile industry. There were 164 pumping engines, 24 blowing engines and 308 engines driving machinery.[3] This shows that Boulton had been correct when he advocated concentrating on the rotative type. One of the earliest, and possibly the most important, of the early rotative engines was ordered by Samuel Wyatt in July 1783 to drive the machinery in the Albion corn mill on the banks of the Thames near Blackfriars bridge. In the previous year, Boulton had been in touch with Wyatt's elder brother, William, about building a steam mill in London, obviously as an advertisement for their engine. Samuel Wyatt was the architect of the Albion Mill[4] and, in June 1783, went to Birmingham where he saw the 18 inch rotative engine at work. He was convinced that it ran smoothly enough for grinding corn and that a larger one could be applied to drive a commercial flour mill.

Both Boulton and Watt took part shares in the Albion Mill which was to be on a scale hitherto never contemplated. The largest mill in London at that time had four pairs of stones but it was proposed that there should be 30 at the Albion Mill. To drive them, three engines with cylinders 34 in diam. would be required. The first engine began regular work driving six pairs of stones in April 1786. Early drawings show that it was designed with the rack and sector arrangement to connect the piston rod to the beam. However, in September 1784, a scheme was prepared incorporating the original form of Watt's parallel motion. Whether this engine or one in Hull was the first fitted with the parallel motion has been disputed.[5]

The person responsible for the erection of the Albion Mill engine and the millwright work connected with it was John Rennie. He had studied at Edinburgh University under Robison and Black, both of whom had been associates of Watt at Glasgow. Rennie was trained as a millwright by Andrew

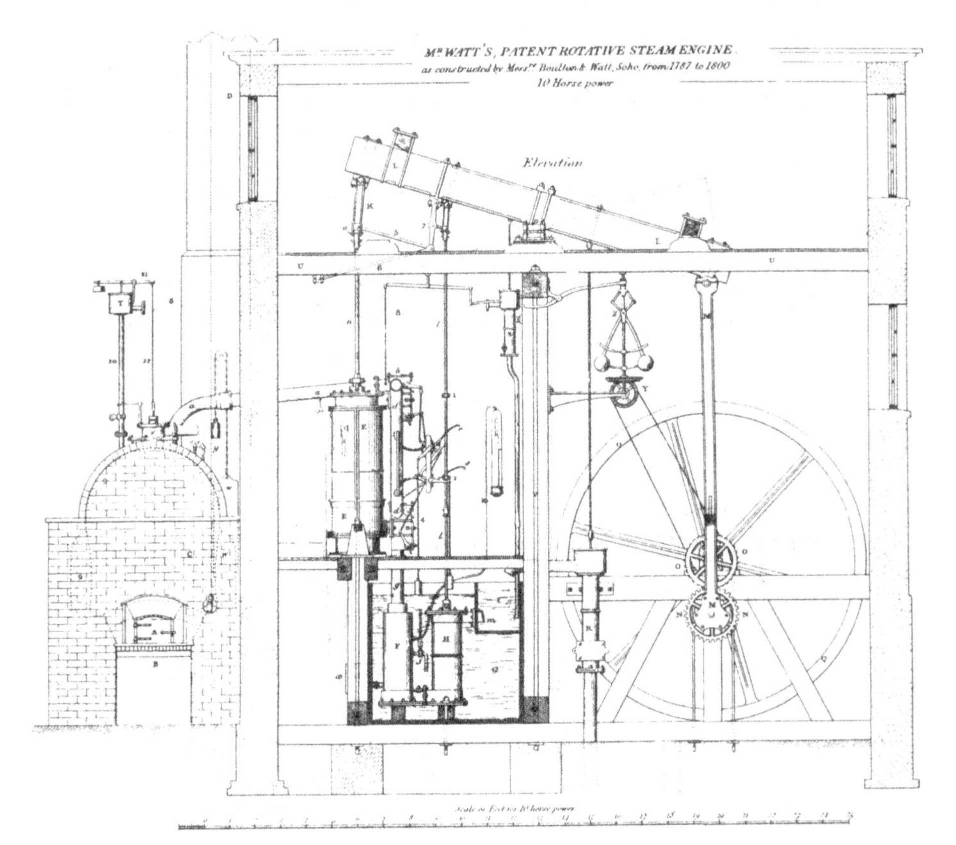

Figure 20 Early Boulton & Watt engine with wooden framing, sun and planet gear, governor and wagon boiler with automatic feed water controls. (Farey, *Steam Engine*.)

Meikle in Scotland. He was employed by Watt at the Albion Mill and it was largely due to his stubborn persistence that the machinery was set to work. He introduced cast-iron gearing as well as shafting, an important step in the history of engineering. The formal opening ceremony was held on 9 March 1786, in front of a large crowd of spectators, and the mill at once became a rendezvous for the curious and serious alike. A second engine was erected in 1788 but the whole venture ended in a spectacular fire during the night of 2 March 1791, when the mill was totally destroyed. It had not been a successful financial venture, but did establish the reputation of the Boulton & Watt rotative engine.

Rennie was involved also with the first Boulton & Watt rotative engine which powered a textile mill. In June 1785, the Robinsons of Papplewick, just to the north of Nottingham, sent Boulton & Watt details and dimensions of the water-wheel in one of their mills because they wanted a steam engine to help power the mill. During the early 1780s, the Robinsons built a series of seven cotton spinning mills, with a complex arrangement of dams and aqueducts to supply them. The valley here is relatively flat and the River Leen quite small so the highest mill, Castle Mill, was supplied directly from a reservoir. The Robinsons had the highest mills on the river, so they might have been expected to have had an uninterrupted flow of water, but, above them, lived Lord Byron in his estate of Newstead Priory.

This Lord Byron loved making ornamental ponds on his lands where he held mock naval battles with his servants. The Robinsons found that Lord Byron was tampering with their water supplies so much that they were not able to work their mills. They ordered their steam engine, one of 10 h.p., in case the pending law suit went against them. A letter in August 1785 from the sons James and John, reveals their predicament.

> We can not nor will not set about any of the Buildings etc until we hear from you & the time is going on & we are all asleep, I beg you will give due attention to our situation & lett Lord Byron see that we can do without him.[6]

The reply from Boulton & Watt shows that they were trying to hurry things along and had put Rennie in charge of the designs and plans.

> My absence from home prevented me from writing to you on Mr. Rennies letter & plan coming here . . . You may immediately set about the little additional Building marked out by Mr. Rennie . . .
>
> I have repeatedly pressed Mr. Wilkinson's people to give your things all the dispatch at the foundry which they could, but I have not yet heard of their being sent off – the parts which we are making here we are forwarding with all the dispatch we can, but . . . things take time.[7]

Ill fate seems to have dogged this engine from the start, but its misfortunes do show some of the problems Boulton & Watt had to overcome to establish the rotative engine and their steam engine business. As no parts had arrived at Papplewick by the beginning of October, the father, George, wrote to Birmingham,

> We have got everything done here to your directions & we have been in full expectation for some time past of hearing from you & that the weighty parts of the Engine had been sent off & to our great Disappointment hear nothing about them from you, I am in some doubt whether you have not forgote our situation, as we expect next month Lord Byron intends to give us some trouble.[8]

Watt replied two days later (a good comment on the postal service in those days),

> I have not forgot you; but have done all I could to push your work forward. I hear from Bersham that your cyl^{d}. & condenser were sent off to the canal a week ago and I hope are now arrived with you. The things here are going on as fast as they can. The matter which is most behind is the boiler, in plates for which we were long disappointed & afterwards found some difficulty in procuring a man to make it such workmen being very scarce here. We have at last got one, he is at work upon it and will not be long about it.[9]

The castings from Bersham failed to arrive and it was not until the middle of December that they were located at Manchester.[10] The engine erector did not arrive until after that Christmas to start installing the engine.

However, by the end of February 1786, the engine was finished at last.

> This morning R^{d} Cartwright left us – the engine being finished & set to work on Saturday [trouble with the boiler flues delayed its regular working] . . . It is only doing justice to R^{d} Cartwright to inform you that on the whole he has behaved himself very well & stuck to business & appear^{d} to be anxious for the work being well done.[11]

The immediate need for the engine was soon removed because the law suit against Lord Byron 'was happily given in our favour last week at Nottingham'.[12] The engine was not used very much for it was not properly aligned with the rest of the mill.

> We have waited some time in writing to you in expectation of the recovery of our Clerk and Engine Smith – who are now both Dead – wch will oblige us to send you another Smith for your instructions – as soon as we can meet with one . . . for by the illness of our two managers of the Engine she has stood still since August.
>
> It is unfortunate for us that she is not square with our works wch occasions a considerable degree of friction and some danger by the

> wheels working uneasy & requires a skilful hand to manage her. From the correctness of the Plans sent by you and Mr Renie setting out the foundation of the works with our Builder & Millwright – we had every reason to expect a most compleat piece of work – but it was not till Mr Harrison came over that it was discovered to be abt 7 in out of the square of our works . . . we fear no effectual remedy can be apply'd without removing the whole Foundation & Engine.[13]

Possibly, as a result of this experience, when the Robinsons found they needed to augment the power of the waterwheel at their Lower Mill, they had an atmospheric engine built at a cost, according to the insurance valuations, of £400.[14]

The subsequent history of this first Boulton & Watt textile mill engine suggests that, by 1790, people had begun to realise that this type was the best. After the death of his father, John Robinson inherited the Castle Mill with the original steam engine and prospects of increasing trade made him write to Boulton & Watt in that year,

> The Steam Engine that was erected here a few years since when we apprehended a stoppage of the Water I wish to apply again that less water may be used and to turn an Increase of Works . . . My brother . . . has some Intention at erecting one at another of the Mills.[15]

His brother James did install another 10 h.p. Boulton & Watt engine in the same year, also to supplement his waterwheel, but both engines may have ceased working during the slump of 1793–4 because in 1797, James wrote,

> Our trade appears to be in a reviving state and if we should succeed in obtaining peace I have little doubt of an increased demand & encouragement to carry on my works to their former Extent.[16]

He wanted to know what Boulton & Watt would charge if he and his brother commuted the unexpired portion of the patent, because their engines would work for only about twelve months during the remaining three years.

From this, it would seem that the Robinsons were reasonably satisfied with the engines, but a letter from Southern shows that Boulton & Watt did not consider that they had benefitted much from their first engine in a textile mill.

> They have rather to regret the erection of your first engine than to acknowledge the benefits they have hitherto received from it, if their time and judgement are worth a valuable consideration; and feeling the terms on which you had it granted to be unfavourable to their just interests, resolved soon after the experiment of yours, to make it a general rule in future to propose terms, such as have been lately handed to you, and which your contract is the only exception to in your whole country.[17]

One clue to the reason for Boulton & Watt's attitude may be found in a letter from Henry Pearson, another textile millowner, who wrote,

> I am told Mr. Robinson wont let anybody see his Engine, should not like to solicit him on that subject, please to say where you can recommend me to see one in some other place.[18]

A successful experiment in one mill would be copied by the managers of others in the area, and Boulton & Watt may have been disappointed that they could not use this engine as an advertisement for their rotative steam engines.

This account of the erection of the engine for the Robinsons has been covered at some length beause it shows the problems that faced Boulton & Watt, and indeed other engineering firms which followed in their footsteps. Boulton had envisaged making engines for the whole world, probably without fully realising all the problems in design, in control over manufacturing particularly when sub-contracting was involved, in erecting complex machinery far away from the parent company, in training skilled men to both erect and then to run the machines, in the continued maintenance and in an after-sales service. The world of mechanical engineering was then in its infancy and the Boulton & Watt rotative engine demanded higher standards in its manufacture and maintenance than anything else in its scale at that time. It was, after all, the most advanced piece of engineering and technology of its day.

At first, most people were totally unacquainted with any steam engines, let alone those of Boulton & Watt. John Kendrew, who had built a mill at Darlington, wrote in 1791,

> As we are totally unacquainted with Engine work in this part of the country, we hope you will be particular in your directions.[19]

Drinkwater, who was going to build a mill close to the middle of Manchester, had found himself in almost the same situation.

> As on the one hand I am almost an entire stranger to the Nature of your, or any other Engine, of this sort, & on the other hand the intended establishment taken altogether – will be rather expensive – you will excuse me being a little tedious and inquisitive at the outset.[20]

Other letters in the Boulton & Watt Collection at Birmingham show that some people had begun building and even equipping their mills before considering how to drive them.[21]

Anyone purchasing a Boulton & Watt engine had to meet three different installation charges before he could run his engine. First, the parts of the engine itself had to be paid for; then there was the engine house; and finally, there was the premium for the use of their patent charged by Boulton & Watt. The customer was left to arrange a large part of the installation himself and many may not have realised what was involved when Boulton & Watt stated,

> Our terms are to provide materials & Workmanship at the expense of our Employers, to furnish drawings & directions without any specific charge.[22]

Although Boulton & Watt did not charge for their advice or their drawings, the expenses of their men to erect the engine had to be met by the purchaser. Because of the demands of the purchaser and the fact that the situation of the engine house might vary so much from one site to another, Boulton & Watt never included the building in their estimates, nor any 'masonry nor the carpenter work of the house nor any part of the mill beyond the first rotative shaft'.[23]

Basically, Boulton & Watt did the calculations and sent the drawings. A letter to Drinkwater gave details of the work which Boulton & Watt would do for the 8 h.p. engine they were erecting for him during the later part of 1789.

> Messrs Boulton and Watt will undertake to furnish all the cast iron, hammered iron, brass, & copper work of your engine including the screw bolts of the cistern, fire door, grate barrs & other mounting of the boiler – the flywheel with its shaft (the latter not exceeding 10 feet in length) – but excluding the boiler itself, the hammered iron straps & screws for the framing & all brickwork – for the sum of Two hundred & fifty seven pounds delivered at Birmingham – They can furnish the boiler for a certain sum, but if Mr. Drinkwater chuses, they will procure it made as cheap as they can – Nor can they undertake the carpenter work nor agree for the expense of the workmen putting the whole together – But in their opinion these articles may be done within the prices below.

Copper boiler about	45
Woodwork of Engine & Iron framing	60
Workmanship in fitting up	45
	145
B. & W. as above	257
	£402[24]

What Drinkwater felt about having to provide so much of the engine himself is not recorded. Timothy Harris of Nottingham was told to send his millwright or engineer to Birmingham to have the drawings explained to him if he did not understand them,[25] but later his engine broke down and it appeared that some of the wooden framing had been erected too low.[26] The millowner had to supply not only the engine house but also the wooden framing and even the main wooden beam itself.

Up to 1795, Boulton & Watt normally had the cast-iron parts supplied by Wilkinson at Bersham. He could bore the cylinders more accurately than anyone else but the delays experienced by the Robinsons proved to be all too typical and this was one of the reasons why Boulton's dream of a steam engine manufactory at Soho became a reality in 1795. Land was purchased in Smethick, conveniently situated for transport by the side of the Birmingham and Wolverhampton Canal, at a spot about a mile distant from the original Soho Manufactory. The buildings erected comprised forge, smithy, boring mill, turning, fitting and carpenter's shops, drying kiln, foundry and air furnace. A wet dock from the canal was excavated for loading and unloading barges. The money for building and equipping Soho Foundry was advanced by Boulton and Watt jointly.[27]

In the days before the Foundry, the regulators or valves were always made at Soho and so were the other small iron parts. Transport of heavy bulky goods was difficult because the canal system was still in embryo, and the railway system non-existent. The parts of Robinson's engine being sent to Manchester instead of Nottingham were not the only instances of delay or loss in transit. Sometimes Boulton & Watt recommended that the less critical castings, such as the flywheel, be ordered from a local foundry, if a good one were close at hand. Boilers for the new engine at the Shudehill Mill were too large to be sent by canal through the Harecastle tunnel. It seems that they had been made by Wilkinson but had to be cut in pieces again so they could be sent by land more easily.[28] They leaked badly when reassembled in Manchester.

When the engine house had been completed and all the parts of the engine had arrived, the task of erection followed. This was the responsibility of Boulton & Watt who had their own engine erectors. Such skilled men were difficult to find and had to be trained since the rotative engines needed much greater accuracy than the earlier pumping ones. To help these men, in 1779, Boulton & Watt had printed *Directions for Erecting and Working the Newly-Invented Steam Engines*, complete with illustrations, and in about 1784 issued a supplement giving *Directions for Working Rotative Engines*.[29] Delays in installation were liable to occur when many engines were ready for erection at the same time through a shortage of trained erectors.

> I do not know how we shall be able to manage for Engineers to the Engines now nearly ready for putting up viz. Mr. Allms, Messrs Robinsons, the Chester engine, and one in this neighbourhood. If Allms materials were come to hand it is possible J. Law might do it & be time enough for Messrs Robinsons, though that is barely possible. B. Muir is in Yorkshire where he can not be done in time and does not understand rotatives. Could you light on some clever young fellow at London that would engage for moderate wages for 5 years if liked upon trial.[30]

These men had to be highly skilled because all the various parts of the engines were sent direct to the sites without any preliminary trial erection in Birmingham. The basic alignment was quite difficult because there were three main sections, the base of the cylinder, the bottom of the pillar supporting the main beam and also the bearings for the crankshaft. All these might be on different levels. The crankshaft had to be set at right angles to the main beam and dead level at the right position beneath it so the crankpin would run true in the big end bearing of the connecting rod. Boulton & Watt supplied a rotative engine in 1806 to William Balston for his papermill at Maidstone.[31] Its cast-iron beam has been preserved and has a spigot at the connecting rod end which allows the bearings for the rod itself to take up slight misalignment. The plummer blocks for the bearings of the crankshaft, main beam and parallel motion had two sets of holding-down bolts. One set secured a plate which gave rough alignment while the second set bolted the plummer block itself to the plate and adjustment could be made by wedges and wooden packing pieces. The cylinder was almost detached from the rest of the engine. Careful measurement and alignment were necessary to place it at the right height, so the piston would not touch either the top or the bottom cylinder cover and also so it was set vertically beneath the end of the beam. The parallel motion links would be set by gauges which again had to be aligned accurately to ensure that the piston rod ran smoothly through the stuffing gland in the top of the cylinder. Alignment of the cylinder was not quite so critical on the earlier pumping engines which had chains running over arch heads.

Not only the engine erector but also the man who subsequently looked after and ran the engine had to be skilled. The misfortunes which happened to Richard Gorton's engine at Cuckney near Mansfield show how much the mill-owners depended upon reliable enginemen. His 14 h.p. engine started running in June 1788 but gave continual trouble. Very soon afterwards, Gorton wrote,

> The Sun Wheels are constantly out of repair & do us much harm. Mr. Law [one of Boulton & Watt's erectors] authorises me to say that the segments are not as they ought to be and I think the maker should find me better.[32]

The trouble continued,

> It is much our misfortune if my Engine Man does not understand the Engine, for the Sun Wheels do their work badly. This day we were obliged to turn all our hands off while one of the segments was pieced which was broken, scarce a day passes without their breaking or threatening to break. Mr. Law has been several times – these frequent misfortunes are both alarming and very prejudicial.[33]

A new pair was despatched and Southern agreed that it was not the fault of the wheels but rather the man who managed the engine. He may well have been right, for, three months later, the engine suffered a serious breakdown.

> Our engine goes worse & worse every day, we are never sure of being able to proceed with our weaving – This week the working beam

> slip'd out of its centre. This morning we can hardly keep the steam up & it goes so irregular, that the weavers can do nothing.[34]

To this, Southern was able to reply, somewhat gleefully,

> I am extremely sorry to hear of the accidents you have been so frequently tormented with, and of the irregularity in the motion of your engine. They are such as very seldom occur anywhere else, and I am persuaded that they, in general, have proceeded from negligence of your engineer, if not from intention. The accident at the beam slipping out of the centre is such a one as I can not conceive to have happened, without the most gross neglect. I am sorry to impute a fault to any man, that I can not be positively certain he is guilty of, but from the disposition he has, which you informed me he has, of making alterations about the working gear, and other irish improvements (much for the worse) I must honestly say that I suspect he is the man that causes the mischief. Try to procure another, and let him not be suffered to come near the engine.[35]

From the subsequent history of this engine, Southern could have been wrong, for there was further trouble in 1791 when Richard Cartwright went to look at it.

> The cylinder is cracked 14 inches long, about 1½ inches below the top, the Cylinder cover has 3 cracks in it, the Piston is worn too small for the Cylinder. The 4 spindles [of the valves] are all worn too small and need a great deal of packing.[36]

Cartwright was told to put the engine in order, but he disgraced himself, for 'he minds his drinking more than his work and cares not what expense he puts us to'.[37] In spite of, or perhaps because of, Cartwright's drunken administrations, the engine went no better, and, in August four months later, Gorton wrote again,

> Our engine must be some way out of order, tho' it seems to do its work well, but it takes so much coal & so much packing of the Piston that we are sure an Engine to raise water would be infinitely cheaper to us. We have given over working in the night, because we thought it would save the packing & give us more time to keep the engine in repair but we are no better for it. If the piston be not packed every 3 days, it goes so heavy that it takes all the coal we can feed it with. We burn 2½ cwt an hour after it has been packed one day, something must be done, but with your Cart Man we had not patience, he certainly spent more time in the Ale house than at the engine, and we thought did more harm than good perhaps we did not give him time.[38]

Joseph Varley was sent to inspect the engine and he found that the cylinder was at least half an inch out of the perpendicular,[39] but this did not solve the problem for the piston still needed packing every day and a half. Apparently the engine was not overloaded for

> When the engine is freshly packed, we have plenty of Power to spare, what is the fault we can not tell, but a terrible fault it is. We have ordered the Piston to be packed every day – Pray send us a good Doctor, Mr. Southern should see it. At present we are paying a premium for using more Coal than the old Engine instead of paying for saving coal.[40]

A new cylinder cover was ordered and other repairs must have been made as the engine worked satisfactorily until Gorton decided to sell it in 1793 when he said that scarcity of coal deprived him of any profit.[41] Similar tales about incompetent enginemen and engines that were 'rogues' were related in the Lancashire mills for the next 150 years.

Boulton & Watt failed to recognise the need for an 'after-sales service'. Later, this became the mainstay of many a mill engine builder to keep the workforce occupied when orders for engines were scarce. Peter Ewart had been trained as a millwright in Scotland before moving south to join Boulton & Watt. Then he went to Manchester to establish his own business and also to look after the affairs of Boulton & Watt there but he relinquished this when he was offered a partnership in Samuel Oldknow's cotton mill in 1792. His final letter to Birmingham reveals the problems he encountered with their engines. Other engine erectors had to overcome the same sort of difficulties right up to the end of the steam era. Working all through the weekend including the nights was standard practice in order to have the mill running again for the Monday morning.

> I have had a great deal of trouble here with Engine Tenders [the men who ran the engines], finding it very difficult to make them keep the Engines in good order; and when anything is to be done about them, I find it much less trouble to do it myself than to give them directions or to stand over them while they do it and I think, that when the Engines now ordered for this part of the country are put up; any one person will find ample employment in keeping them in tune; not to mention the time & attention for the new erections; for which reason I think it would be much better for you to have some person to devote the whole of his attention to the Engines without being obliged to depend upon support from any business of his own, and I make no doubt that the proprietors of the engines would be willing to contribute considerably towards his support, tho' he must spend a good deal of his time in doing things which they think ought to be done at

> your expense; for they argue that if you undertake to supply them with a perfect machine, they are not to be at the expense of repairing and adjusting it soon after it is erected, which is always necessary – they employ such a number of people in their factories (which they must keep in pay whether the Engine works or not) that if it stops for ever so short a time it is immediately known everywhere to the prejudice of the character of the engine. For which reason, he must take the opportunity of Sundays and nights to make the necessary adjustments and he must certainly do it with his own hands for some time till the people become more accustomed to it, then he may perhaps get the business done by standing over them and giving directions – I must set off for Scotland tomorrow morning.[42]

The greater part of this letter could be written even today about the trials and tribulations of commissioning new machinery and training new staff.

Once Watt had begun to produce steam engines, modifications were made continually to improve them. One of the first changes was in the valve gear. The swivelling plate used by Newcomen and Smeaton as the steam inlet valve was replaced by a conical plug, shaped exactly like a bath plug, which was lifted up and down. These were some of the parts manufactured by the skilled craftsmen at the Soho Manufactory because great accuracy was required to ensure that they seated evenly and were steam-tight. These valves remained the standard type on all Boulton & Watt engines until about 1800 and probably on those built by other manufacturers too. Their method of operation was altered by Matthew Murray of Leeds when he used concentric valve stems, again around 1800, but these did not find much favour.[43] The great disadvantage of these valves was the force needed to lift them off their seats because they had to be raised against the pressure of the steam which could be considerable if the area of the valve were large.

At the turn of the century, both Matthew Murray and William Murdock, Boulton & Watt's 'chief engineer', invented types of slide valves.[44] Murray used a short valve with long passages to the cylinder ends. He invented a planing machine to smooth the flat surfaces of the valve face and the ports on the cylinder casting. This type became the most popular on small engines and on railway locomotives. The disadvantage was that the hot incoming steam had to pass along passages which had been cooled by the exhaust steam. In 1799, Murdock developed the long slide valve driven by an eccentric on the crankshaft. Ports with flat faces were cast at either end of the cylinder in order to keep the steam passages into and out of the cylinder as short as possible. The valve stretched between the ports for the whole length of the cylinder and was flat on the side next to the ports and semi-circular on its back. The exhaust passage was cast into the whole length of the valve to carry the steam away to the condenser. In the form originally pioneered by Murdock, it was extremely heavy and was

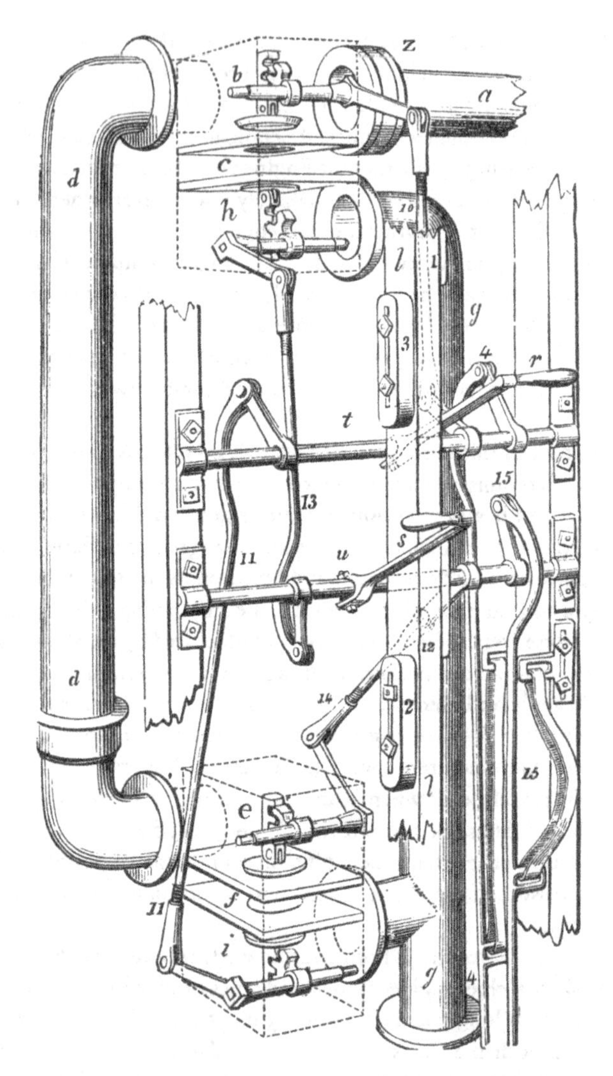

Figure 21 The plug valves and gearing of one of the Albion Mill engines. (Bourne, *Steam Engine*.)

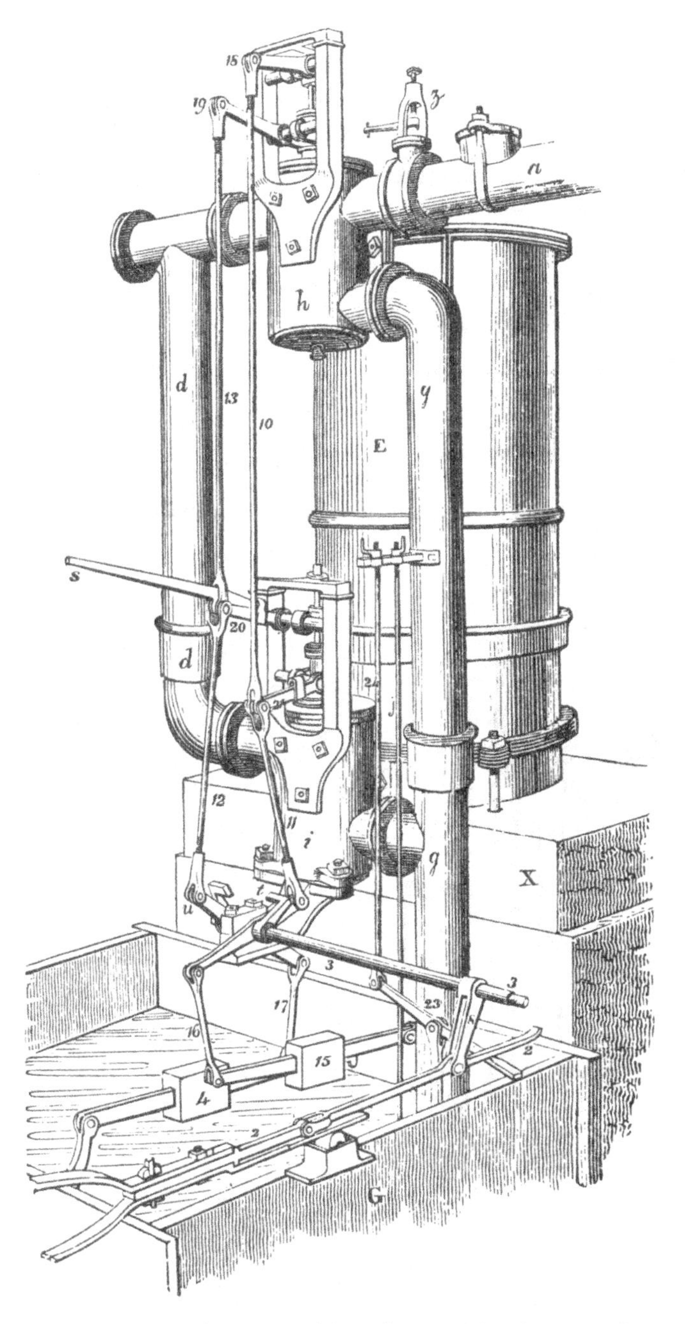

Figure 22 Plug valves operated through concentric valve stems. (Bourne, *Steam Engine.*)

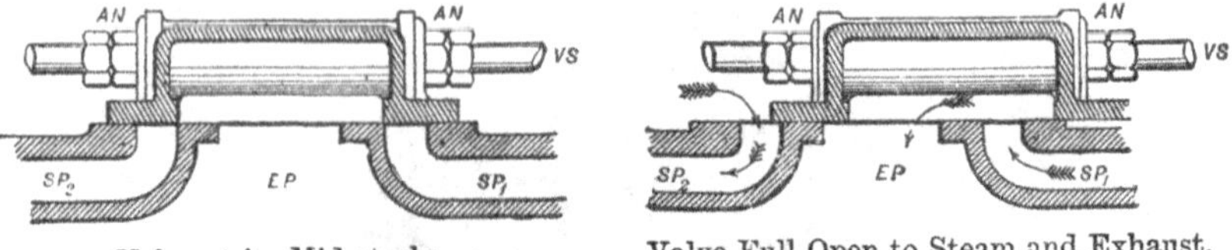

Valve at its Mid-stroke.

Valve Full Open to Steam and Exhaust.

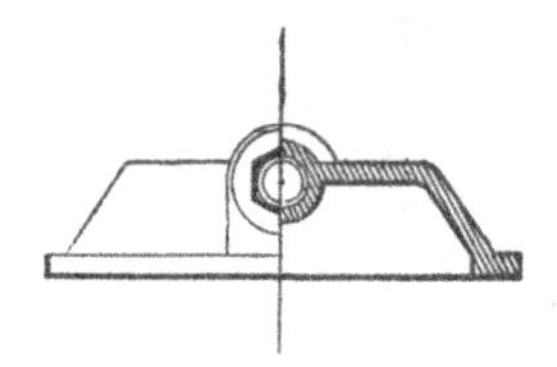

End View—Cross Section.

Figure 23 Murray's short slide valve. (Jamieson, *Elementary Manual.*)

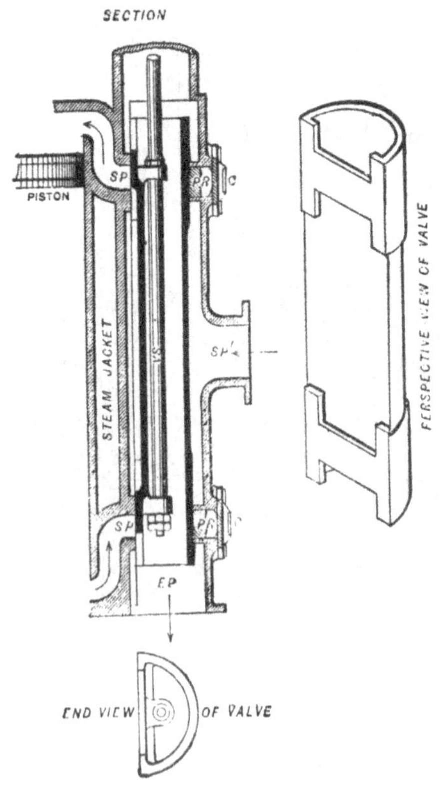

Figure 24 Murdock's long 'D' slide valve. (Jamieson, *Elementary Manual.*)

poor thermodynamically too because the exhaust steam inside the valve passage was heated by the incoming fresh steam. The other problem was keeping the packing round the semi-circular back tight.

A modification of Murdock's valve became very popular for mill engines and on engines for ships. The middle section of the valve was replaced by a rod or rods joining the two ends. The valve faces were flat, and, if the rod had a pin-joint, the alignment of the surfaces on which they worked was not so critical. This was an advantage before planing machines were generally available. On the beam engine from Haydock dating from about 1830 and now reerected in the Manchester science museum, the castings for the valve ports were bolted onto the cylinder against narrow chipping surfaces from which metal could be removed quickly to allow for lining up. The loss of Murdock's middle section made the valve lighter but then both the semi-circular backs had to be packed, a cause of constant leakage. The pressure of the steam on any slide valve caused the rubbing surfaces to wear, which, on the Haydock engine, was well over half an inch. The friction caused loss of power which was one reason for the popularity of Murray's smaller valves. On the low-pressure cylinders of later mill engines, a valve similar in form to Murray's was used, which was split in two, one part for each end of the cylinder. In this way, the short steam passages were retained and the packing round the back became unnecessary. Simplicity was achieved at the expense of economy for slide valves did not allow the steam to exhaust as quickly as the older plug valves. So there was loss of power with the consumption of a greater quantity of fuel.[45]

For spinning, it was essential that the number of revolutions per minute should be kept as constant as possible. It may well have been Rennie with his experience in millwrighting and corn mills who adapted the 'Whirling Regulator' or governor to the steam engine. The first engines to be fitted were those at the Mint in 1788. Pearson's engine did not have one when it was built in 1787, for, in 1791, Southern regretted they did not have one already made but recommended a throttle valve which could be sent by the next coach.[46] Drinkwater realised the value of the governor and inquired about it in 1789.

> Among those inventions one I understand is of a Nature soly calculated to secure more effectually an equal motion under different degrees of heat from the fire – a Property so extremely essential in preparing cotton to work into fine yarn – that I wou'd on no act. have you deny the use of this instrument.[47]

Boulton & Watt replied,

> The only new invention of any consequence we have lately added to the Engine is the regulator or Governor you mention & which is not absolutely necessary, though it certainly when kept in order itself gives greater steadiness & regularity, especially in the cases of taking

> off or putting on work, as it will not permit 2 strokes per minute of increase of velocity though all the work were taken away at once.[48]

We shall see how the governor was improved later to give much closer control of the speed than this. It consisted of a pair of arms, pivoted at the top, with weights in the form of balls at their lower ends. On rotation, centrifugal force caused the balls to fly outwards, and, through the position of the pivots, upwards. The arms were joined to a linkage so that a throttle valve in the main steam inlet pipe could be opened or closed as the balls moved up and down according to speed. If the speed were reduced, the balls moved in and down. This caused the throttle valve to be opened and so more steam passed to the engine which meant the speed increased again. Even if the boiler pressure remained constant, when some spinning machines were stopped in the mill, the engine would settle down to run at a slightly higher speed under the control of an ordinary pendulum governor because the angle of the governor arms and the throttle valve opening were directly linked. At the Etruscan Bone Mill in Etruria, Stoke on Trent, the Bateman & Sherratt beam engine would be adjusted manually from time to time to keep the same speed as the load gradually became less and less. The engine driver twisted a screwed knuckle on one of the rods linking the governor to the throttle valve so that the engine and governor rotated at the same speed but the throttle valve opening varied according to the load.

Another important change in the design and construction of engines was the replacement of wooden parts by cast iron. Smeaton had used cast iron for a

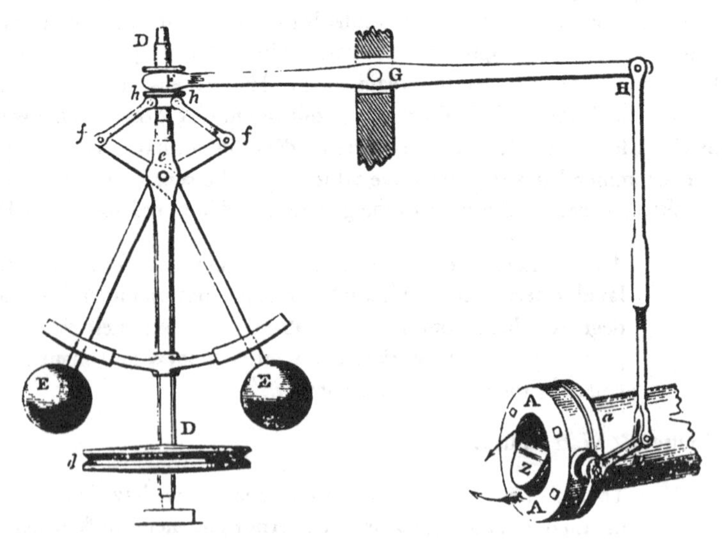

Figure 25 The 'Whirling Regulator' or governor developed for steam engines. (Farey, *Steam Engine*.)

windmill shaft in 1754 and Rennie had used it for the gearing at the Albion Mill. Owing to a series of fires in cotton spinning mills, William Strutt had experimented with one type of fire-proof construction for a mill at Derby in 1792 and followed it by others. His friend, Charles Bage who was an associate of Boulton & Watt, went further when he designed a flax spinning mill at Shrewsbury in 1796.[49] Cast-iron columns were used, as in Strutt's mills, but the novel feature was that the main floor beams were cast iron as well. Other people were experimenting with the structural use of cast iron too. An itinerant lecturer on mechanics and 'natural philosophy', John Banks, had carried out experiments to determine the breaking load on bars of cast iron at Coalbrookdale during April 1795.[50] He found that, in general, cast iron was more uniform than wood, and that it was 3⅓–4½ times stronger than oak of the same dimensions and from 5 to 6½ times stronger than deal. He also found that he could make a beam stronger by casting it in the form of a parabola with a saving in weight. It was advisable 'to put a moulding round them, 3 or 4 inches broader than the thickness of the beam by which any twist to a side will be guarded against'.[51] Banks seems to have acted as adviser to the little-known Yorkshire firm of Aydon & Elwell who pioneered the use of cast iron for the main beams of engines around 1796.[52]

Boulton & Watt started using cast iron for the framing of their engines around 1800, when they were lagging behind other engine builders, notably Fenton & Murray of Leeds.[53] In 1799, Murdock was sent from Birmingham to Leeds to try and discover the secret of Murray's much better castings. M. R. Boulton wrote gleefully to James Watt Junior that Murdock 'by plentiful doses of ale succeeded in extracting from him [one of Murray's foundry men] the arcane and mysteries of his superior performance'.[54] In desperation to try and improve the quality of their castings, Boulton & Watt even ordered a boat-load of moulding sand from the same source as Murray. Industrial espionage continued but three years later, they still had not succeeded in emulating him.[55]

One of the earliest engines built by Boulton & Watt entirely from cast iron was that for the Salford Twist Company. This mill was the first in the Manchester area to have a fire-proof structure similar to that at Shrewsbury and later was one of the first to be lit by gas. Drawings for the main beam of the engine show that a sort of 'space-frame' was proposed. Another suggestion was for a parabolic shape but without the strengthening moulding round the edge as Banks recommended. This engine was fitted with the older form of plug valves and the tests which were carried out on it by George Lee will be described later. Therefore, by 1800, the rotative beam engine had been developed to a form that it would retain for nearly another fifty years without any major modifications.

It is now necessary to retrace our steps back to when Boulton & Watt first started selling their new steam engines. Boulton had insisted on the prolongation of the 1769 patent so they could recover their development costs; a problem still confronting inventors today. Because they did not sell complete machines, they had to find another way of charging for the use of their patent. We have seen how their pumping engines were much more efficient and consumed less coal than the earlier Newcomen types. Boulton & Watt decided that their engine should be paid for by a royalty, or 'premium', based on the saving of fuel effected by their engines compared with the consumption of an atmospheric engine doing the same work. They stipulated that they should receive one-third of the value of the fuel saved, which in fact was determined by a formula that gave a figure less than the actual saving.

Watt made elaborate tests on existing pumping engines, both in Cornwall and other places, to establish a relationship between his engine and the others. Based on these trials, he concluded that the effective pressure load for the average atmospheric engine was 7 p.s.i. and for his own 10.5 p.s.i. He worked out a standard figure for his engines based on diameter and stroke. To keep a record of the number of strokes made, he developed a 'counter' which logged the see-sawing of the main beam. In this way, the amount of work the engine did could be recorded.[56]

Such a method could not be applied on the rotative engines because the load could vary so much and therefore a basic sum was charged. This was £5, or in the London area £6, per horsepower per annum until the patent expired in 1800. If the engine remained out of use for over a year, the payments were discontinued 'from the day on which the said engine shall be discontinued to be used or worked, until it shall again recommence to be used'.[57] Instead of paying the premium each year, an annuity to cover the whole period up to 1800 could be purchased, but William Douglas, who did this, found himself caught out.

> Our engine had stood still for 12 months, & we set it to work again 5 weeks since, only, so that our bargain with you was rather a dear one.[58]

There were many complaints about this system of payment.[59] Where an engine was intended to supplement the power of a waterwheel in periods of drought, it would be run intermittently and the charges per horsepower actually used would be very high. Where a mill was being established, production would build up slowly. The Shudehill Mill took seven years before it was filled with machinery. In such a case, the millowner wanted an engine powerful enough to drive the whole mill but felt it unfair that he should have to pay for excess capacity at the beginning. Then textile machinery was being invented so quickly that it was difficult to know how much power might be required to drive a mill with new machines. John Cartwright was faced with this dilemma when he was building a mill at Retford in 1788 in which to install his brother Edmund's power looms. There was no precedent or basis on which the power needed to drive such looms could be judged and, exceptionally, Boulton & Watt remitted a half year's premium.[60] Determining the power of an engine became even more difficult when an element of expansive working was introduced. Usually the power was given in terms of horsepower, but this became widely inaccurate when the pressure began to fall part way through the stroke.

Comparison with horses had been made since the very first invention of the steam engine.[61] Savery had claimed that his pumping engine could replace ten or twelve horses because it could raise 'as much water as two Horses working together at one time . . . and for which there must be constantly kept ten or twelve Horses'.[62] Triewald had described how five 'fire-and-air machines' were to be installed at Konigsberg (Hungary) to replace 500 horses then being used for pumping water out of the mines.[63] Smeaton sometimes stated the power of engines that he designed in terms of the number of horses they might replace. For example, he claimed that the Chacewater engine erected in 1775 was equivalent to 150 horses, or to 450 horses working in shifts for continuous operation.[64]

The difficulty was that there was no agreed definition of the power of a horse at this period. Smeaton made one horsepower equal five men, but French writers thought six or seven.[65] Desaguliers adopted the figure of 27,500 lb and later Smeaton took the figure of a weight of 22,916 lb raised one foot high in one minute as a standard for one horsepower. When Watt had to determine the power of his rotative engines, he decided to take the figure of 33,000 ft lb, which has been accepted ever since. This concept of horsepower could be applied easily to pumping engines because the volume and weight of water raised could be calculated from the size of the pumps, the number of strokes and the height the water was raised. The power of a waterwheel could be determined in the reverse way by finding out how much water flowed through the sluice gate and the distance it fell round the wheel.

The power of an engine, measured by the number of foot-pounds expended in a given time, was a concept which could be applied to any machine which lifted a weight through a vertical height. But the performance of a cotton mill could not easily be considered in similar terms because this could not be calculated as a weight. One way to determine the power was to compare a mill driven by a steam engine with one actually driven by horses. For example, Arkwright owned a cotton mill in Nottingham where there was a horsewheel of 27 ft in diameter. A note on the plan says '6 strong horses at a time, Mr Arkwright said 9 horses'.[66] Boulton & Watt decided to erect a 12 h.p. engine because these horses were worked very hard and were changed frequently. Another Nottingham mill owner, John James, wrote to Boulton & Watt about an engine.

> We now make use of Eight Horsers but shold we agment our macheanery to take 10 Horsers to work the said mill I shold be glad to know the difference betweene a mill that wourks Eight and that of 10 and the Quantity of Cole it will take to worke the said mill with a day reconing twelve howers for each day.[67]

In this case, Boulton & Watt supplied an 8 h.p. engine.

There seems to have been no way of estimating accurately the power required to drive any one machine. This was a problem which had been faced already by people who experimented with earlier forms of spinning machines. Among the Paul & Wyatt papers at Birmingham there is an *Essay on Friction*, written about 1750, in which the authors were trying to establish what power was lost through friction in one of Paul & Wyatt's horse-driven spinning machines. Watt recorded another experiment, not made by him, on the power required for spinning when horses were used.

> When the machinery was turning *with & without* cotton in it; the difference in power was *barely if at all sensible*. Hence it appears that the mere extension of the cotton wool requires but a small portion of power; and that the power is chiefly absorbed by friction – The latter deduction points at another viz. that a good deal depends on the construction.[68]

Although Watt soon learned by experience to allow about 1 h.p. for every 100 spindles, yet he admitted that he could not judge accurately.

> We have erected several engines for Cotton Mills, but the machinery differs so much in different places that we can not direct you as to power. At Nottingham it takes a horses power to drive 100 spindles & carding & Roving for them, but Mr. Peels engine we are informed works twice that number or more pr horse having only 14 horses power. It seems to depend on the perfection of the machinery & the velocity with which the spindles turn, fine twist also takes less carding pr spindle than coarse.[69]

The figure of 100 spindles to each horsepower was probably widely taken as a rule of thumb by the turn of the century in mills where the spinning was based on Arkwright's water frame. Robertson Buchanan, who quoted this figure in 1808, also noted that when flax was being spun rather than cotton, 75 spindles could be driven by 1 h.p.; but when cotton was being spun on a mule, 1,000 spindles were equivalent to 1 h.p.[70] This large figure may indicate that Buchanan was referring to mills where mule spinning was not fully mechanised. In every case, the figures refer not only to the power needed for driving the spinning machines but also the carding and roving machines that necessarily accompanied them.

But even if these rough figures for cotton spinning were widely used, when a new type of machine appeared, only trial and error could determine how many could be driven by any particular steam engine. This is clear from Watt's correspondence with John Cartwright. Watt wrote that it was impossible to be certain

Table 2. *Results from Lee's indicator experiments, 8–9 December 1795*[61]

Experiment progress no.	Load		Indicator in cylinder top		Effective pressure
	Preparation	Spinning	Depression	Elevation	
46	22	21.5	13.85	2.50	11.35
47	11	21.5	14.00	3.90	10.10
48	—	21.5	14.00	5.10	8.90
49	—	11.5	14.20	8.80	5.40
50	—	—	14.20	12.50	1.70

> what number of looms will be equal to the *whole* power of the Engine . . . At the same time we can not entertain a wish that you should encumber yourself with an engine too large for your use, so heap expenses on an infant manufactory.[71]

However, rough figures begin to be available for determining the load on an engine.

Matching an engine to its load was a problem which could be examined in two stages, the first being to measure the power absorbed by the machinery being driven, and the second to calculate the dimensions of an engine that would supply this power. But data relevant to both points could be obtained if the power developed by an engine could be measured as it drove sample batches of machinery. This approach was adopted in the 1790s, when pressure gauges were attached to engine cylinders. One such gauge which was employed by Banks took the form of an enclosed mercury barometer. Another type of gauge or 'indicator' had a pointer which moved against a scale. This was actuated by a spring-loaded piston. Both types of instrument were difficult to operate because, if an engine were running at 15 or 20 strokes per minute, there would be some inertia in the instrument and it would be difficult to take readings by eye. Even then only maximum and minimum pressures could be recorded.

It was about 1790 that Boulton & Watt introduced an 'indicator', but some of the best evidence for its use dates from 1793, when Watt's friend, George Lee, began a long series of experiments in the famous Philips & Lee cotton spinning concern which bought a Watt rotative engine in 1792. In a typical series of experiments, Lee started with the engine working under full load, which he expressed as 22 units of cotton spinning and 22 units of preparation machinery. Then progressively he reduced the load, disconnecting first the preparation and then the spinning machinery until the engine was driving simply the line shafting and gearing. Highest and lowest cylinder pressures were measured in each instance, so that Lee could estimate how much power was needed by each group of machines (see Table 2). It is not clear how Lee interpreted his results, but if he assumed that 'effective pressure' was proportional to the power developed, the figures in his table would have told him that, when his engine was driving all the machinery, the spinning took 64 per cent of the power, the preparation 21 per cent and the 'Geer without machinery' 15 per cent.

Now, although Lee used the indicator to measure the highest and lowest pressures reached in the cylinder, and although he probably took the difference between these two readings to estimate the power, any such estimates were invariably very approximate. The pressure difference acting on the piston of an engine was not constant throughout the stroke, and, if Lee were to get an exact measure of the power of the engine, the variations in pressure would have to be recorded continuously. A method of doing this was evolved by Southern in 1796, when he invented a device whereby a pencil replaced the pointer on the indicator and traced a line, recording the pressure, on a paper which moved backward and forward in pace with the piston.[72] The line traced out was effectively a graph, representing a plot of the pressure against the displacement of the piston. The average pressure available for driving the engine was given by the average distance between the top and bottom lines of the figure, once the scale had been determined. Southern evidently got the device working in April 1796. On 12 April, after Lee had heard about this innovation, he wrote,

> I am like a man parch'd with thirst in the Expectation of relief, or a Woman dying to hear (or tell) a Secret – to know Southern's Mode of determining power.[73]

A week later, he had been told more about the invention, and wrote,

> Southern's scheme is highly scientific & Ingenious but will require nicety in the Execution – a Good Expt. upon a pump Engine wod. not be amiss.[74]

Lee had used the word 'power' to describe the measurements made on the new principle, but Southern's device was probably at first conceived as a means of measuring the average pressure. This was implied by the name 'indicator diagram' which was given to it, and the first published account of the device described it as a means of measuring pressure and not power.[75] However, Watt and Southern quickly realised that the indicator diagram could be used to give a direct measure of an engine's power. For, provided the engine was always run at the same speed, the area of the figure on the diagram was directly proportional to the power being developed by the engine at the time when the diagram was taken.

In 1798, Lee used the indicator to check that the valve settings on an engine were arranged to give maximum power,[76] and indicator diagrams were used for the same purpose in 1803 when a new engine was installed in Thomas Houldsworth's cotton mill on Little Lever Street, Manchester. A copy of these later diagrams shows that, when the engine was first started, its valve settings were a long way from the optimum. The indicator diagram must have been extremely valuable in adjusting the valves of a new rotative engine. This particular example may represent the earliest indicator diagram to survive, and its interest is enhanced by the fact that Houldsworth's engine was roughly the same size as the one on which Lee had carried out his experiments, and, like Lee's engine, it operated throughout the stroke at pressures well below that of the atmosphere. It had a 29 in diameter cylinder with a 6 ft stroke and was rated at 32 h.p.

This diagram serves well to illustrate the inaccuracies involved in the experiments that Lee was doing with his indicator. One of these earlier instruments fitted to Houldsworth's engine might have shown a maximum pressure of about 6 in of mercury below atmospheric pressure and a minimum pressure of about 12.2. Thus the 'effective pressure' would appear to be 6.2 in of mercury. But for the diagram shown as the full line in the figure, the average pressure, given by the average distance between the top and bottom lines, was

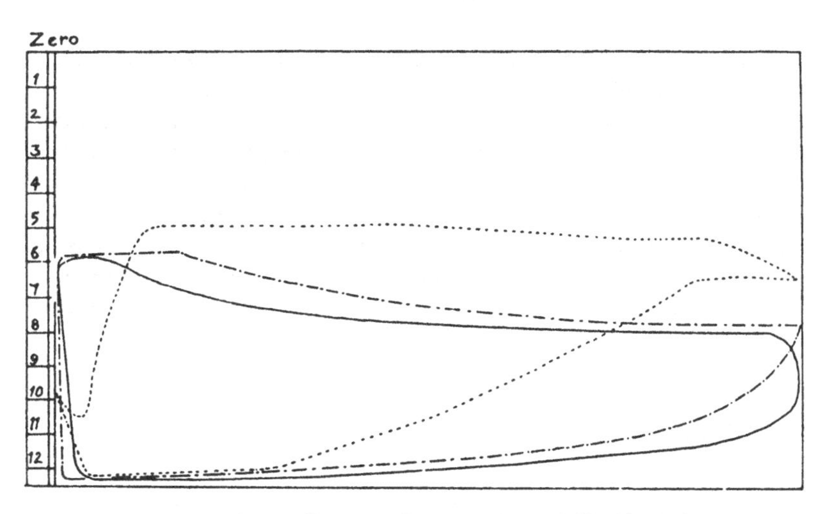

Figure 26 Indicator diagrams taken in 1803 on 'Mr Houldsworth's Engine' in Manchester. The diagrams have been redrawn, the lines on the original being in colour. The significance of the three diagrams shown here is explained as follows:

Dotted line: 'Diagram made with the Valves opening only 1¾ in or less, the Working Gear according to the first construction... 19 March 1803.'

Chain line: '[Diagram] made with the Valves opening a *little* in advance. 22d. March 1803.'

Full line: 'Diagram made with the Eccentric Circle advance 15°... 25th. March 1803.'

There follows a remark that the difference in area enclosed by the last line and the first 'shews the quantity of steam saved.' Assuming that the engine was running at the same speed as Lee's engine, the area enclosed by the last line shows the power developed to be 9.4 horsepower. (Boulton & Watt Collection.)

only 4.1 in. Thus Lee's method of measurement might overestimate the pressure available to drive the engine and might have exaggerated the power developed.

Lee was well aware of the defect in his earlier measurements and was quick to repeat the experiments with the new technique. Influenced, perhaps, by his French books on steam and water power, he called Southern's device a 'dynamigraph' when he described his most recent investigations to Strutt in 1803.

> In the Experiments which I made by an instrument which I ventur'd to call a Dynamigraph I detach'd every kind of Resistance from the Engine and having determined by the figure describ'd the power absorb'd by it, I successively but separately added the Mill Work, the Carding, the Drawing, the Roving & the Spinning.[77]

With each set of machines successively connected up, Lee measured the power delivered to the mill by the engine. It is clear that, once his experimental procedure had been modified to employ the indicator with diagrams, the first satisfactory answer had been found to the problem of matching an engine to the machines it was supposed to drive. When allowance had been made for power absorbed by the mechanical parts of the engine, the indicator diagram provided accurate data about the power needed to run different types of machines and about the capability of different engines. The indicator was developed later into a vital piece of apparatus for checking the valve setting and performance of steam engines. It was as if people could see what was actually happening to the steam inside the cylinder. The next stages in the development of the rotative engine with higher pressures and higher speeds would have been impossible without it.[78]

6
Such unbounded power

In 1834, on the occasion of an appeal for a memorial to Watt which would be erected in Westminster Abbey, the effect of his steam engine was summed up in the following words:

> A time will come when the science of destruction shall bend before the arts of peace; when the genius which multiplies our powers, which creates new products, which diffuses comfort and happiness among the great mass of people, shall occupy, in the general estimation of mankind, that rank which reason and common sense now assign it.
>
> Then Watt will appear before the grand jury of the inhabitants of the two worlds. Everyone will behold him, with the help of his steam engine, penetrating in a few weeks into the bowels of the earth, to depths which, before his time, could not have been reached without an age of the most toilsome labour, excavating vast mines, clearing them in a few minutes of the immense volume of water which daily inundates them, and extracting from a virgin soil the inexhaustible mineral treasures which nature has there deposited.
>
> Combining delicacy with power, Watt will twist, with equal success, the huge ropes of the gigantic cable by which the man-of-war rides at anchor in the midst of the raging ocean, and the microscopic filaments of the aerial gauze and lace of which fashionable dresses are so principally formed.
>
> A few strokes of the same engine will bring vast swamps into cultivation; and fertile countries will also thus be spared the periodical returns of deadly pestilential fevers, caused in those places by the heat of the summer sun.
>
> The great mechanical powers which had formerly to be sought for in mountainous districts, at the foot of rapid cascades, will, thanks to Watt's invention, readily and easily arise in the midst of towns, on any storey of a house.
>
> The extent of these powers will vary at the will of the mechanician; it will no longer depend, as heretofore, on the most inconstant of natural causes, on atmospheric influences.
>
> The common branches of each manufacture may be carried on in

> one common space, under the same roof; and their products, as they are perfectioned, will diminish in price.
>
> The population well supplied in food, with clothing, and with fuel, will rapidly increase; it will, by degrees, cover with elegant mansions, every part of the earth; even those which might justly have been termed the Steppes of Europe, and which the barrenness of ages seemed to condemn to be, forever, the exclusive domain of wild beasts.
>
> In a few years, hamlets will become great towns; in a few years, boroughs, such as Birmingham, where there could hardly be counted thirty streets, will take their place among the largest, the handsomest, and the richest cities of a mighty kingdom.
>
> Installed in ships, the steam engine will exercise a power a hundredfold greater than the triple and quadruple ranks of rowers, of whom our forefathers were wont to exact a labour which is deemed a punishment for the most atrocious criminals.
>
> By the help of a few bushels of coal, man will vanquish the elements; he will play with calms and contrary winds and storms. The passage from one place to another will be much more speedily accomplished; the moment of arrival of the packets may be known beforehand, like that of the public coaches; no one will any longer wander on the shore for whole weeks and months, with a heart tortured with anguish, watching with restless eye the horizon for the dim outline of the vessel, which is to restore a father, a mother, a brother, or a friend.
>
> Lastly: the steam engine drawing in its train thousands of travellers, will run on railroads with far greater speed than the swiftest racehorse, carrying only his light jockey.
>
> Such is a very brief sketch of the benefits which have been bequeathed to the world by that machine.[1]

Yet, even at the time these words were written, the supremacy of the Watt rotative beam engine had been challenged in many different ways. We shall follow the changes in design which were brought about through the many different applications mentioned above. A great many people thought that Watt had brought the steam engine to such a state of perfection that the chance of any further improvement was excluded. These views were held not only by people averse to innovation but also by scientists. As Thomas Tredgold put it,

> The idea that Watt had done everything possible to be done reflecting the power of steam had stopped inquiry among men of science, and left the manufacturers and capitalists of the country, who were wishful to encourage improvement, to be guided by vain and ignorant projectors or ruined by pretending knavery.[2]

One of the challenges that Watt himself had to face came through high-pressure steam. This was, of course, nothing new, and may be traced back in one way to the aeolipile made by Hero of Alexandria. Even Watt himself had stated in his 1769 patent that

> I intend in many cases to employ the expansive force of steam to press on the pistons . . . In cases where cold water cannot be had in plenty, the engines may be wrought by this force of steam only, by discharging the steam into the open air after it has done its office.[3]

His 1784 patent included a specification for a steam road carriage with two steam cylinders in which again he would

> employ the elastic force of steam to give motion to these pistons, and after it has performed its office, I discharge it into the atmosphere by a proper regulating valve.[4]

Condensing apparatus might be fitted also on these carriages. The idea of a steam carriage seems to have been Murdock's. The story is that Watt only included it to pacify Murdock who had actually set off from Cornwall for London to take out a patent but was met by Boulton on the way. Murdock was far too valuable to Boulton & Watt for keeping their pumping engines running in Cornwall and nothing more was heard of the steam carriage.[5] Watt himself always advocated low-pressure steam, no doubt partly for safety reasons through fear of boiler explosions and partly because high-pressure steam could avoid the necessity of using the separate condenser.

The obvious advantage of high-pressure steam was that smaller engines could be made to produce the same power. Then, dispensing with the separate condenser again reduced the size and complexity of engines, reduced the power wasted through friction working the air pump and avoided dependence on cooling water. Of course, high-pressure engines could be fitted with condensers and people were puzzled to find that both condensing and non-condensing high-pressure engines might use less coal than equivalent low pressure ones. The American, Oliver Evans, was not only one of the first people to pioneer high-pressure reciprocating steam engines, but he commented on their economy too.

Evans had described a boiler that was suitable for higher-pressure steam in 1792, but lack of funds prevented his introducing it. So it was not until 1801–2 that his first experimental steam engine and boiler were completed.[6] This engine had a single cylinder set vertically which exhausted directly into the atmosphere. A flywheel, driven by a crank, was set above a wooden frame. The steam valves were oscillating three-way cocks and the engine rotated at 30 r.p.m. A later engine had a vertical cylinder connected to one end of a beam in the 'grasshopper' arrangement. Evans fitted it into an amphibious machine called the 'Orukter Amphibolos' or 'Amphibious Digger', which was meant to be a self-propelled steam dredger that was launched in 1805.[7]

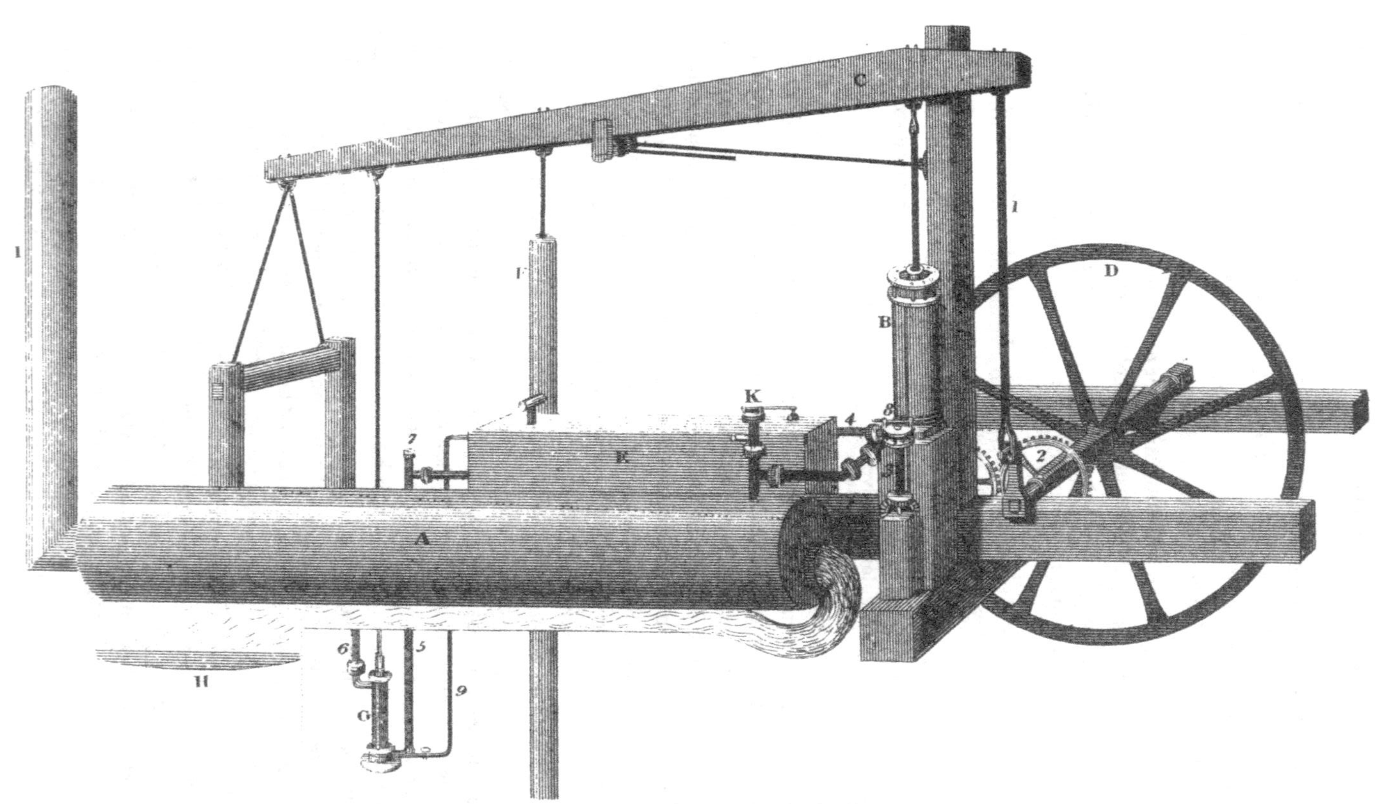

Figure 27 Evans's 'grasshopper' high-pressure engine and boiler. (Smithsonian Institution.)

The steam pressure in these first engines was 50 p.s.i. but, in later ones, this was increased to 100 or 150 p.s.i. Evans found that,

> If the steam be confined by the load on the safety-valve, to raise its power to 100 pounds to the inch . . . the power will be equal to twenty horses hitched . . . If the steam be confined by 150 pounds, the power of the engine will be equal to 30 horses, when the steam is shut off at one third of the stroke, and striking 36 strokes per minute.
>
> The more the steam is confined, and the shorter it be shut off by the regulator, the greater will be the power obtained by the fuel. For every addition of 30 degrees of heat to the water doubles the power. So that doubling the heat of the water increases the power 100 times. On these principles fuel may be lessened to one third part consumed by other engines. This engine is not more than one fourth the weight of others; is more simple, durable, and cheap, and more suitable for every purpose; especially for propelling boats and land carriages.
>
> The power of my engine rises in geometrical proportion, while the consumption of fuel has only an arithmetical ratio; in such proportion that every time I add one-fourth more to the consumption of fuel, the powers of the engine were doubled; and that twice the quantity of fuel required to drive one saw would drive 16 saws, at least . . . so that the more we resist the steam the greater is the effect of the engine.[8]

It was to be another 40 years before the true answer to the economy of Evans's engines was discovered, but the practical advantages of using higher-pressure steam and, in particular expansive working, were quickly realised.

Watt had discovered the economy of expansive working, but had not pursued it due to the low pressure of steam used and probably also due to the need for an even pressure on the piston to promote smooth running. It was in Cornwall, where the need for engines with the greatest possible economy was so important, that expansive working was first introduced generally, but, before discussing the introduction of high-pressure steam into this region, there was one other change in the pumping engines there which really paved the way. The pump normally used in the older atmospheric engines was the bucket type which has been described earlier. This needed a cylinder with a smooth interior bore, something not easy to achieve with the machine tools of that date. When the leather flap valves wanted replacing, the pump rods had to be pulled up to lift the bucket out of the barrel, and the bucket also needed regular packing as well. The plunger pump had been known for many years,[9] but does not seem to have been used much in Cornwall until shortly before 1800. It consisted of a rod or solid plunger moving up and down in a casing. As it moved down into the casing, it displaced water through one set of valves and as it moved up, more water was drawn in through another set. Its advantages were that the plunger could be turned smooth easily in a lathe while the casing did not need boring if

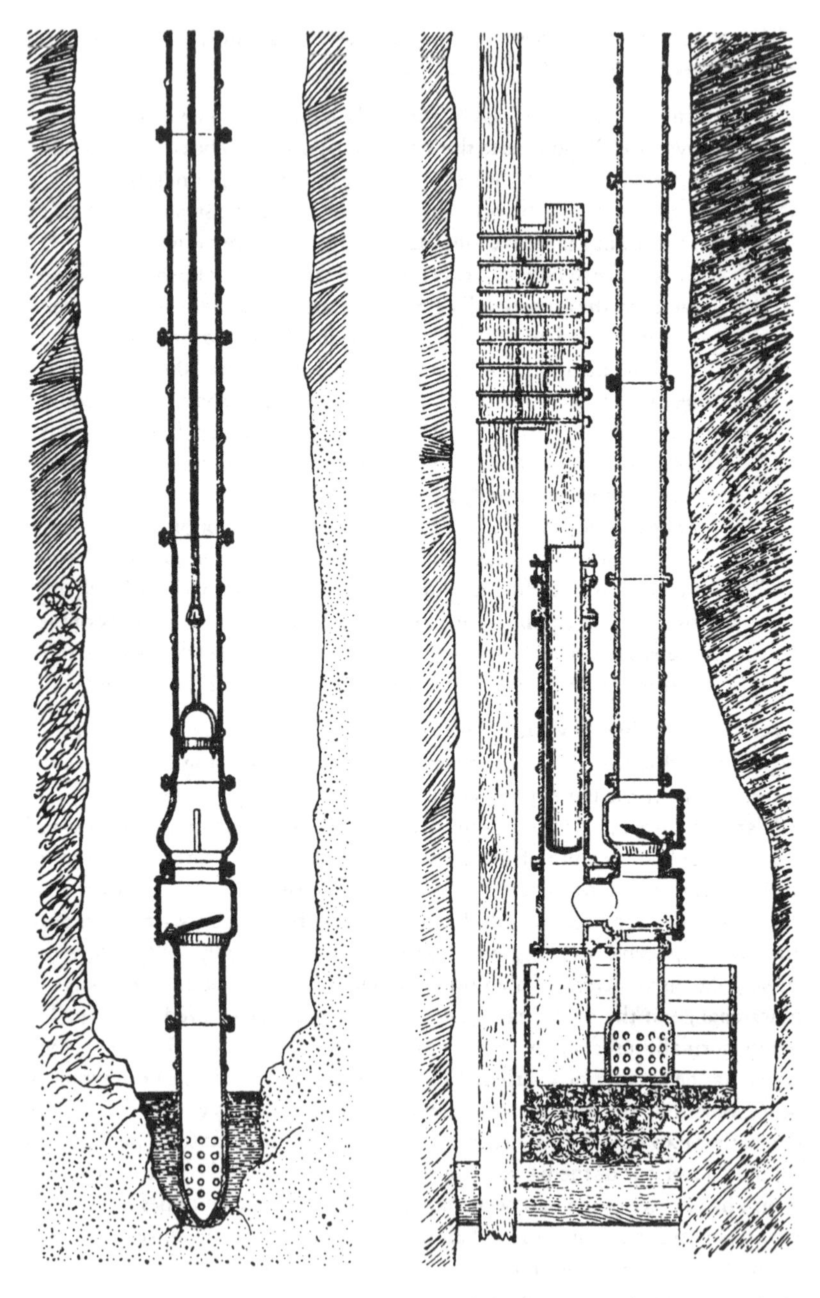

Figure 28 Cornish mine pumps. On the left is the earlier bucket pump which was superseded by the plunger type on the right. (Barton, *Cornish Beam Engine*.)

there were a packing round the top. This packing was readily accessible for maintenance, as were the other valves which could be reached through cover plates. Another advantage was that the plunger pump was less liable to obstruction by grit or dirt in the water.[10]

The introduction of the plunger pump has been ascribed to both Richard Trevithick and William Murdock in 1797.[11] Put succinctly, with the plunger pump, the engine pumped the rods and the rods pumped the water, reversing the earlier practice. It was the weight of the plungers and the rods which forced up the water so, while the steam was being transferred from the top of the piston to the bottom in the steam cylinder, the water was being forced up in a steady stream through the valves. On the power stroke, the steam pressure in the cylinder raised the plunger. It was of no great consequence if the plunger were accelerated too fast at the beginning of the stroke because a vacuum would be formed under it which would be filled with water when the engine slowed down at the end of the stroke. If the steam pressure had been pushing the plunger to raise the water, the speed variation could not have been so great as the water would reach a maximum velocity through the valves. This reversal of effort in the pumping action was vital for the later successful development of the Cornish engine.

Interest in high-pressure steam may have been aroused in Trevithick through his work on water-pressure engines. Just as power applied to a plunger can raise water, so the reverse, with water pushing a plunger or piston, can be used to generate power. Trevithick's first water-pressure engine was erected at Prince William (Roskear) Mine in 1798.[12] Several others followed. That at Wheal Druid had a fall of water of 204 ft which, together with the tail pipe, gave a pressure of about 100 p.s.i., or much higher than anything in steam engine design at that period.[13] These engines had much smaller moving parts than their low-pressure rivals. Problems with water-hammer in the pipes and the lack of any expansion which carried a steam piston on a little at the end of the stroke and so completely shut one set of valves and opened another, were all difficulties that Trevithick had to overcome. Dealing with water at high pressure must have helped him when he switched to steam at similar pressures.

A little after his first water-pressure engine, Trevithick turned his attention to the high-pressure steam engine. In 1800, a portable one was sent to Wheal Hope and another was despatched to London in charge of Woolf who was later to raise the efficiency of the Cornish engine to unprecedented heights. The engines became nick-named 'puffers' because they exhausted directly into the atmosphere. One was used as a whim or winding engine at the Cook's Kitchen Mine. 'I recollect the valley engine, because she was a puffer, and you could hear her for miles'.[14] In all probability, the steam pressure on this engine was 25 p.s.i., and it would seem from the previous comment that no attempt was made to use the steam expansively (just as this was impossible in the water-pressure engine).

For the next few years, Trevithick was deeply involved trying to promote his high-pressure engines. There was his Camborne road locomotive which was tried out on Christmas Eve 1801. He placed one in a dredging machine in 1803. He used others to pump water from the tunnel which he was building below the Thames. More famous perhaps were his railway locomotives and then the 'Catch-me-who-can', demonstrated near Euston in 1808. People's distrust of the high-pressure engine was confirmed when the boiler of a stationary engine exploded at Greenwich on 8 September 1803. It was the usual tale; the boy who had been trained to work the engine went off to catch eels and a labourer stopped the engine without releasing the safety valve. Trevithick wrote,

> I believe that Mr B. & Watt is abt to do mee every engurey in their power for the have don their outemost to repoart the exploseion both in the newspapers and private letters very different to what it really is.[15]

In spite of this set-back, Trevithick persevered because the compactness of high-pressure engines, as well as their economy, made them attractive to many potential users.[16] One engine erected at Tredegar Iron Works in 1801 for driving the puddling mills there, was being run at pressures from 50 to 100 p.s.i. On a slightly later engine at Penydarren, there was an attempt to use the steam expansively when driving a tilt-hammer, which had a more regular load than the rolling mill. When Trevithick tried expansive working on the rolling mill, the engine would stop through the carelessness of the workmen. Trevithick's comment was,

> When the cylinder was full of steam the rollers could not stop it; and as coal is not an object here, Mr. Homfray wished the engine might be worked to its full power. The saving of coal would be very great by working expansively.[17]

Trevithick ought to have fitted a heavier flywheel to even out the irregularities in the power output when working expansively. In this instance, the need for power won over economy, a choice that frequently had to be made later.

The economy of these engines was illustrated in 1806 when one was tested against a Boulton & Watt engine.

> Their steam was not above 4 lbs. to the inch: *mine was near 40 lbs. to the inch*; yet I raised my steam of near 40 lbs. with a third of the coals by which they got their of 4 lbs. to the inch. This is what I cannot account for, unless it is by getting the fire very small and extremely hot. Another advantage I have is that there is no smoke that goes off from my fire to clog the sides of the boiler while the common boilers soon get soot half an inch thick, and the mud falls on the bottom of the boiler, where the fire ought to act; but in these new boilers the mud falls to the bottom, where there is no fire, and both the inside and

the outside of the tubes are clean and exposed to both fire and water . . . Then the coal did do above one-seventh of the duty that it now does.[18]

The economy of an engine was determined partly by the design of its boilers so the whole question of boiler design and the best ways of firing will be discussed in the next chapter. Here it is sufficient to note the bewilderment felt at the economy of these engines. In 1804, another engine in Wales was connected to a pump to ascertain its powers for raising water. It worked at an average pressure on the piston of 65 p.s.i. and presumably was running non-condensing. If so, the duty it returned of 17.5 million pounds was quite reasonable, especially if it were a rotative engine.[19]

Trevithick's many ventures, including the Thames Tunnel, were financial failures and he returned to Cornwall in 1811 bankrupt. He set himself up as a consulting engineer and, in about 1812, installed new boilers at the Dolcoath Mine. These were the type later called the 'Cornish' boiler, cylindrical in shape with a central flue running through the middle which had the fire inside at one end. These new boilers, working with higher-pressure steam and other improvements instituted by Trevithick, are claimed to have raised the duty to 40 million pounds from the usual 18 to 20 million of Boulton & Watt's day. Extensive trials had been carried out on the engine at Wheal Harland in 1798 when a duty of 27.50 million pounds was achieved. Watt stated at the time that it was so perfect that further improvement could not be expected.[20] Trevithick's figure of 40 million pounds at this date is almost certainly too high, but, in course of time, his boiler superseded all others in Cornwall.

The year 1811 was when Captain John Davey of Gwinear started to report the duty of his engines at Wheal Alfred, which were considered to be the best in the county and returned about 20 million pounds duty. The rivalry between the different mines, caused by having their duty reported as this custom spread, soon showed a dramatic improvement in the monthly figures. In August 1811, eight engines were reported with an average duty of 15.7 million and, in December, the number had increased to twelve, averaging 17 million.[21] Captain Joel Lean took over the reporting which continued for almost a century. In 1814, the three Dolcoath engines returned duties of 21,445,912, 26,756,063 and 32,027,842 lb.[22]

A third event at this same time foreshadowed the pattern of the next 40 years of Cornish engine development. It was the starting of a small engine at Wheal Prosper by Trevithick in the summer of 1812. This was a single-acting engine, but the cylinder was inverted. The boiler pressure was 40 p.s.i., which was exceptionally high for the period. The important fact was that the steam was used expansively and this was the first instance of the true cycle of the Cornish engine.[23] Although recording a duty of 26.7 million pounds in January 1813, such high figures soon decreased and, by the end of that year, it was doing a mere 15.4 million. In the Cornish cycle, steam was admitted through the inlet valve to the top of the piston. After the piston had moved down the cylinder for a pre-set part of the stroke, the valve was closed and the steam inside the cylinder expanded. The cylinder was jacketed with a separate steam casing so there would have been little condensation when the fresh steam entered the cylinder. As this steam expanded, condensation would have been kept low by the steam receiving heat from the cylinder walls. At the termination of the stroke, the equilibrium valve was opened so that the steam inside the top of the cylinder could pass from above to below the piston and the piston rose on a dead stroke. This steam must have been heated by the cylinder walls as the piston was rising which would have been a waste of heat but this was preferred to having the bottom of the cylinder open to the condenser all the time. On the working stroke, a third valve was opened after the equilibrium valve had closed. A communication was opened to the condenser and the steam passed from the bottom of the piston into the vacuum of the condenser. The pressure of the steam below the piston needed to be kept as low as possible after expansion to avoid a pressure drop and so a temperature fall in the exhaust steam. The working stroke as the piston moved down the cylinder was made with the pressure of the fresh steam on top against a vacuum underneath. There may have been a slight thermodynamic advantage in the single-acting cylinder when set up in the usual vertical position because the top with the inlet was always the hottest and the bottom with the exhaust cooler.

Shortly after the erection of the Wheal Prosper engine, Trevithick experimented with a 'plunger pole' engine. In this design, the piston was replaced by a plunger similar to those on the mine pumps, but, in this case, it was pushed upwards by the pressure of the steam. A large engine of this type was erected at the Herland Mine in 1816 and worked non-condensing.[24] A description of it when it had first started working was given by James Banfield in 1871.

> When a young man, living on a farm at Gurlyn, I was sent to Gwinear to bring home six or seven bullocks. Herland Mine was not much out of my way, so I drove the bullocks across Herland Common toward the engine-house. Just as the bullocks came near the engine-house the engine was put to work. The steam roared like thunder through an underground pipe about 50 feet long, and then went off like a gun every stroke of the engine. The bullocks galloped off – some one way and some another. I went into the engine-house.[25]

If this account is true, it would appear, through the sound of the exhaust, that Trevithick had not learnt properly the lessons of expansive working.

The case for expansive working was discussed by Davis Giddy, a close friend of Trevithick, in a letter written to him in February 1816.

> I am of opinion that the stronger the steam is used, the more advantageous it will be found. To what degree it must be applied expansively must be determined by experience in different cases. It will depend on the rate at which the engine requires to be worked, and on the quantity of matter put into motion, so that as large a portion as possible of the inertia given in the beginning of the stroke may be taken out of it at the end.
>
> Some recent experiments made in France prove, as I am told, for I have not seen them, that very little heat is consumed in raising the temperature of steam. And if this be true, of course there must be a great saving of fuel by using steam of several atmospheres' strength, and working expansive through a large portion of the cylinder.[26]

After some initial trouble with badly made castings, the boiler pressure on the Herland engine was raised to 150 p.s.i. Working at five-sixths of the stroke expansively, with a speed of 20 strokes a minute or double the speed of a Watt engine, a duty of 48 million pounds was returned. Trevithick claimed 57 million[27] and this promise of highly economical performance led to orders for many more engines of this type on the understanding that they would cost much less than the Watt engine of equal power and perform the same work with one-third of the coal. Soon faults in design became apparent, for the plunger wore more quickly in the middle and packing became difficult. Although it is not stated, with steam at this high pressure, there must have been problems with the oil and packing burning. The worst defect was masked at first by the high pressure of the steam, for the plunger cooled quickly when it was exposed to the air as it was pushed out of the cylinder and this lost valuable heat. The plunger pole engine soon disappeared after Trevithick had left for South America in 1816.

In the meantime, Woolf had returned to Cornwall from London. In that city, he had successfully obtained employment with Joseph Bramah where, already a skilled workman in wood, he became equally proficient in metal. He followed this by becoming engineer at the Griffin Brewery. Here, in 1803, he patented a new type of boiler which later he used successfully for high-pressure steam and which will be described in the next chapter.[28] Through testing this boiler, he evolved ideas about the expansion of steam which he patented in the following year.[29] One aspect of this patent was that he envisaged using the steam first in a small cylinder to drive a piston in that one, and then employing the steam again in a second, larger cylinder. The principles and advantages of compounding will be discussed later. While his theory of the expansion of steam proved to be wrong, he did realise that, if the steam were expanded adiabatically, it would still be

> capable of producing a sufficient action against the piston of a steam engine to cause the same to rise in the old engine (with a counterpoise)

> of Newcomen, or to be carried into the vacuous parts of the cylinder in the improved engines first brought into effect by Boulton and Watt.[30]

In other words, having used the steam once, Woolf realised that he could use it again and so obtain more power.

In 1804, he altered a 6 h.p. engine made by Fenton & Murray for the Griffin Brewery to work on his new principle. It was a failure because the ratio of the cylinder proportions was wrong. Even with a steam pressure of 40 p.s.i., it was incapable of performing the work it was appointed to do, but the consumption of fuel was found to be small which encouraged Woolf to persevere.[31] Owing to the time it took Woolf to make his engine work properly, the brewery owners called in Rennie who reported so adversely that it was replaced by a Boulton & Watt rotative engine which was one of their earliest with 'D' slide valves and a pair of steam pipes in an architectural pattern similar to the Haydock engine. Woolf continued to improve his engines so that soon they performed the same work as the Watt rotatives with half the fuel consumption.[32] He achieved this by better design, by a higher standard of engineering and by using steam at higher pressures. He formed a partnership with Humphrey Edwards but demand for his engines in London proved to be insufficient to sustain a manufactory there. By May 1811, the partnership had been terminated. Edwards, after continuing alone in London for a while, moved to France where he achieved success by building Woolf's engines and boilers which became very popular there.[33] Woolf returned to Cornwall.

Farey held a high opinion of Woolf's engineering ability and undoubtedly his training under Bramah bore its fruit in Cornwall. Higher standards of manufacture and fitting were essential for the high-pressure engines and this Woolf was able to provide.

> In Mr. Woolf's workshops, he adopted the use of improved turning lathes made of iron, with slide-rests to hold the turning tools. Such lathes were used by accurate mechanicians on a small scale, but had not been adopted by steam-engine makers. He also made a slide lathe for turning piston rods and other cylindrical work, which required very little skill or exertion of the workman, and paved the way for the more modern self-acting slide lathe. By these means the expense of turning and boring iron work was reduced, so as to admit of giving extreme accuracy of fitting to all parts of the engines.[34]

In 1816, he joined Henry Harvey of Hayle to help develop that foundry which became so famous for its Cornish engines all over the world. Woolf continued with this association for many years and the renown in which he was held may be judged by the remark that his engines 'were more like ornaments for a showroom than machines for draining a mine'.[35]

Meanwhile, on 17 May 1811, an advertisement under Woolf's name appeared in the *West Briton* setting out terms for his engine.

> This engine is now brought to such a high degree of perfection as to require not more than one-third part of the Fuel employed in working engines on Messrs Boulton & Watt's construction . . . The Patentee engages to warrant a saving of at least one-half of the usual quantity of coals; and as a further inducement for their adoption, agrees to allow those who shall employ his Patent Engine, one half of the actual saving in fuel.[36]

Woolf's first double-cylinder engine in Cornwall was a small one for winding at Wheal Fortune. The duty in May 1813 was given as 5.3 million.[37] His first double-cylinder engine designed for pumping water at a Cornish mine was erected at Wheal Abraham and was the largest made up to that time. The high-pressure cylinder was 24 in diam., 4.3 ft stroke and the low-pressure cylinder 45 in diam. and 7 ft stroke. When first reported in October 1814, its duty was 34 million pounds. Soon afterwards, a slight defect was discovered in some part of the castings. When this had been rectified, the duty advanced in 1815 to a staggering 52.2 million pounds while the highest duty of 55.9 million pounds was recorded in May 1816.[38]

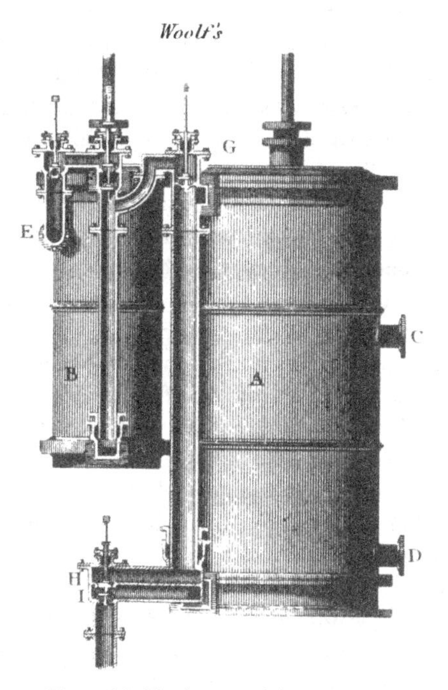

Figure 29 The layout of the two cylinders on Woolf's engine. (Rees, *Cyclopaedia*.)

In April 1815, Woolf's second and slightly larger compound was reported at Wheal Vor, which did an amazing 45 million pound duty in its initial month. In June, it passed the 50 million mark, the first Cornish engine to do so.[39] Thirty-five engines were reported during that year with an average duty of 20.5 million pounds.[40] The best Dolcoath engine was returning a duty of 35 million so Woolf's engines far outstripped the performance of all the others. The new Dolcoath Great Engine, erected in 1816, cylinder 76 in diam., 9 ft stroke, attained 40.7 million pounds soon after starting and was the first single engine to have passed the 40 million mark, but it still fell short of the best compounds. In 1815, it was said that 'Woolf's engine performed the same duty by consuming 34 bushels of coal as the engine of Boulton and Watt does on the consumption of 116'.[41] However, Woolf's engines, after a period in service, began to give trouble. The cast-iron boilers were prone to cracking so the pressures had to be lowered. In 1816, Wheal Abraham was loaded from 62 to 68 p.s.i. on the safety valve but a more usual pressure was probably around 45 p.s.i. The complexity of a compound and the problem of keeping the pistons steam-tight made people revert to the single cylinder engines in the early 1820s but Woolf still considered the compound engine preferable.

To settle the compound versus single engine controversy once for all, John Taylor permitted the erection of two comparable engines in 1824 for the re-working of Wheal Alfred. Both had pumps with 7 ft 6 in stroke. The single, 90 in. diam., 10 ft stroke, was made by Neath Abbey, while Harvey & Co. built the compound with cylinders 40 in diam., 6 ft 6 in stroke and 70 in diam. and 10 ft stroke. Steam for the compound was raised in Woolf's cast-iron boilers while that for the simple was generated in the Cornish type. Both engines returned similar duties, around 42 million pounds. The single engine was lightly loaded and so was worked with a high degree of expansion, while the compound was raising much more water although both were proportioned to exert equal power. This degree of expansion was probably the decisive factor and will be examined further later. Also the steam pressures used in those days were too low for economical working in a compound. Simple engines cost less initially and were easier to maintain subsequently. As the compound at Wheal Alfred showed little better economy, it was the last one on Woolf's principles erected in Cornwall and was soon dismantled. Compound engines were not tried again until James Sims rebuilt an old engine at Carn Brea Mines in 1839 on his system. This one returned 95 million pounds in one month in 1840 and averaged 90 million that year. At Wheal Alfred, Woolf's cast-iron boilers on his compound gave more trouble than Trevithick's Cornish type on the other engine so he never used them again either.[42]

Another reason for the improved performance almost certainly was better proportioning of the steam passages and ports. On the Haydock beam engine and on the Stretham engine in the Fens near Ely, both rotative engines built around 1830, the ports into the cylinders were too small and restricted the flow of steam. Larger steam pipes too helped to improve the flow of steam. If we look at the valve gear on the standard Boulton & Watt rotative engines, the duration of the time that the inlet port is open is the same as that of the exhaust, both on the earlier plug valve engines and the later ones with slide valves. The introduction of valve 'lap', that is extending the face of the slide valve so that the inlet is closed sooner than the exhaust, was not introduced until the later 1830s, almost certainly through experience gained on railway locomotives.[43] Up to this time, designers of the standard rotative beam engines faced the choice of either expansive working with inlet valves cutting off early, which meant that the exhaust valves shut early too giving excessive back-pressure, or no back-pressure and no expansive working.

The power output had to be controlled by the throttle valve in the steam pipe. This is quite clear from the figures quoted by Lee from his experiments with early indicators where the highest pressures in the cylinder were well below atmospheric pressure.[44] The diagrams taken on Houldsworth's engine show very little expansion indeed that can be attributed to the closing of the valve and also a high terminal pressure at the end of the stroke. While this gave an equal force on the piston throughout the stroke, it was bad for economy because the steam left the cylinder at too high a pressure and too high a temperature. This is confirmed by Farey's comments on the Watt engine worked non-expansively and the Woolf engine worked expansively. Terminal pressures in Woolf's rotative engines were probably much lower than in contemporary types at that time which helped to account for the saving in condensing water.

> Respecting the advantages of employing Mr. Woolf's engines instead of Mr. Watt's engines, it should be kept in mind, that the quantity of cold water required for condensation by Mr. Woolf's engines is less than is required for Mr. Watt's engines, in the same proportion as the consumption of fuel is less by one than by the other. In manufacturing towns there is often great difficulty in procuring an adequate supply of cold water for Mr. Watt's engines, and in consequence of a deficiency of cold water, the condensation of the steam is very imperfectly performed, and the engines work under a great disadvantage; whereas if Mr. Woolf's engines were used in the same places, the quantity of cold water that is procured would be sufficient to effect a complete condensation.[45]

The valve gear was different on the pumping engines. There were three valves operated independently, so the steam inlet could be adjusted to work the engine at the best rate of expansion. Woolf is credited with introducing one of the most important modifications to the details of the Cornish engine when, in 1823, he adopted the 'double-beat' drop valve. The valve itself had been thought out by Joe Hornblower but never actually used by him.[46] The important features of this valve were that it had two seating surfaces which gave maximum passages to the steam, and, because it was hollow, minimum pressure was required to raise it off the seatings. Its design was so good that it continued to be used until the last reciprocating uniflow steam engines were built in the 1950s. Such valves gave improved steam flow and could be timed individually so that, on the Cornish engines, the steam could be expanded in the cylinder from full boiler pressure right down to that approaching the vacuum in the condenser which gave excellent thermal efficiency.

Expansive working, of necessity, brought about unequal and diminishing

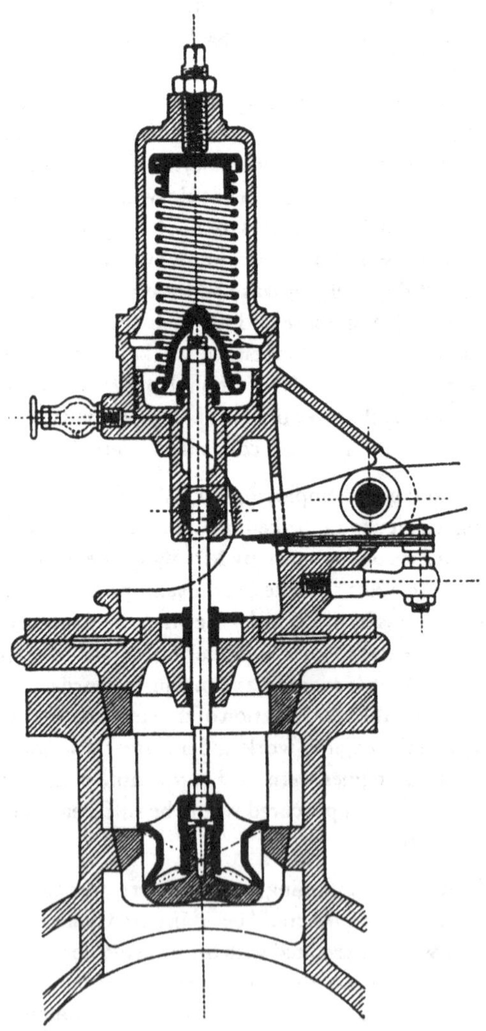

Figure 30 Double-beat drop valve as used on later horizontal engines. (Ewing, *Steam Engine*.)

pressures during the stroke, an evil Watt had tried to cure in his 1782 patent. This is where the change from the bucket to the plunger pump became so crucial in the Cornish pumping engines. The weight of the plungers gave a steady force for raising the water. The force of the steam lifted the unbalanced weight of the plungers and pump rods and 'then the variable force with which the high-pressure steam operates during its expansion becomes an unimportant circumstance'.[47] In fact, the weight of the plungers, pump rods and reciprocating mass of the main beam became an essential feature of the Cornish engine. At the beginning of the stroke, the full force of the high-pressure steam was used to accelerate this mass. Towards the end of the stroke, the inertia of this mass carried the engine through, gradually slowing down as the steam pressure fell as a result of expansion. Judged properly, the valves would be tripped as the momentum died away and no power was wasted stopping the engine ready for the next stroke.[48] In this way, the inertia of the mass of the moving parts counteracted the excessive variation in steam pressure caused by expansive working and helped the engines to work more smoothly and efficiently.

However, what was gained in one way was lost in another, because the parts of the engines had to be made very strong to withstand the initial pressure and shock of the incoming steam.[49] The working of Taylor's 85 in engine at the United Mines in the 1840s at a mere tenth cut-off and 50 p.s.i. gave the engine, and with it the pitwork, a very severe shock each time the steam valve opened because the initial pressure produced a force on the piston of nearly 100 tons. To avoid this, Woolf recommended opening the inlet valve slowly at first to wire-draw the steam. Then on some engines a small protuberance was cast on top of the piston which partially occupied the steam port and thus slightly lessened the initial force of the steam.[50] Nevertheless, great strength was necessary in both engines and pitwork for, when something broke, the consequences were disastrous.

The duty figures continued to climb during the 1820s. By paying attention to details such as improved insulation of boilers and steam pipes, size of steam pipes, etc., one of the leading engineers, Samuel Grose, increased the duty at Wheal Towan to no less than 87 million pounds in April 1828. The boilers might be covered with brickwork or wooden lagging and the pipes were surrounded by wooden casings filled with ashes.[51] In 1835, the hitherto steady increase in duty was shattered by the starting of Austen's 80 in engine at Fowey Consols erected by the relatively unknown engineer, William West. That July it averaged 90 million pounds and in September 97.8 million. The goal of 100 million must have seemed attainable at last. This engine had a public trial on 22–3 October 1835 when an incredible 125 million pounds duty was recorded. This caused a furore because it was felt that the tests should have covered a longer period.[52] Yet, through further minor improvements, Taylor's 85 in engine gave figures of over 100 million pounds duty consistently between April and September 1843, with the best at 105 million. At this time, the general practice was to use steam at 30 p.s.i. above atmospheric pressure with the cut-off one quarter or one fifth of the stroke.[53] However, the days of glory of the Cornish engine were beginning to come to an end, for such high rates of economical working were only reached by superb maintenance. Soon it was thought to be safer to reduce the efficiency of the engine and lessen the working loads.

In any engine, there could be only one rate of expansion which would give the most economical working, which also implies that it worked best at only one load. Once the mine had been drained down to its working level, the volume of water pumped each stroke and the height to which it was lifted, remained constant and so these engines could be run, under normal conditions, at an optimum setting. However, the duty could be affected by the engine having to work faster or slower through the need to pump out more or less water according to weather conditions. The mine might be deepened, which increased the load. The duty would also be affected by varying qualities of coal, but, at this period, there was no way of determining the calorific value of the coal. Even allowing for these factors, the increase in performance of the Cornish engines was quite remarkable and aroused wide interest in other parts of the country and abroad.

In 1835, Thomas Lean calculated that the improved engines were saving nearly 100,000 tons of coal a year valued at £84,300, when compared with 1814.

> Wheal Towan was working in 1814, and also in 1835: the mine was drained by two engines at both periods. In 1814, the depth of the mine was 79 fathoms, and 328 million gallons of water were pumped out by the consumption of 49000 bushels of coal. In 1835, the consumption was 2000 bushels less; yet the mine was 50 fathoms deeper, and the water discharged 53 million gallons more. Or to put it in another form; in 1814 the greatest quantity of water pumped 100 fathoms deep in Wheal Towan, by the consumption of a ton of coal, was 66000 gallons; while in 1835, 269000 gallons were drawn from the same depth for the same expenditure.[54]

It was probably the general decline of mining in Cornwall as much as anything that discouraged further development of the Cornish pumping engine. In other spheres, the way ahead lay, ironically, in the use of compounding to give a smoother running engine, and higher boiler pressures to give a more economical engine, both of which had been pioneered by Arthur Woolf but which had to be discarded through the limitations of technology available at that time. Also, in the background, lay the problem of the theoretical knowledge of the properties of steam and the best way of using it. This may be summed up by questioning whether there was any inherent advantage in using a small amount of high-pressure steam with all the dangers that involved, or whether the same result could be achieved more safely by using a larger volume of low-pressure steam. If the steam engine is a pressure engine, the answer lies with the latter, but if the steam engine is a heat engine, then the hotter the steam, and so the higher the pressure, the better.

The lead of the Cornish engineers with their more efficient engines was not followed by the manufacturing districts. One reason was that Watt had laid down a rule which others copied for years afterwards without troubling themselves to inquire, or even think, whether it could be improved upon or not.

> The matter before them was all-sufficient because it answered up to a certain point of working duty, and thus mutual contentment reigned where an equal desire for further knowledge ought to have been.[55]

The textile manufacturer was concerned primarily with meeting production schedules. He also had to consider the employment of his workforce. For him, it was better to have a less efficient engine which could be guaranteed not to break down rather than have his people being paid for idleness.

In the period between 1800 and 1830, there was one major advance in textile mill design. This concerned the line-shafting which distributed the power from the waterwheel or steam engine to the various machines. William Fairbairn claimed that he introduced this in 1815.[56] At the beginning of the nineteenth century, the earlier wooden shafting had been replaced by lengths of cast iron, square in section. Often these were badly aligned and jointed so that, in some cases, the power to set them in motion was almost the same as that required by the machinery they drove. Running at a speed of 50 r.p.m., they had to be made very heavy to transmit the power, and had to have large diameter pullies fitted on them to give the correct speed to the machinery. Not only did they take up a great deal of space, but they also blocked out the light which was a great disadvantage in spinning mills.[57]

Fairbairn realised that, if the speed could be increased, the power transmitted could also be increased. A shaft running at 80 r.p.m. will transmit 100 h.p. with no more stress than 50 h.p. at 40 revolutions. Where the machinery was running at high speed, it was advantageous to have the shafting running at a proportional velocity. If the machinery turned at 500 r.p.m., then Fairbairn recommended that the line-shafting should rotate at half that speed, i.e. 250 r.p.m. Therefore, the line-shafting had to be entirely redesigned and Fairbairn changed to wrought iron which could be turned accurately in a lathe and made to run true. These new shafts were a third of the old weight, so there was a saving in the initial cost, in the power absorbed and in maintenance. Their lightness meant they could be hung from beams in the ceilings, well out of the way, where they obstructed neither the light nor the other machinery. Work or energy which was previously absorbed in transmission was now conveyed to the machinery. Dr Andrew Ure commented, 'The method of increased velocities in the driving arms of factories is undoubtedly one of the most remarkable

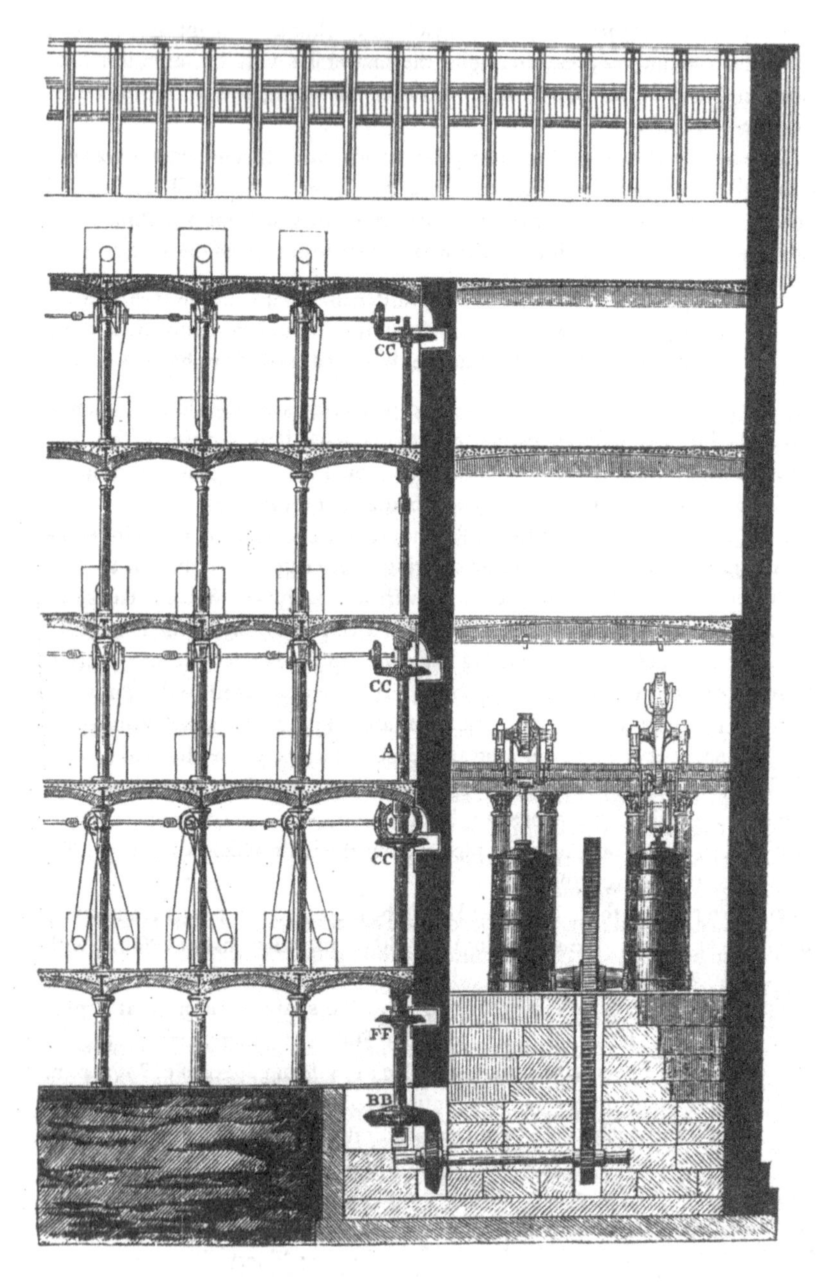

Figure 31 Section through a textile mill showing the mill gearing and shafting taking the drive from the engine to the machines on the different floors.

improvements in practical dynamics.[58] Fairbairn ascribed his own success in life to the saving of power he thus introduced and it established an entirely new system of operation in transmission techniques.

There was no matching improvement in mill engine design for a long time. Very few details have survived about what was actually built, but the single cylinder beam engine, such as the one installed at the Albion Mill in Manchester around 1818, was probably typical for its period.[59] It would have been a low-pressure condensing engine, for right up to 1840, the maximum steam pressure used in most textile mills was only 10 p.s.i.[60] The first Order Book of Benjamin Hick covers the period from October 1833 to March 1836.[61] It is difficult to judge how typical it is because, while Hick was well known from his days at Rothwells, he may not have had the capacity for building large engines when he first established his Soho Foundry in Bolton. In 1835, Hick sent the following price list to Messrs Sharp & Co. in Manchester.

> *List of prices furnished to Messrs Sharp & Co. Manchester*
> Steam engines fitted up in the best style of workmanship includg. one wrought iron boiler of sufficient size to work the Engine, (say five superficial feet area at surface of water to each Horse power) with all its Grate bars, Bearers, Fire Door, Damper etc. etc. deld. free of expense (except a small charge for packing cases) within a days journey of Bolton, or at Liverpool if for exportation, payable in a 3 mns. Bill, one half the amount when ready for setting to work & the remainder in 3 mns. after, for foreign orders one half on dely. & remainder in 3 mns. We include requisite plans & instructions for the Engine House, Boiler seating, etc. etc. the time of an experienced workman to erect the Engine if for this country.

10 Horse Engine & Boiler		£ 520.0.0
12		570.0.0
16		660.0.0
20		750.0.0
25		860.0.0
30		950.0.0
40		1250.0.0
50	with 2 25 horse boilers	1650.0.0
60	with 2 30 horse boilers	1850.0.0

> For coupled engines with their cranks at rightangles including the pair of first motion spur wheels, double price.
>
> We are making many 'Independent Frame' Engines with short strokes as substitutes for Marine Engines where room is an object, the cost of which is the same as for house-built engines with long strokes.[62]

This list is interesting because it suggests that the single-cylinder beam engine with one wagon boiler was still the standard type. Other types were beginning to make their appearance, possibly as the result of the influence of steamships.

The number of engines actually ordered from Hick's during the period of the Order Book was 39. In the two full years covered, there were 11 in 1834 and 23 in 1835. The power ranged from one at 12 h.p. to a double one at 160. The average size was 48 h.p. and the most popular type was a pair of 30 horse engines coupled together to give a 60 horse. Five of these were ordered. Out of the 39 engines, 26 were single and thirteen coupled. This in itself shows a trend in engine design. The idea of using two cylinders to give a more even turning force originated with the various types of atmospheric engines driving textile mills but the idea was never employed much with the double-acting Watt engines. Watt had included two cylinders on the steam carriage he patented in 1784 and so did Richard Trevithick and Andrew Vivian in 1802 for the same purpose. They specified that the cranks should be set 'at one quarter turn asunder'. In 1802, Murray also patented 'new combined steam engines for producing a circular power, and for certain machinery thereunto belonging . . . for spinning cotton, flax, tow and wool, or for any purpose requiring circular power',[63] in which two double-acting cylinders drove a crankshaft with two cranks at right angles to each other. Sometimes pairs of cylinders like this were referred to as two engines. Various two-cylinder layouts were designed for ships and, of course, railway locomotives. Millowners, after seeing that such engines actually gave a smoother turning force, may have been persuaded to begin to change to this type.

The average power of the engines built by Hick at 48 h.p. shows the size of a typical mill engine at that date and corresponds with figures mentioned by other authors too.[64] When James Montgomery was writing his book on cotton spinning at the same time, he described the layout of a textile mill, six storeys high, capable of containing 23,000 spindles. It would require an engine of from 40 to 50 h.p. He recommended that,

> Every Spinning Factory ought to have a little more power than is merely necessary to drive it; because it is well known that the weight of the machinery will often vary with the weather, the quality of the oil used, etc; consequently, when there is barely a sufficiency of power, the engine will be frequently so overburdened, as to render it incapable of driving the machinery at a regular speed, thus requiring more trouble and expense for fuel etc.[65]

This size of engine was still less than the largest waterwheels for, in 1820, a 100 h.p. wheel was installed at Quarry Bank Mill, Styal, and Fairbairn was building waterwheels of 200 h.p. capacity in the 1830s. But the size of steam engines in textile mills was about to increase dramatically and Hick's order in 1836 for a 160 h.p. double engine for Horrocks & Nuttall, Preston, may reflect this trend.[66]

Table 3. Number of power looms in Britain 1803–57[68]

Year	1803	1820	1829	1833	1857
Number of looms					
in England		12,150	45,500	85,000	
in Scotland		2,500	10,000	15,000	
Total	2,400	14,650	55,500	100,000	250,000

Fairbairn's improved line-shafting not only allowed the same power to be transmitted through lighter shafting but it also allowed more power to be transmitted through heavier shafting and this may have been one factor increasing the size of mills. But there were also improvements to textile machinery. From a slow start by Edmund Cartwright in 1785, the power loom had been gradually developed into a reliable machine in the 1820s. One person who has been credited with its eventual success is Richard Roberts, who in 1822 constructed a loom largely of iron with various other improvements.[67] The growth of weaving by power may be judged by Table 3.

This increase in numbers alone masks the demand for power which power weaving placed on the steam engines because there was a great increase in productivity too on each power loom.

> A very good *hand weaver*, 25 or 30 years of age, will weave *two* pieces of 9-8th shirting per week, each 24 yards long, containing 100 shoots of weft in an inch . . .
>
> In 1823, a *steam-loom* weaver, about 15 years of age, attending two looms, could weave *seven* similar pieces in a week.
>
> In 1826, a steam-loom weaver, about 15 years of age, attending to two looms, could weave *twelve* similar pieces in a week; some could weave fifteen pieces.
>
> In 1833, a steam-loom weaver, from 15 to 20 years of age, assisted by a girl about 12 years of age, attending to four looms, can weave *eighteen* similar pieces in a week, some can weave twenty pieces.[69]

Shortly before 1850, each loom had a capacity of around 25 yards per day or six pieces of cloth each 28 yards long per week. At this period, there was a fashion for building vertically integrated spinning and weaving mills where both operations were driven by the same steam engine. For this, much more powerful engines were needed.

On the spinning side, Richard Roberts solved the complex problem of making the mule 'self-acting'. On earlier mules, while from the 1790s the actual spinning was done by power, the winding on of the yarn still had to be controlled by the spinner. A strike of mule spinners at Stalybridge for higher wages in 1824 resulted in the employers there approaching leading engineers of the district to see if they could make the mule self-acting (i.e. make it wind on automatically), so less skilled men could be employed in order to break the strike. Roberts took out one patent in 1825, but did not find an effective solution until he patented his quadrant winding mechanism in 1830.[70] The size of the semi-powered mules probably had not exceeded 400 spindles with 250 to 300 being more normal. With the self-actors, the number soon increased to 600 and even 1,000 spindles in the cotton mills.[71] By 1837, there were over half a million spindles in use worked on the basis of Roberts's self-acting principle and the number continued to grow rapidly. These new mules helped to extend the demand for power.

The larger textile mills needed larger steam engines and this was one reason for the introduction of coupled engines which were also found to drive the mills much more smoothly because the cranks were set at right angles. Attempts were made to increase the speed of existing engines. Watt had limited the piston speed to 240 ft/min but this was raised to 320 to 350 ft/min. Shortening the piston stroke permitted engine speeds to rise. Where attempts were made to increase speeds of older engines, numerous failures followed. Attempts to obtain more power by raising the boiler pressure also led to disasters and explosions so mill owners were very wary of making any changes to their engines and plant.[72]

We can see the beginning of a new era in engine design in the notebooks of Charles Beyer, who became head of the Drawing Office and then Chief Engineer at Sharp Roberts & Co. There was some sort of link between Hick and Sharp Roberts. The probability is that they combined to equip textile mills, with Hick supplying the engines and Sharp Roberts the textile machinery. In 1842, Beyer visited the mill of Messrs Sharp & Murray in Bradford (probably the one near Manchester), where he saw the latest trends in both mill and engine design. The mill building itself was four storeys and alongside it was a weaving shed, with traditional north light roofing, capable of housing 550 looms. Gas lighting was installed.

> The mill is engaged in the fancy business, weaving a great variety of *coloured goods* . . . The engine is nominal *40 HP* and Mr. Hick warrants it capable to exert 90 indicated horsepower. It is on Wolfe's principle.[73]

Hick appears to have pioneered both the high-pressure and the compound engine in textile mills. In October 1842, he sent another price list of his engines to Sharp Roberts, offering engines on the Common, High Pressure, Condensing and also their 'Compound two cylinder engines working both High and Low Pressure', see Table 4. It is obvious from the letter quoted above, and from comments by Farey,[74] that Hick's compounds followed Woolf's designs. Unfortunately no figures are given for the boiler pressures but judging by the extra

Table 4. *Prices of steam engines supplied by Benjamin Hick in about 1841*[75]

	Compound high-pressure		Condensing		Compound with two cylinders working both high and low pressure	
Horse power	Engine & 1 boiler	Extra boiler	Engine & 1 boiler	Extra boiler	Engine & 1 boiler	Extra boiler
6	£ 270	£ 50	£ 380	£ 50	£ 400	£ 65
10	370	65	500	65	550	100
12	420	80	560	80	630	120
16	500	100	650	100	770	150
20	600	120	750	120	900	180
25	700	150	860	150	1,020	220
30	800	180	950	180	1,130	260
40	1,070	240	1,200	240	1,450	320
50	1,300	280	1,450	280	1,750	400

charge for the boilers of the compound engines, these must have been worked at a higher pressure than the others. An estimate of fuel consumption was given. For the common high pressure engine, he estimated 18 lb of coal per horse power per hour, for the condensing, 14 lb and for the compound 5 lb.[75] With such economy, it is obvious to us looking back with hindsight where the future of the steam engine lay.

7
Good servants but bad masters

> The boiler is, in fact, to the steam engine what the living principle is to animated existence. Like the stomach, it requires food to maintain the temperature, circulation and constant action, which constitute the energy of the steam engine as motive power. To keep up the temperature we have to feed, stoke and replenish the furnace with fuel, and we may safely consider it a large digester, endowed with the functions of producing that supply of force required in the maintenance of the action of the steam engine.[1]

So wrote Fairbairn in 1861. He had good reason to value the importance of boilers. He had built many steam engines and carried out experiments to improve their efficiency. He had also experimented with different types of boilers and tried various ways of burning coal in their furnaces as efficiently as possible, with the result that he developed the 'Lancashire' boiler, which was the most usual type in cotton mills from roughly the 1850s until the final abandonment of steam power.

Yet boilers have been too often the forgotten and most neglected parts of steam engines, with sometimes literally fatal results. This was because most people lacked even a basic understanding about how they should function. First, it was necessary to learn about the nature of coal and the different properties of the various types. Then the furnace had to be designed to burn the particular coal in the best way. After that, the heat from the burning coal and gases had to be transferred to the water inside the boiler with as little loss as possible. From the boiler, the steam had to be taken to the engine with the minimum waste of heat and pressure. The boiler had to be supplied with water; the water level had to be monitored inside the boiler; provision had to be made for excess steam to escape through safety valves. All this took many years of trial and error. In fact, it was as late as 1888 that R. H. Thurston summed up the objectives of good boiler design,

> 1st. To secure complete combustion of the fuel without permitting dilution of the products of combustion by excess of air.
> 2nd. To secure as high temperature of furnace as possible.
> 3rd. To so arrange heating-surfaces that, without checking draught,

> the available heat shall be most completely taken up and utilized.
> 4th. To make the form of the boiler such that it shall be constructed without difficulty or excessive expense.[2]

Early fanciful pictures of boilers for say de Caus's steam fountain in 1615 or even Savery's engine show a spherical boiler sitting on top of an open fire.[3] The open fire was probably a figment of the imagination, for an original drawing of Savery's engine dated to 1699 depicts his boiler placed in a brick setting.[4] Brick settings were standard for all boilers in textile mills until a few package boilers started to be introduced in the 1970s. A spherical boiler was the best shape to withstand the pressures of 150 p.s.i. which Savery used according to Desaguliers and, in 1717 or 1718, Desaguliers built such a boiler for the Savery engine he sent to Czar Peter I for his gardens at St Petersburg.[5] However, spherical boilers were not as efficient as vertical cylindrical ones.[6]

The problems Savery faced illustrate well the difficulties which confronted steam engine builders for another 150 years in the construction of boilers. Savery used copper sheets riveted and soldered together. His boilers did not exceed 30 in in diameter and so were quite small, holding 5 or 6 hogsheads (or gallons according to another account) of water.[7] Soft soldering was satisfactory for the boilers Desaguliers sent to Russia where the water was forced up only about 11 ft, but, when Savery tried much higher lifts, the steam became so hot that this type of solder turned pasty and then melted.[8] Savery found that in practice he had to go to the great expense of having the joints soldered with spelter or hard solder[9] and also to limit the pressure to three atmospheres which gave a steam temperature of 135 °C (274 °F). This meant that he could raise water only about 100 ft. How to make joints and seams of boilers steam-tight, particularly close to the fire, remained a problem for very many years. Internal stays helped to give strength and Savery resorted to them as well.[10]

It is therefore no wonder that Newcomen tried to avoid all these problems by using low-pressure steam because the engine near Dudley Castle lifted the water over 150 ft, which, in a Savery engine, would have necessitated steam at 153 °C (307 °F). Newcomen based the design of his boilers on those in brewhouses. They were described by the term 'haystack' because they had domed tops mounted on a cylinder which was closed at the lower end by a dished bottom.[11] The boiler of the Dudley Castle engine was 5 ft 6 in diam., 6 ft 1 in high and contained 13 hogsheads. The domed top was made of lead and the side and the bottom of copper sheets. The design may have been modified on subsequent engines where the diameter of the dome was made greater than that of the cylinder which was flanged out at the top where it joined the lead.[12] The fire was situated under the concave dished bottom. The edge of the dish rested on the brickwork of the furnace. The hot gases passed in a circular flue round the side of the boiler before going up the chimney. The edge of the upper flange also rested on the brick setting of the flue and there were various accidents on later engines when the stoker allowed the water level to fall and uncover the flange.

Very quickly wrought iron was substituted for copper. The first copper boiler supplied with the 1715 engine at Whitehaven burnt out at the flange and in 1717 was replaced by a wrought-iron one.[13] Various accounts for building boilers show that the lead dome was retained until at least 1733[14] and it is claimed that it was Smeaton who first built a boiler completely of iron at Long Benton in 1772.[15] John Spedding experienced severe corrosion on the boiler at Whitehaven for the mixture of lead and iron, together with impure water from the mines as the boiler feed, must have created ideal conditions for electrolytic action, something not understood at that period.

Triewald was highly critical of the design of the early boilers which could not supply enough steam.

> The cause of this conclusion was the false principles concerning the steam which the inventors harboured in their minds, thinking that the steam rises from or is generated by the boiling water in proportion to the quantities of water. In consequence of these false principles, they made their boilers very high, as can easily be seen from the Stafford machine, the boiler of which they made of greater height than width, thus not knowing that they should give the boiler a suitable shape. Neither did they possess the knowledge of the great importance of letting the fire play all around the sides as well as at the bottom of the boiler – not to mention many other improvements which a sound theory concerning the fire-machine seems to suggest and demand.[16]

It is easy to understand how Newcomen could have thought that it was the volume of water and not the heating surface which produced the steam. The large steam space was probably essential, not only to act as a reservoir for the steam but to prevent priming as well, a problem that will be discussed later. C. F. Partington, writing in 1822, commented that,

> It is of considerable importance that this part of the boiler be accurately proportioned to the power of the engine. If the boiler-top be too small, it requires the steam to be heated to a greater degree to increase its elastic force sufficiently to work the engine, and then the condensate on entering the cylinder will be greater.[17]

What Partington does not say is that the increase in temperature implies an increase in pressure which would have upset the balance of the engine, even if the early boilers could have withstood it safely. The higher pressure might have caused the piston to rise quicker than the pump buckets descended and so jerked the suspension chains.

Starting in 1770, Smeaton made extensive tests on an experimental atmospheric engine which he built in the grounds of his house at Austhorpe near Leeds. He seems to have made few changes to the basic shape of the boiler which by this time had lost the flanged portion. He supplied three haystack boilers of this later shape made of cast iron, 10 ft in diameter and 16 ft 4 in high to steam an engine with a 66 in diam. cylinder and 8 ft 6 in stroke for pumping out the docks at Cronstadt in Russia around 1775.[18] He took care to use good quality water or rain water wherever possible and warmed it before sending it into the boiler. One thing he discovered in his experiments was that the greatest amount of steam was raised with a thin, clear fire, evenly spread over the grate, with a bright flame, and stirred regularly.

> When the fire was made with the same coals, heaped thick on the grate, to keep up a large body of fire, and stirring it but very rarely, the performance or mechanical power exerted by the engine, with a given allowance of fuel, was only five-sixths of what it was with a clear bright fire; though all the other circumstances were kept as nearly as possible the same in both cases.[19]

As the call arose for more powerful engines, so the boilers had to be built larger and larger. Leupold, writing in 1725, mentions one of 7 ft diam.[20] Those Smeaton sent to Russia were 10 ft, and John Curr, writing in 1797, advised that no boiler should be constructed with a diameter larger than 17 ft.[21] Curr's advice was not heeded, for Tredgold recounts the tale, but does not give the date, of a spectacular boiler explosion at Cyrfartha Iron Works in South Wales.

> The boiler was constructed of the old spherical form, twenty feet in diameter; the thickness of the plates when new was, top plates a full quarter of an inch, bottom plates half an inch; load on the safety valve 7 lbs. per circular inch. Many lives were lost by this explosion; and the boiler was thrown to a distance of 150 feet, to a place 30 feet above the level of its former seat. The upper plates were undoubtedly too weak.[22]

Boilers of this size could not be fitted conveniently inside the engine house, and in any case it began to be realised that, if an engine had to work continuously, two boilers might be needed so one could be cleaned out while the other was still at work. But this presented a problem because Newcomen placed the main steam valve in the top of the boiler immediately under the cylinder. Smeaton retained this form of construction in the engine he built for Long Benton in 1772. With a cylinder height of 10 ft, on top of the same amount for the boiler, the height and width of the engine house became excessive when the heights of the fireplace, valve gear, piston rod and arch head were added too.

By this time, other engine builders had moved the boiler away from the cylinder to the side of the engine. This was the position of the boiler on the

Fairbottom Bobs engine which has been dated to 1760, although the boiler that survived with the engine was much later. In the North East of England, William Brown erected an engine at Benwell Colliery in 1762 which had three boilers and, in the following year, one which was amongst the largest ever built at the Walker Colliery which had four boilers.[23] The idea soon spread for the detail on a map of 1769 of the Persberg iron ore mine in Sweden shows an atmospheric engine in one building with the haystack boiler beside it. This engine was delivered from the Carron Company in Scotland. It was erected by the Scotsman, Ebenezer Grieve, with two of his workmen in 1765–6 and worked until 1772.[24] While this arrangement may not have reduced the total height required because the boiler still had to be sited below the level of the cylinder to allow for drainage of condensate in the pipes, the engine house could be narrowed and the beams supporting the cylinder made shorter.

The haystack boiler was the type that would almost certainly have been used for the first atmospheric engines in the cotton mills. Farey has a picture of such an engine with a crank and a connecting rod while the boiler is still placed underneath the cylinder.[25] This type remained popular in colliery districts for many years into the nineteenth century. Some of the illustrations in Thomas Hair's *Views of the Collieries . . . in Northumberland and Durham* drawn in the late 1830s show haystack boilers at various collieries. Hartley Colliery has three in a row working and a fourth spare.[26] Robert Armstrong, writing soon afterwards, said that such boilers were still being used in the Staffordshire collieries but were being replaced by the egg-ended type.[27]

For his first engines, Watt used haystack boilers, but he soon found that Cornish engineers preferred a long type called the 'wagon' boiler from its shape.[28] This had a rectangular box for the bottom with the top shaped from half a cylinder. The bottom and two longer sides were concave to help withstand the pressure better and to provide improved heating surfaces. The two short sides, or ends, were left flat. The fire was situated beneath one end and the gases passed underneath the boiler before being taken in a flue right round the sides and passing to the chimney. The tops might be covered with wooden lagging or a layer of bricks to help retain the heat. The drawings for the second rotative engine erected by Boulton & Watt in a textile mill, that for Timothy Harris of Nottingham which started running in 1786, show a plan for a wagon boiler sketched over one for a haystack boiler.[29] The wagon boiler soon became the more popular type in cotton mills.

Watt favoured copper for boilers. Not only was the art of the coppersmith more advanced even in the 1770s than that of the blacksmith, but he must have realised that copper conducted heat much better than iron. Farey instances a 10 h.p. engine for a starch manufactory in Lambeth which Boulton & Watt erected in 1795 with a copper haystack boiler.[30] Watt took great care to arrive at the best proportions for his boilers and found that he must allow a greater heating surface when using iron. His early engines were under-boilered and this defect led him to determine, for the first time, a desirable ratio between boiler heating surface and cylinder volume. He specified 4 sq. ft of heating surface per cubic foot of cylinder volume.[31] He found it needed 8 sq. ft of surface area to boil a cubic foot of water in an hour and he allowed 12 sq. ft for each horse-power.[32]

To supply feed water to the boiler, Boulton & Watt developed a system where the water flowed in by gravity under the control of a float inside the boiler. When the water level in the boiler fell, the float also fell and so opened a valve in the bottom of a cistern on the top of a stand-pipe. The cistern was filled with hot water from the overflow of the condensate and injection water. The stand-pipe finished nearly at the bottom of the boiler, well below the normal water

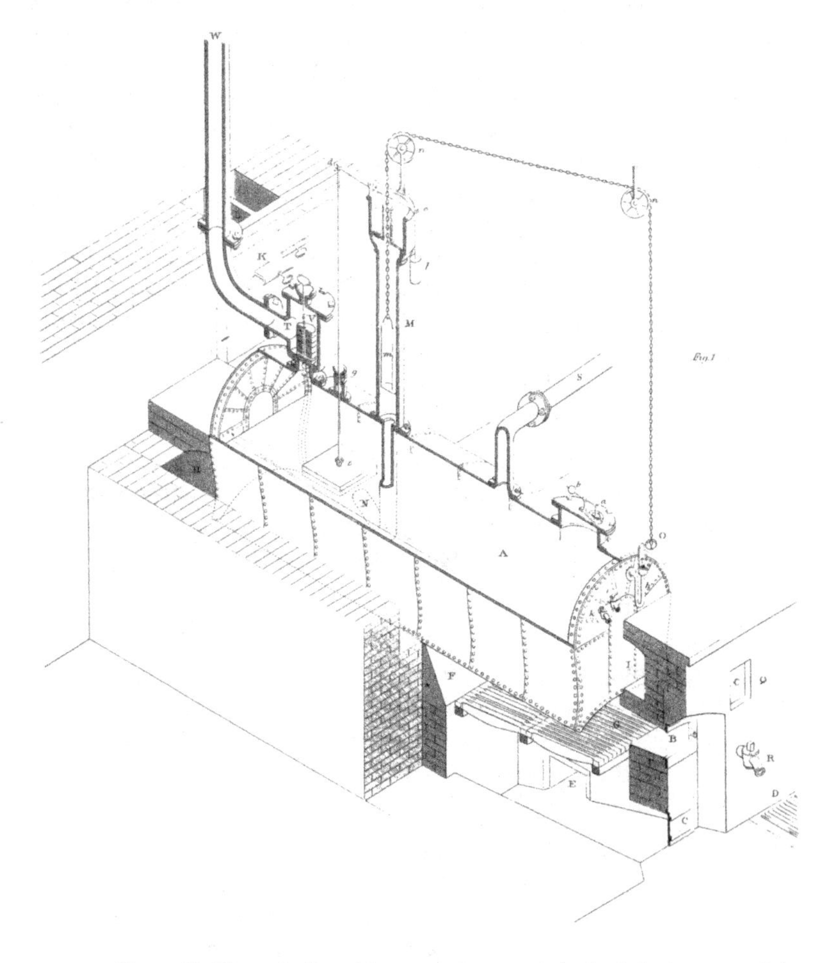

Figure 32 Wagon boiler with automatic controls for both feed water and the flue damper. (Tredgold, *Steam Engine.*)

level. As steam pressure rose, so the water was forced up the stand-pipe which had to be made tall enough to cope with the proper working pressure. This arrangement is shown on the drawings for the engine supplied to Oldknow in 1791 for his cotton mill in Stockport,[33] and became standard on both haystack and wagon boilers. In 1799, Murray connected the damper at the bottom of the chimney to another float which he placed inside the stand-pipe. When the steam pressure, and so the water in the stand-pipe, fell, the float fell too, opening the damper automatically to give the fire more air to make it burn brighter. The opposite happened when the pressure rose.[34] The fact that both the water level and the action of the fire could be controlled automatically was seen as a great achievement in the development of the steam engine.

Cotton millowners seem to have been very conservative over their steam engines and preferred to retain well-tried designs. This applied to their boilers too. Rothwell, Hick & Rothwell, of Bolton, increased the heating surface to 15 sq. ft per horsepower in the wagon boilers they supplied with their steam engines in the middle 1820s.[35] When Hick established his own foundry in 1833, he was offering wrought-iron wagon boilers as the normal type. One was 24 ft long by 8 ft wide,[36] and had an internal flue running throughout its length. Many were also stayed to help withstand higher pressures. The internal flue also acted as a stay strengthening the flat ends as well as increasing the heating surface. Boulton & Watt continued to build such boilers until 1842 at least, but, in their case, the internal flue did not run right through from end to end. It bent at right angles near the back and was joined to the bottom.[37] This could not have created such a strong construction.

Armstrong wrote in 1838,

> Within the last eight or ten years, the waggon boiler may be said to have been almost universal in the cotton district around Manchester; and it is believed that four fifths of all the boilers at work in this neighbourhood at the present time are of this kind.[38]

Two years later in 1840, he carried out a survey and ascertained that about three-quarters of all the boilers in the neighbourhood of Manchester were of the wagon shape.[39] But by that time, in spite of the extra strength given by the internal flue and also by stays, the wagon boiler had become outmoded through developments elsewhere in steam engine design which called for higher boiler pressures. From the 4 or 5 p.s.i. which was all that Watt had considered safe, during the 1830s pressure for the Lancashire mill engines nudged up to 12 p.s.i. and was beginning to increase steadily.

The craft of the boilermaker, and the design of boilers, had to be drastically changed to enable steam to be generated safely at higher pressures. The manufacture of wrought iron had been cheapened when Henry Cort patented his puddling process in 1783.[40] Up to that time, wrought iron had been produced by the bloomery process in small pieces. Boiler plates were made by hammering these together but the welding might not always be sound and the thickness was liable to vary. Cort's process yielded a larger lump of wrought iron which was hammered and then rolled, giving more uniform sheets. Even so, the sheets were still very small. Boiler plates about 4 ft by 8 in by ½ in were being rolled at the Coalbrookdale Company's works at Horsehay in 1790 but no other works in Shropshire made them that size.[41] Even in 1797, Curr's book, which is the earliest account of boilermaking, gives the largest plate as 5 ft 9½ in by 16¾ in by probably ⅜ in.[42]

The rivet holes in these plates had to be punched by hand, which might cause distortions. When the plates were lined up, the holes might not match properly so drifts were used to try and make the holes true but this often stretched the plates and weakened them. The rivets were hammered tight by hand, but if the holes were not true, they might not fit properly and might leave a gap through which the steam or water escaped. Between the layers of the plates, some form of caulking had to be inserted, such as rope fibres or paper smeared with white lead. Boulton & Watt recommended hammering the edges of the plate with a blunt chisel to try and caulk it and also painting every joint and rivet above the waterline with thin putty. If all this failed, resort was had to pouring dung or oatmeal into the boiler while the engine was working.

The pioneering work on high-pressure steam was carried out by Cornish engineers, in particular Richard Trevithick and Arthur Woolf. Both at first turned to cast iron because round 1800 the iron founder's art was more highly developed than smithing. Here it is interesting to note that the Select Committee of the House of Commons, set up in 1817 to investigate the cause of the boiler explosion in the steam boat at Norwich, found many engineers who still favoured cast-iron boilers. For example, John Hall of Dartford, Alexander Tilloch of Islington and even Thomas Lean of Cornwall, all considered that cast-iron boilers could be made stronger than wrought-iron ones. They believed that wrought-iron boilers, being riveted together, could not be as strong as those cast in a solid mass,[43] but others strongly disagreed with them. At this time, it was still not possible to make wrought-iron plates as thick as cast-iron ones, so, for high pressures, there was no alternative but to use cast iron.

Trevithick had built his first cast-iron boiler in 1803.[44] It had the traditional haystack shape but by 1806 he had designed a cylindrical boiler which had a horizontal barrel cast integrally with one slightly dished end. The other end was closed with a bolted-on end plate. On this were mounted the fire grate and a wrought-iron return 'U' tube taking the gases to the chimney which was also mounted on this end plate.[45] He generated steam at pressures ranging from 50 to 90 or 100 p.s.i. in these early engines which were quite small. At this period, Trevithick constructed an even larger return tube boiler. In a letter of 1804, Samuel Homfray of Penydarren Place, South Wales, wrote,

> We are now so thoroughly convinced of the superiority of these engines that I have just begun another of larger size. The boiler is to be 24 or 26 feet long, 7 feet diameter, fire tube at wide end 4 feet 4 inches and at narrow end, where it takes the chimney, 21 inches, steam cylinder 23 inches diameter. This boiler, on account of its tube withinside, will, I have no doubt, get steam in proportion, and work the engine with much less coals than our present one.[46]

The shell of this boiler also was made of cast iron. These boilers firmly established the advantages of placing the furnace inside the fire tube of the boiler.

The introduction of the 'Cornish' boiler is usually given as 1811 or 1812 when Trevithick replaced the Boulton & Watt wagon boilers at the Dolcoath mine in Cornwall with this type.[47] These boilers were cylindrical with flat ends. A fire tube ran through the middle and helped to stay the ends. The furnace was in one end of the tube. The gases passed to the back through the tube where they

Figure 33 Trevithick's high-pressure engine and return flue boiler. (Rees, *Cyclopaedia.*)

divided before passing outside along either side to the front. There they joined in a central flue under the boiler and then went out to the chimney. The idea was that, with such long flues, the maximum amount of heat could be extracted.

Trevithick never patented this type of boiler, probably because other people had built examples before him. As early as 1792 in the United States of America, Evans had constructed a cylindrical boiler with a fire placed under one end in a brick setting so the gases passed underneath and then back along a tubular flue through the boiler to a chimney at the firebox end. By 1804, he had moved the fire to inside the tube and was using steam at 100–150 p.s.i.[48] In England, Ralph Dodd patented in 1806 a cylindrical boiler with a central flue but it did not have the return passage underneath.[49]

Trevithick constructed his boilers with wrought-iron plates but, when those at Dolcoath were first installed, they leaked 'very much; we could hardly keep the fire up sometimes'.[50] The high-pressure steam blew out the caulking in the joints because the standard of manufacture was inadequate and also the heat burnt it. It was his fellow countryman, Woolf, who solved this problem with improved machine tools. Trevithick gave his boilers an hydraulic test to four times their working pressure[51] which ought to have ensured their safety. However, no boiler could be made strong enough to withstand people tying down or adding weights onto the safety valve to increase the pressure, and Trevithick's smaller high-pressure engines acquired an unenviable reputation for boiler explosions through this cause.

The advantages of the Cornish boiler were many. Its basic design was sounder than either the haystack or wagon types for withstanding higher pressures and the end plates could be reinforced with stays. Its long flues permitted more heat to pass into the water to raise steam, but its greatest advantage was the position of the fire. Being placed inside the boiler, all the heat was radiated directly to the water and not to the brick setting. In older types, the fire heated the bottom plates where all the impurities and sediment had settled. This could cause serious overheating of the plates, particularly if the sediment had formed hard layers. Loose particles were carried up by the ascending bubbles of steam and formed a scum on the surface. This could cause severe priming when water was carried up into the steam pipe and into the engine. Priming was caused also by dirty feed water. Beside the River Medlock in Manchester stood more steam engines than on any river of equal length in England. This water was highly polluted and caused severe priming.[52] The higher the pressure of the steam, the more these problems increased. In Cornish boilers, the sediment fell to and remained at the bottom from where it could be easily washed out through the draining-down pipe.

Cornish boilers had two defects. The internal tube had to be quite large so the water level on top of it was rather meagre and had to be watched carefully; something rather difficult before the days of gauge glasses. Also the tube could collapse unless it were strengthened. Yet the Cornish boiler remained a popular type to the end of the steam era but was not employed much in the textile industry for reasons which will be explained later.

In order to increase the heating surface exposed to the direct action of the fire, Woolf patented a totally different type of boiler in 1803.[53] He placed a series of small tubes side by side horizontally which were connected together by a larger cylinder placed at right angles above them. The lower tubes and half the cylinder were filled with water and the boiler setting was constructed so that the heat from the fire, which was below the first of the small tubes, passed alternately over and under the others in a waving course.[54] He proposed other ways of arranging these tubes too. It is interesting to note that Woolf recommended short tubes

> to prevent any derangement taking place in the furnace or in the tubes, by the expansion and contraction by changes of temperature, which would be much more considerable in one tube of the whole length of the furnace than when divided into three portions.[55]

While these boilers could be made from any material, Woolf preferred cast iron because he considered that he 'could make a cast-iron boiler stronger and more to be depended on for great pressure than wrought-iron'.[56] He designed his boilers to stand from 14 to 20 times the pressure at which they were intended to be worked and fitted two safety valves.[57]

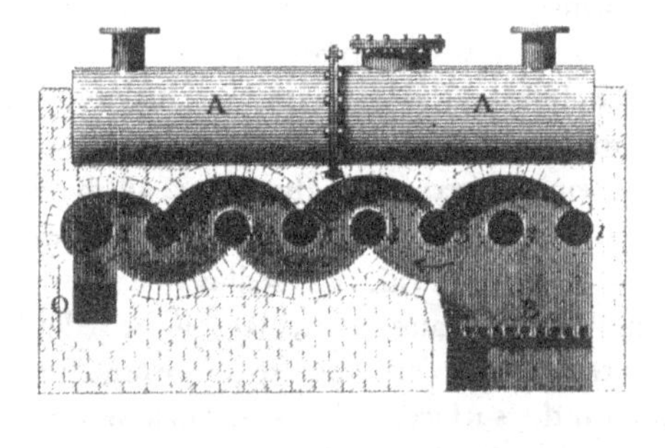

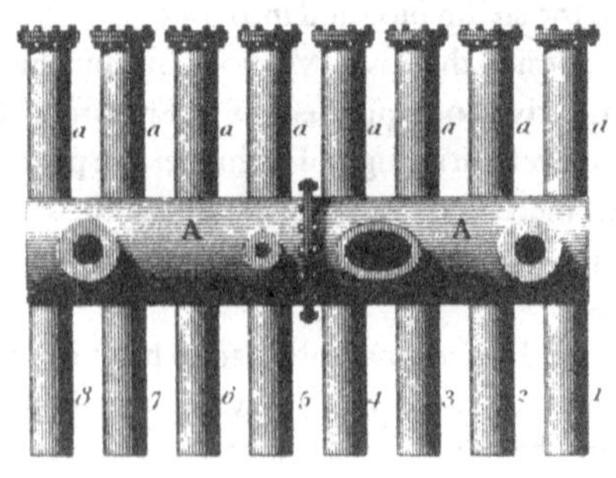

Figure 34 Woolf's cast-iron boiler. (Rees, *Cyclopaedia*.)

Table 5. *Improvement in efficiency of pumping engines, 1769–1859*[61]

	Duty in lbs
1769 Average of old atmospheric engines	5,590,000
1772 Smeaton's atmospheric engine	9,450,000
1776 Watt's improved engine	21,600,000
1779–1788 Watt's engine working expansively	26,600,000
1820 Engine improved by Cornish engineers	28,000,000
1830 Average duty of Cornish engines	43,350,000
1859 Average duty of Cornish engines	54,000,000
1859 Extreme duty of best engine	80,000,000

Woolf had built his first boilers in London, but in 1811 he returned to Cornwall and started working on the large mine pumping engines there. He fitted several with his boilers but it was found that the small tubes burnt out through incrustation on the inside from minerals in the water. Also, with the repeated heating and cooling of the cast iron, cracks tended to develop so that the water began to leak out and the boilers had to be worked at reduced pressure.[58] In fact, cast iron was not a suitable material for boilers where it could be subjected to unequal heating and at this period there was no way of discovering whether there were any flaws in the castings which might fatally weaken them.

Woolf's Cornish practice involved him in constructing during 1818 a large wrought-iron boiler of Trevithick's design which he succeeded in making steam-tight without placing rope yarn between the laps of the plates. He achieved this by strict attention to detail, such as the correct formation of the rivet holes by means of a specially designed cylindrical punch and making all the rivets to a uniform gauge. Later he advanced the art of boilermaking still further by designing a 'holing engine' to facilitate punching the rivet holes in boiler plates.[59]

As has been seen earlier, the work of the Cornish engineers dramatically improved the efficiency of the steam engine. This was recorded by Thomas Lean who, starting in 1811, published monthly reports on the performance of each engine.[60] The results were briefly summarised by W. S. Jevons in 1865, see Table 5, although with different figures from those discussed earlier. At various times it was suggested that a similar system of reporting ought to be arranged in the textile areas but nothing ever happened.[62] Some of the Cornish engines had returned much better figures than Jevons shows. For example, Woolf's second compound engine at Wheal Vor did a staggering 45 million pounds duty in 1815. In 1828, Grose had improved this to 87 million on the Wheal Town engine, with a boiler pressure of 40 p.s.i.[63] and figures of over 100 million pounds were reached eventually (see p. 111).

This increase in performance was achieved partly by better boiler design, such as the Cornish boiler, partly by attention to details such as more thorough insulation; partly due to improvements in the engine themselves such as using steam expansively; and partly through using higher steam pressures. In 1830, most of the cotton mills were still using steam at about 6 or 7 p.s.i.,[64] but the millowners began to take notice of the potential savings which could be attributed only to high pressure. The fact that this could be done safely was demonstrated on their own doorsteps by the locomotives on the Liverpool & Manchester Railway. From its opening on 15 September 1830, locomotives were making regular journeys hauling passengers in safety with steam pressures of 50 p.s.i. in their boilers. The fears of repeated boiler explosions proved unfounded.[65]

Seeing railway locomotives at work, something which people did not do with the engines hidden away in textile mills, began to give confidence in the use of high-pressure steam. In the 1840s, Ernst Alban wrote,

> The high-pressure engine has generally been conceded only a very limited field of application, and considered as only applicable to a range of objects. The late introduction of railways, and the great interest excited by them in all quarters, seems, however, now about to place the principle of the high-pressure engine in a higher point of view. It is generally found, that a subject which has lain for a time dormant, has on its revival been taken up with greater zeal than before; and thus it is we now find that in England, so long exclusively the country of the WATT engine, the high-pressure plan is occupying the attention of engineers, and furnishing employment for the workshops throughout the land . . . Thus England appears again about to become the mart for the high-pressure engine, and all now look to that enlightened nation for the perfect dispersion of the obscurity in which the subject has heretofore been enveloped.[66]

As the art of boiler making improved, so steam pressures rose. Francis Trevithick described how, in 1843, to cope with trains lengthened to fifteen carriages, he was allowed to risk an increase in boiler pressure on the Crewe locomotives from 50 to 60 p.s.i. Then, in 1845 when the line over Shap was opened, steam pressure was raised to 75 p.s.i.[67] This was increased to 90 or 100 p.s.i. around 1852 and reached 120 p.s.i. only when the original Crewe type of locomotive was given up for new designs in 1857–8.[68] David Joy used 100 p.s.i. on the 'Jenny Lind' locomotive in 1847.

In order to be able to increase the pressures for their stationary engines, Lancashire millowners turned to a variety of boiler designs. The simplest was the cylindrical boiler with hemispherical ends, often called the 'egg-ended' boiler, which made its appearance around 1814.[69] This had a similar setting to the wagon boiler with a fire below it at one end and the flue passing underneath to the back and then all round at a higher level before reaching the chimney. A variation was the cylindrical boiler with flat ends and a tube for the gases passing through the middle. Both these suffered from the same defect as the wagon boiler for they were liable to prime badly because the sludge fell to the bottom where the action of the fire stirred it up. These boilers tended to be bulky because the heating surface was comparatively small.[70] However, their advantage was that the fire grate was not restricted in size, as was the case with the Cornish boiler.

A large fire grate was necessary, particularly in the Lancashire area where the quality of coal was inferior to the South Wales steam coal sent to Cornwall. Here it is interesting to note that the Great Western Railway, which used Welsh coal in its locomotives, retained a narrow firebox for its most powerful 4–6–0 King class engines in 1927 whereas the London & North Eastern Railway employed a wide firebox for the 4–6–2 *Flying Scotsman* in 1923 because the calorific value of its coal was not so high. In the cotton mills, a type of boiler, developed by the Butterley Co. of Derbyshire around 1811, became popular for a short time. It was called the 'whistle-mouth' boiler because the Cornish boiler was taken and a portion of the lower part cut away at one end. The fire was placed here so that the grate could be made the full width of the boiler. The gases still passed away through a flue inside the boiler in the same manner as the Cornish type.[71]

The best solution was eventually found in the 'Lancashire' boiler. It has not been possible to discover who actually developed this type which was similar to the Cornish except that it had twin fire tubes. A variation known as the 'breeches' boiler, in which the fire tubes joined together halfway through the boiler, existed in 1820. Robert Stevenson & Co. built a locomotive, the *Lancashire Witch*, in 1828 for the Liverpool & Manchester Railway with twin fireboxes inside a cylindrical boiler.[72] In his essay written in 1838, Armstrong says that the ordinary Cornish boilers did not

> answer at all for the factories in this district. Those which have been found to answer best, are much wider, in proportion to their length, than in Cornwall, and with two or more flues placed as low down in the boiler as possible, so as to leave sufficient steam room.[73]

John Bourne's *Treatise on the Steam Engine*, published in 1846, has an illustration of three 'Cornish' boilers erected by Messrs Maudsley & Field at the London terminus of the London & Blackwall Railway for operating the stationary haulage engines on that line which opened in 1840.[74] These boilers had twin fire tubes. The advantages were a greater depth of water over the tubes and an increased grate area compared with the ordinary Cornish boiler.

William Fairbairn and John Hetherington patented the 'Lancashire' boiler in 1844,[75] when they patented not so much the type as the way the boiler was to be fired. In all probability too, Fairbairn derived the idea of the twin fires and thc method of firing from Charles Williams who had been experimenting with smoke consumption on board ships by dividing large furnaces with an internal partition and firing each side at regular intervals.[76] Their method of firing, which will be discussed later, helped to reduce the smoke pollution and, for this reason, it seems that the Lancashire boiler became the most popular type in the textile areas, see Table 6. It was also an economical steam raiser and could be strengthened to take higher pressures. Fairbairn proposed using two fire tubes which tapered to a smaller diameter beyond the firebox. His boiler was supported on a single row of fire-bricks in the middle so the gases passed down one side and back along the other to the chimney. Later there were two rows of fire-bricks and the gases passed down the centre between them and then back along the sides, or in the opposite direction. It was claimed that, with the same thickness of plates, these boilers could withstand four times the pressure of older types.[77]

The idea of twin fireboxes was copied by other builders. Some joined the flues together beyond the grates to form a sort of combustion chamber where the gases from both fires could mix. One design passed the gases through a series of small diameter horizontal tubes beyond the combustion chamber, like those on a railway locomotive boiler. Boilers of this design were installed by Fairbairn in the mill at Saltaire in 1851.[78] The more successful type was patented by William and John Galloway in 1851.[79] In their design, the two fire tubes united into a single long kidney-shaped tube for the rest of the length of the boiler beyond the furnaces. The flat top and bottom portions of this tube were united and strengthened by a series of vertical conical tubes riveted into the main one. These conical tubes improved the circulation of the water and increased the heating surface as the hot gases passed around them. They were made conical

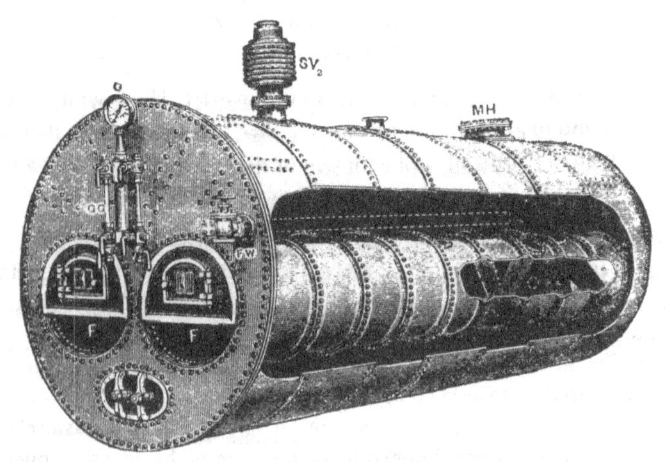

Figure 35 A Lancashire boiler of the 1880s with cross tubes in the fire tube. (Jamieson, *Elementary Manual.*)

Table 6. *Types and pressures of boilers in the Manchester area, 1859*[a]

	Pressure in lbs per square inch					
Type of boiler	−15	15–30	30–45	45–60	60+	Total
Cylindrical with flues (Lancashire)	119	406	326	253	73	1,177
Cylindrical, no flues	36	21	15	16	6	94
Galloway	4	29	36	33	1	103
Water-tube	3	14	24	43	35	119
Multiflued	—	19	19	5	1	44
Butterley	24	1	1	—	—	26
Waggon	7	—	—	—	—	7
Total	193	490	421	350	116	1,570

[a]From Kanefsky, 'Power Technology', p. 126

with the greater diameter at the top to assist the escape of the steam. Galloway invented special machine tools for forming these conical tubes and this boiler remained a popular type until the firm closed in about 1930.

In the 1830s and 1840s, there were attempts to save fuel by heating the boiler feed water with the exhaust gases before they passed up the chimney. In one experiment, an old boiler was placed in the chimney flue, but the improvement was negligible.[80] Edward Green patented a successful design of 'economiser' in 1845. The first one was installed that year in David Illingworth's Providence Mill, Bradford, Yorks. It consisted of rows of vertical cast-iron tubes, connected at top and bottom to headers. The design was modified over the years so that it became possible to take out any tube individually through the top header should one need replacing. Special machine tools were designed to bore out the seatings in the headers and turn the matching tapers on the ends of the tubes. The important feature of Green's economisers, which made them successful, was the scraper apparatus. This was moved up and down the tubes.

> with a continuous and alternating motion so as to keep the pipes continually free from any deposit of soot, thus permitting always the full action of the heated air and gases, which otherwise would soon be impaired or destroyed . . . the gear be made to alternate three or four times an hour.[81]

It was important that the economisers be supplied with hot water, otherwise the sulphur from the coal would condense on them and cause corrosion. Likewise the water in the economisers should not become hot enough to boil, or steam formed in the top header which could become overheated. Therefore

Figure 36 Green's economiser sectioned to show the scraping apparatus on the vertical water tubes. (Nasmith, Cotton Mill Construction.)

by-pass dampers were fitted to divert the hot gases directly up the chimney when the engine was stationary and the boilers did not need water. Normally the economisers were filled with water at full boiler pressure. Fairbairn said,

> It is found that when the waste gases escape at a temperature of 400° to 500°, the feed water can be heated to an average of 225°, the temperature of the gases after leaving the pipes being reduced to 250°. To produce this effect, 10 square feet of the heating surface are provided for each horsepower.[82]

Eventually economisers were fitted in practically every textile mill. They absorbed heat which would otherwise have been wasted but, more important, the temperature of the feed water was raised by about 60 °C (140 °F) before it entered the boilers so the chance of cold water straining them was reduced.

Another small but important safety feature which became essential on high-pressure boilers was the fusible plug. The idea was that, if the water level fell and uncovered the metal shell near the fire, a plug made of a lead alloy would melt and allow the steam to blow out into the fire, dampening it, before any excessive pressure was reached in the boiler. Trevithick fitted a lead plug in his boilers in 1803.[83] Early forms of plugs were not entirely reliable and scale sometimes formed over them so they did not work. They were continually improved and became an essential requirement for any boilers insured by the Association for the Prevention of Steam Boiler Explosions, later known as the Manchester Steam Users Association, founded in 1854.[84] It is interesting to note that high pressure meant something very different from what it does today, because, at its foundation, the Association would not insure boilers worked at over 60 p.s.i.

Behind the development of boilers to withstand higher pressures lay two other questions. The first was how to burn the fuel efficiently and the second was how to transfer the heat thus generated into the water. For burning the fuel, the first necessity was to understand the chemical reactions that occurred. It must be remembered that modern knowledge of chemistry is very recent. To sum this up very briefly, it was only in the 1770s that Joseph Priestley discovered that the air we breathe consisted of a mixture of gases. Among them he identified 'fixed air', carbon dioxide, and in 1774 'dephlogisticated air', oxygen, but there was still no proper comprehension of their role in the process of combustion. Cavendish isolated 'inflammable air', hydrogen, in 1783 but the composition of water was not determined until Friedrich Humbolt and J.-L. Gay-Lussac found that it consisted of two parts of hydrogen and one part of oxygen in 1805. John Dalton's publication in 1803 of his *New System of Chemical Philosophy* launched the start of the theory of atomic proportions as we know it today, but it took some years before the correct quantities of air needed for burning with different types of coal could be determined. The work of Sir Humphrey Davy in 1812, when he was advising on the problems of explosions in coal mines, was the first time that any systematic study had been carried out on the process of combustion.

Linked with the best method of combustion was the problem of smoke abatement. Although Watt had taken out a patent in 1785[85] for an improved way of burning coal, pollution increased with the growing number of textile mills and associated industries. One reason was almost certainly the method generally followed in firing. In spite of Smeaton discovering the importance of having a clear, bright fire, it would seem that most stokers shovelled on a large quantity of coal and shut the furnace door.[86] Today we are familiar with the bunsen burner and can see that the amount of air controls both the quality and the temperature of the flame, but this was not realised until the chemical properties were understood. The coal newly shovelled on had first to be 'coked', that is to have the volatile gases driven off. This absorbed heat and deadened the fire so that, in an extreme case, there might be insufficient heat above the fire to sustain the chemical reaction necessary to ignite the gases. Also there might not be enough oxygen left in the air which had been drawn through the fire to sustain combustion above it. Therefore firing these early boilers could cause the emission of vast amounts of smoke which sooted up the boiler surfaces as well as the creation of carbon monoxide which was a further source of heat loss.[87]

All sorts of smoke consuming apparatus and methods were suggested and the arguments raged furiously up to the early 1840s.[88] What evolved was the realisation that it was essential to introduce a secondary source of air, either above or beyond the fire. Fairbairn records that it was John Wakefield of Manchester who first turned his attention to admitting air beyond the fire which he did in the early 1820s.[89] Williams continued with the principle in the 1830s and this system achieved considerable popularity.[90] Another method was to introduce supplementary air through the firing door. When the fire door was opened for shovelling in another charge of coal, the volume of air drawn in was too large and cooled the furnace. But, by making in the firing door holes which could be regulated in size, the stoker could control this extra draught to produce the best burning conditions. The effect of this extra air supply was shown in experiments by Houldsworth in 1842 at Manchester where the fire became hotter by about 10 per cent and there was about a third saving of fuel.[91]

These changes began to reduce pollution.

> Dense black columns [of smoke] poisoning the air of the surrounding neighbourhood . . . are prevented now, however, by a moderate enlargement of the fire-beds and flues, and the introduction of air to the surface of the fire through perforated doors, and plates placed between them and the fire. The furnace itself being constructed to admit the quantity of air required for perfect combustion, the perforated plates secure such a mechanical division and distribution of the common atmosphere as to ensure its becoming instantaneously

> heated, and promoting, instead of retarding, as a column of cold air does, in a great degree the combustion sought. The whole mystery of all the smoke preventing apparatus now in public favour lies in this very simple secret.[92]

It was not until 1858 that roughly similar principles were successfully applied for burning coal in railway locomotive fireboxes.[93]

Now it is possible to understand the importance of Fairbairn's Lancashire boiler. At first, he used the Williams system of admitting air beyond the fire, but soon others let it in through the fire-hole door. Fairbairn recommended firing each furnace alternately with a little coal at a time. In this way, neither fire was smothered with a fresh charge of fuel and remained burning brightly. While one was coking the fresh fuel and so would be burning at a lower temperature, the other would be burning properly and so the steam supply could be maintained more evenly. All this reduced the smoke, saved fuel and raised steam better.

There still remained the problem of transferring the heat to the water. The Lancashire boiler presented the maximum surface area closest to the fire where the radiant heat could strike it. Haystack and wagon boilers with their dished bottoms were better in this respect than egg-ended boilers. The importance of radiant heat probably was not fully realised until the advent of the railway locomotive. In the 1840s, Daniel Clark carried out experiments on locomotive boilers which showed that most of the evaporation occurred round the firebox and decreased very quickly along the boiler tubes.[94]

Watt and many others felt that, in the ideal boiler, the gases would pass to the chimney at the same temperature as the steam.[95] As boiler pressures rose, this presented a dilemma because the temperature of the exhaust gases rose too and potentially useful heat was being wasted. Long tortuous flues were seen as desirable because in them as much heat as possible could be extracted from the products of combustion.[96] The Cornish engines were seen as examples of good design because their boilers had ample capacity and did not have to be forced.[97] Yet there was a paradox because boilers where the exhaust gases left at 225–260 °C (400–500 °F) gave very good economy too.[98] The answer lay in balancing the length of the flues so that the gases passed through at the optimum speed. If the flues were too long, the gases were too sluggish and, while they lost heat, the optimum conditions for transfer did not occur.[99] This dilemma was removed when economisers were introduced because then the gases could leave the boiler at a reasonably high temperature and transfer this extra heat to the feed water in the economisers.

In order to develop more economical engines using higher-pressure steam, the methods of boiler manufacture had to be improved. To begin with, the wrought-iron plates had to be increased in size and thickness to reduce the number of joints and withstand the higher pressures. The limit in 1831 was about 3 ft by 4 ft by ⅜ in.[100] In 1838, the Coalbrookdale Company had a mill in which plates 5 ft wide could be rolled, but this is said to have been the only one in the world capable of such a dimension. The largest plates ever made at that time measured 10 ft 7 in long, 5 ft 1 in wide and 7/16 in thick,[101] but these were exceptional and would have been welded. Even if the plates could have been made thicker, riveting them by hand would have been impossible because half-inch plate was about the maximum.[102]

Developments in machine tools raised the standard of boiler manufacture. One improvement was planing the edges of the plates so they could be more accurately aligned in the punching machines. Plate rolling machines had been produced before 1840 which made perfect cylinders. Instead of the earlier punching machines with a single punch, double punches began to be used so that one nipple entered the last hole while the other made a new one. This greater accuracy enabled the earlier practice of putting drifts in the holes and forcing the plates into alignment with jacks to be abandoned.[103] Double riveting came into fashion for, although the individual rivets were spaced further apart, the two rows together made a stronger joint without weakening the plates so much. A rivet-making machine was invented by Robert Griffiths in 1835, but most important of all was the riveting machine which Fairbairn developed from his punching and shearing machine in 1837.[104] At last plates up to 3/4 in could be riveted and the rivets closed better than by hand so boilers could be built for higher pressures. There still remained one problem to solve, which was how to rivet the final end onto a Lancashire boiler. This had to be done by hand with one man inside until somebody had the brilliant idea of making the flange on the outside and not the inside. This can be seen on the boilers of the Stretham engine, where the two 1871 boilers have internal flanges but the 1878 replacement has an external one.[105]

There was one last development to take place in boiler design and that was the replacement of wrought iron by the stronger mild steel which permitted pressures to rise still further. The Bessemer tilting converter was patented in 1860 and was followed about three years later by the Siemens open-hearth process. These inaugurated the era of cheap steel. Steel rails were being rolled for railway lines at Crewe in 1863 and, in 1865, steel plates were permitted by the railway authorities and other engineers for boilers in place of wrought iron.[106] At that time, Jevons saw a bright new future with steel.

> Such are the qualities of steel, that, if it were cheap enough, its uses would be infinite. Our engines, machines, vessels, railroads, conveyances, furniture would all be made of it, with an immense improvement in strength, durability, and lightness. Our whole industry would be thrown into a new state of progress. It would be like a repetition of that substitution of iron for wood in mill work which Brindley, and Smeaton, and Rennie brought about.[107]

8
An uncultivated field

> The field of High-pressure Engines is yet so uncultivated, and the state of our knowledge and experience is yet so imperfect with reference to the merits or de-merits of these machines, now taking such an important part in the intercourse of the world, that every voice raised on the subject deserves attention.[1]

Around 1850, the compound engine, which was suitable for higher steam pressures, had been introduced based on Woolf's principles but had not been widely accepted in the textile areas. The Lancashire boiler was gaining in popularity and this again was suitable for higher pressures. In spite of the various pointers to more efficient and economical steam engines, the Lancashire textile millowners continued to prefer the low-pressure beam engine either in its single-cylinder form or with pairs of engines throughout the period up to the 1860s. Such engines were regarded as safer because less damage occurred if there were boiler explosions and their reliability had been proved. Also until about 1850, there was no theoretical reason to consider that an engine using higher pressures would give greater economy and that therefore there could be advantages in adopting high-pressure steam.

It is interesting to note where high-pressure steam was introduced first. It was in cases where a high starting torque was essential. The textile mill engine never started to drive the whole mill at the beginning of the day because, at night just before the mill was due to stop running, a whistle was sounded and the operatives disengaged their machines. So the mill engine only had to overcome its own internal friction and that of the line-shafting when first starting. This might be around 15 per cent of the total power generated. On the other hand, the railway locomotive had to start its train from rest, often with the steam acting in a single cylinder. The need for a compact engine and light weight was another consideration in this instance. In the case of the Cornish pumping engine, all the moving parts and the pumps had to be accelerated from rest at the beginning of each stroke and so steam of a reasonable pressure was an advantage to make a quick start. Once the engine was moving, the airpump would assist in creating a vacuum and there would be no difficulty in finishing the stroke.

There was another application of the steam engine where high pressures were adopted early on and that was in the winding engine. Once again a high starting torque was needed because it was necessary to raise the whole of the load at the commencement of the wind. Just when steam engines began to replace horse whims for winding coal or ore out of mines is unknown but this was an important application in the early development of the steam engine. Ordinary atmospheric and condensing engines were applied originally, much in the same way as the rotative mill engines. Even some of the single-acting Cornish engines were adopted in Cornwall for winding ore out of the mines. These must have been difficult to start unless stopped in the correct position.

Winding railway wagons up inclines with steam engines probably started when the West India Docks were being built in London in 1804 and spoil from the bottom of the dock was excavated. The use of fixed engines to drag trains on railways by ropes was introduced by Cook in 1808.[2] George Stephenson in his youth worked on a winding engine at Willington Quay for dumping ships ballast. He built two two-cylinder low-pressure beam engines for hauling wagons up the inclines on the Stockton & Darlington Railway which were installed ready for the opening in 1825. The order was placed in 1823 and Stephenson undertook to

> erect two 30-horse engines combined on one axle at Brusselton hill-top . . . for £3,482.15.0 and also another engine at Etherley hill-top, two engines each of 15 horse-power combined on one axle for the sum of £1,982.15.0 . . . The two boilers for the first named engine to be 8 feet diameter by 20 feet long and the boiler for the Etherley engine to be of the same dimensions . . . The size of the working cylinders to be 30 inches [60 in stroke] for the Brusselton engine and for the Etherley 22 inches [36 in stroke].[3]

While the engine for the Etherley incline was said to have been originally intended for a steam boat and so was a side-lever type with the reciprocating beam placed below the vertical cylinders, that for Brusselton and those in subsequent orders were conventional beam engines. It is interesting to note that Stephenson was using pairs of cylinders as early as this which seems to precede their use in textile mills. Hick's textile mill engines which have been mentioned earlier were built after 1830 and Fairbairn installed a double engine at Orrell Mill, Stockport, in 1834. At this period the Stephensons were quite active in the stationary engine field and the value of their Forth Street Works stationary engine production, including the winding engines for the Stockton & Darlington Railway inclines, was twice that of its locomotive production up to the end of 1828. Engines varying in size from 8 to 40 h.p. were made for factories, foundries, iron-works and collieries as well as railways. They were employed for pumping, winding, blowing and powering machine tools.

A major advance was made in 1827 when orders were received from the Canterbury & Whitstable Railway for three winding engines using high-pressure steam. Through the financial difficulties which beset that railway, only two 25 h.p. engines were completed at first and these not until well into 1829. The first high-pressure engine actually set to work by the Stephensons was a pumping engine for the Mount Moor Colliery which was running in 1827 with a pressure of 32 p.s.i.[4] The boiler for this engine must have been constructed with experience gained from locomotive boilers, particularly the 'Experiment'. The first high-pressure winding engine to start working was a smaller 10 h.p. one installed on the 1 in 60 Daubhill incline of the Bolton & Leigh Railway. It was running in November 1827 with steam at 30 p.s.i. It provided Stephenson with a demonstration engine to show visiting engineers who were considering such designs for engines of higher pressure.

At this period, the stationary steam engine was still being considered as a serious alternative, if not the preferred alternative, to locomotives for railway motive power. Francis Thompson, engineer to the Brunton & Shields Railway, operated that line in 1828 with six high-pressure engines which ranged from 6 to 24 h.p. working at 30 p.s.i. It was a period of rapid improvement not only in railway locomotive design but also in stationary engine design too. The development of the high-pressure winding engine by Thompson and the Stephensons became one of the main points of contention in the lengthy deliberations about the motive power for the Liverpool & Manchester Railway. In his report of 1829 to the managing committee of that line, James Walker favoured stationary steam engines for haulage and the steam railway locomotive did not gain ascendancy until the Rainhill Trials in October 1829.[5]

On that railway, there was one section which it was accepted would be far too steep for working with locomotives and that was the Wapping tunnel, taking the line down to the docks in Liverpool. The Committee had placed the order for this engine in April 1829, some months before the Rainhill Trials, and it was

> reported that their recommendation that Robt. Stephenson & Co. of Newcastle should furnish the 50 Horse Engine for the mouth of the Tunnel, as since only *one* Engine had been decided on, it did not seem adviseable to advertise for contracts.[6]

This was the largest engine constructed up to that time at Forth Street, although with only one cylinder, but the boilers were made by Isaac Horton, West Bromwich. The engine not only was more powerful than most textile mill engines at that time but the boiler pressure of 30 p.s.i. would have been higher too. A second engine was added in 1831. These engines must have helped textile millowners to become accustomed to the idea of steam at higher pressures.

The Stephensons continued to build high-pressure engines until 1835 when the demands for locomotives supplanted them. In 1834, they built their largest winding engine which was one of 100 h.p. installed at Pontop Ridge on the Stanhope & Tyne Railway. It was a two-cylinder type with dimensions 28 in diam. by 5 ft stroke.[7] While the market for railway winding engines soon disappeared through the extremely rapid development of the railway locomotive, a letter written by Stephenson in 1830 showed the competition in the engine construction business.

> I am afraid we are a great deal too high for the winding engine at St Helens, Auckland, but we really cannot compete with those engine-builders in the neighbourhood of Newcastle, who not only work for nothing, but who make bad workmanship. For the Liverpool engine we had £1,600, but I dare say you will soon have offers for £1,000; but it is useless attempting to make engines for such prices, because I know it is impossible to make a good and substantial job without reasonable prices.[8]

High-pressure engines would have been smaller than their low-pressure equivalents and so the capital investment should have been less. However, the extent of their use in mining, particularly in the important north eastern area, is completely unknown. Their higher-pressure steam would have made starting the wind so much easier and they could well have become quite popular.

While the pressure of 30 p.s.i. may seem low by modern comparisons, it must have seemed dangerously high to the textile millowner used to only 6 or 7 p.s.i. (see p. 132). It must also have seemed dangerously high to the captains of steam ships too where once again steam pressures remained very low. This must have been due partly to the fear of boiler explosions, which was a very real one as the boat at Norwich showed in 1817 when the safety valve was tied down and the boiler exploded. Already by 1819 the marine engine had outstripped the textile mill engine in horsepower for in that year, the *Waterloo* of 200 tons and 60 h.p., the largest steamer of her year, inaugurated the Liverpool to Belfast route. But low pressure meant that engines soon had to become massive in order to develop enough power.

The problems of marine engine design are well illustrated by those installed in Isambard Kingdom Brunel's three famous ships, the *Great Western*, the *Great Britain* and the *Great Eastern*, which were all the largest ships afloat when launched. For the *Great Western*, Maudslays built a pair of enormous side lever engines, 73½ in bore by 7 ft stroke which weighed 310 tons.[9] With a boiler pressure of only 5 p.s.i., these engines produced 750 i.h.p. which enabled the *Great Western* to make the first really successful crossing of the Atlantic by a steam boat in April 1838. The *Great Britain* was fitted with a screw propeller. Its engine had four cylinders 88 in bore by 6 ft stroke and weighed 340 tons.[10] At 18 r.p.m. and with steam at 15 p.s.i. 1500 h.p. was developed. The *Great Britain* sailed on her maiden voyage for New York on 26 August 1845.

For his last and by far the largest ship in the world when it was launched in 1858, the *Great Eastern*, Brunel sought

> tenders of the following dimensions:– Length, 680 feet; beam, 83 feet; mean draught, about 25 feet; screw engine, indicated horse-power, 4,000; nominal horse-power, 1,600; paddle, indicated horse-power, 2,600; nominal horse-power, 1,000; to work with steam 15 lbs. to 25 lbs., speed of screw 45 to 55 revolutions; paddle, 10 to 12.[11]

The screw engines had four horizontal cylinders, 7 ft bore and 4 ft stroke, which were made by Boulton & Watt. Together with their boilers they were estimated to cost £60,000. There were four inclined cylinders for the paddlewheel engines which were a massive 6 ft 2 in bore by no less than 14 ft stroke and were over 40 ft high. The estimate for them, with boilers, was £42,000.[12] The normal working boiler pressure it seems was 15 p.s.i.

When it is realised that the pressure of the sea on the bottom of the hull of these large ships was greater than the pressure of the steam inside the boilers, the question has to be asked why did the designers not raise the steam pressure and reduce the size of these massive engines? One reason was safety and the fear of boiler explosions which on board ship would have fatal results, besides possibly sinking the vessel. Another reason was that all these ships were supplied with salt water in their boilers. As the water boiled and steam was drawn off, the density of the salt rose until it deposited on the bottom and sides of the boiler. To avoid this, the boilers had to be blown down regularly and topped up with more sea water. A great deal of valuable hot water was wasted in this way which must have added to the fuel consumption. It was a dangerous process for often the boiling water was run into the bilges and pumped out from there. At the end of her first trans-Atlantic crossing, one of the engineers was killed in the act of blowing down on the *Great Western*. This process was made less dangerous in 1837 with the invention of the conical Kingston one-way valve which enabled the blow-down water to be discharged below water level. With the larger diameter of the valve outermost, it sealed itself automatically when discharging ceased. At pressures above 25 p.s.i., another problem was encountered with sea water for there is present besides common salt and magnesium sulphate which remain soluble, carbonate of lime which deposits as a scale of a chalky nature at boiling point and even worse sulphate of lime which at pressures above 25 p.s.i. forms a hard rock-like scale which in those days had to be chipped off. The more a boiler became scaled up, the quicker its performance dropped and the greater the risk of an explosion through the plates burning.[13]

The obvious answer to us would have been the use of surface condensers and to recirculate the boiler feed water. After all, Watt had experimented with surface condensers before he took out his patent in 1769. On the *Post Boy*, a vessel of 65 tons and 20 h.p., David Napier in 1820 tried a surface condenser made from small copper tubes through which the steam passed while in 1822, Marc Brunel patented some form of surface condenser. Samuel Hall in 1837 fitted a surface condenser to the steamer *Wilberforce* but the Thames and Humber mud persistently choked it and it was removed after a few years to be replaced by a straight jet condenser. The first successful condenser is said to have been employed in the P & O vessel *Mooltan* in about 1851.[14] The early examples seem to have been bulky, occupied valuable space and also added more parts to the engine as circulating pumps for both condensate and cooling water were needed. The pipes became furred up, particularly with the tallow used for lubrication. Engineers may also have been reluctant to add the extra complexity of the additional water purifiers which would have been necessary to top up the fresh water. In the 1880 edition of his book originally published in 1872, Henry Evers commented that surface condensers were coming more into use and were fitted into many naval vessels. The *Lord Clyde* had 13,000 vertical tubes in its condenser.[15] The difficulty of sealing all these suggests another reason why surface condensers were not adopted until designers were compelled by the advent of higher pressures.

There were other difficulties which militated against the use of high-pressure steam. The earliest pressure gauges were made from a bent tube filled with mercury and open at the top. The mercury could be lost from it easily, by opening the cock too quickly or by the pitching of the ship. The Bourdon type of pressure gauge did not become available until the 1850s. In land boilers, the level of water was indicated by a stone float inside the boiler working a pointer outside. This would not work reliably in a ship pitching about and it remained difficult to keep a check on boiler levels until the advent of gauge glasses and better designs of boilers.

Yet even while Brunel was building his masterpieces with their massive engines, a few people pioneered the use of high-pressure steam and compound engines. For example, the *Cricket*, built in 1848, was fitted with compound engines and plied between Hungerford and London Bridges. Although the steam pressure was only 36 p.s.i., she was very economical until she blew up with the loss of 17 people after the safety valves had been tied down and a boiler stay broke. The first successful sea-going ship with compound engines has always been considered the *Brandon*, launched by the firm of Randolph Elder in 1854.[16] John Elder joined this company in 1852 and quickly put into practice the new principles which were becoming apparent about the steam engine through the work on the science of heat of people like James Joule, William Thomson, John Rankine and Rudolf Clausius. Elder was born in Glasgow in 1824 and attended both the High School and the College of Engineering at the University there. He was a close friend of Rankine. He learnt his trade at the engine-building works of Robert Napier and was put in charge of the drawing office.

Elder quickly adopted the compound engine and coupled his cranks at 180° to reduce the pressures on the bearings of the crankshaft and so to reduce friction, but there were certain disadvantages in this layout, caused by the use of the

Woolf principle, which will be discussed later. Elder was one of the first to point out that the compound engine had proved itself more efficient than the single-cylinder engine only when the pressure of steam exceeded the customary practice of his time. The *Brandon* required but 3¼ lb of coal per horsepower per hour when the usual consumption was one-third more. Five years later, he was building engines which consumed a third less than the *Brandon* at 2¼–2½ lb.[17] In 1857, E. A. Cowper fitted some 60 h.p. compound engines to the *Era* which had a steam-jacketed receiver added to them. The economy was materially improved and it was this type of engine which was favoured for marine use.[18] The acceptance of the compound engine was rapid and sizes soon increased dramatically. In 1862, Clark commented, 'The days of fuel economy, and of compound marine engines have begun'.[19] The *Alaska* launched in the Clyde in 1881 had compound engines with H.P. cylinder 68 in and L.P. 100 in bore by 6 ft stroke. At 100 p.s.i., 10,000 h.p. was developed.[20]

The compound engine

In 1862, Clark could still write,

> But two questions remain to be decided, – Whether the best results are to be had from two or more cylinders? and, Whether the steam should be surcharged with heat?[21]

In the compound engine, the steam first drives one piston in a small cylinder where it expands and then passes to a second larger cylinder where it expands further driving a second piston. The first person to try to follow this principle was Jonathan Hornblower who is said to have taken up the subject as early as 1776.[22] After further unsuccessful trials on a model in 1778, Hornblower took out a patent in 1781. In 1782, he built an engine with cylinders 19 in and 24 in diam. at Radstock colliery for John Winwood, the Bristol ironfounder.[23] However its performance was described as being so wretched that it did not equal an atmospheric engine. Another engine was erected nearby at Timsbury soon afterwards, followed by an experimental model, before Hornblower started erecting engines in Cornwall where he threatened the near monopoly of Boulton & Watt.

Hornblower's first engine in Cornwall was installed in 1791 at the Tincroft Mine with cylinders 21 in diam., 6 ft stroke and 27 in diam., 8 ft stroke, followed by nine others between then and 1794.[24] Drawings of these engines show two vertical cylinders placed at one end of the beam of a pumping engine so that the larger cylinder was at the end furthest from the beam pivot where its stroke was longer. These engines showed little improvement in economy over Watt's engines in spite of Hornblower's claims. The pipe connecting the valve chest with the bottom of the small high-pressure cylinder seems to have too small a diameter and the condenser again seems to be limited in capacity, probably to try and avoid Watt's patent. Yet the engine at Tincroft Mine was sufficiently successful for Hornblower to petition Parliament for an extension of his patent.

Boulton & Watt opposed Hornblower's Bill and then, after the Bill was thrown out, made the owners of Hornblower engines pay compensation for using the separate condenser. It is interesting to note that, while Robison showed mathematically that there was no difference in performance between Watt's single-cylinder engines and Hornblower's compound, there was an advantage in practice

> because the combined effect of the two pistons, approaches more nearly to a uniform action, than could be done by the same extent of expansive action of the steam, when operating in only one cylinder on Mr. Watt's system.[25]

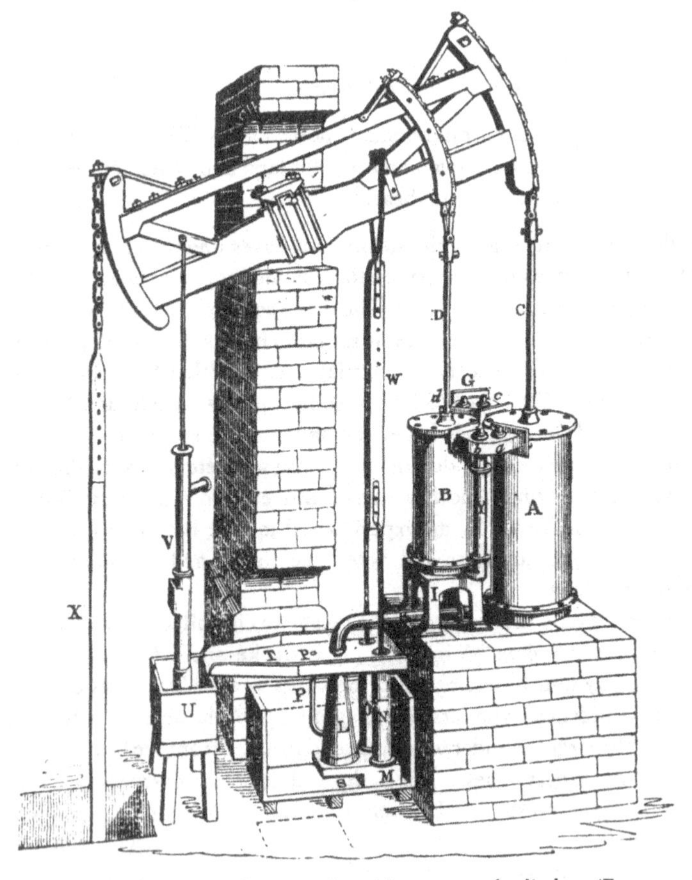

Figure 37 Hornblower's beam engine with compound cylinders. (Farey, *Steam Engine*.)

In reality, the pressure at which it was possible to generate steam at this period was far too low for there to be any thermodynamic savings in compounding and the friction of the two pistons would have been greater. However, it is interesting to note how early the greater smoothness gained by compounding was recognised.

In Rees's Cyclopaedia, which must have been written soon after 1810, it was stated,

> We have been thus full upon this subject, because the gaining more power by the expansion of air or steam acting in double cylinders, has been a favourite idea with many, and there are no less than five different patents for it, but several of these have been upon mistaken notions . . .
>
> The advantage of the expansive principle arises wholly from a peculiar property of steam, by which, when suffered to expand itself to fill a greater space, it decreases in pressure or elastic force by a certain law, which is not fully laid down; that is the relationship between its expansive force and the space which it occupies is not clearly decided.[26]

Of these people who had taken out patents, Woolf was the most important with the one he was granted in 1804.[27] He explained his theory of the expansion of steam, which he considered to be in direct relationship to the pressure.

> I have ascertained by actual experiments . . . that steam, acting with the expansive force of four pounds the square inch against the safety valve exposed to the atmosphere, is capable of expanding itself to four times the volume it occupies, and still to be equal to the pressure of the atmosphere . . . This ratio is progressive, and, nearly if not entirely uniform, so that steam of the expansive force of twenty, thirty, forty, or fifty pounds . . . will expand itself to twenty, thirty, forty, fifty times its volume.[28]

Woolf's fallacious law was rejected by the scientific and engineering communities of his day because it implied that at 1 p.s.i. excess pressure the steam has not expanded at all and begins to do so only after this.

Woolf also had ascertained the temperature of steam at different pressures (see Table 7), and what became apparent to him was that the pressure of the steam rose more quickly than its temperature. He stated,

> By small additions of temperature, an expansive power may be given to steam to enable it to expand to fifty, sixty . . . three hundred, or more, times its volume, without any limitation but what is imposed by the frangible nature of every material of which boilers and the other parts of steam engines have been or can be made.[28]

Table 7. *Woolf's tables relating temperature to pressure and expansion*[28]

	Pounds per square inch.		Degrees of Heat.		Expansibility	
Steam of an elastic force predominating over the pressure of the atmosphere upon a safety valve	5	requires to be maintained by a temperature equal to about	227½	and at these respective degrees of heat, steam can expand itself to about	5	times its volume, and continue equal in elasticity to the pressure of the atmosphere;
	6		230¼		6	
	7		232¾		7	
	8		235¼		8	
	9		237½		9	
	10		239½		10	
	15		250½		15	
	20		259½		20	
	25		267		25	
	30		273		30	
	35		278		35	
	40		282		40	

What is evident is that Woolf was more interested in working his engines at a high pressure to obtain the greatest possible expansion from the steam rather than the greatest heat. This is a point which will be considered again in the next chapter.

In fact, Woolf was one of those who had other 'mistaken notions' as well. He placed his high-pressure cylinder alongside the low-pressure one in a similar position to Hornblower. He also realised that one of his high-pressure cylinders could be added to an existing beam engine of Boulton & Watt's design to increase its power and efficiency. However, when he tried to set the first engine to work on his principle, it did not perform as well as he hoped (see p. 106). Woolf was employed at the Griffin Brewery in London where he had been carrying out a variety of experiments with boilers and using steam for heating purposes.[29] He altered an existing engine and fitted it with cylinders H.P. 8 in diam. by 3 ft stroke and L.P. 30 in by 5 ft, using steam at 40 p.s.i. When set to work in September 1805, although not as powerful as expected, its steam consumption was very small. At first, Woolf did not doubt the correctness of his theory and looked elsewhere for the faults. Eventually a new H.P. cylinder was fitted, 12 in bore and 40 in stroke which reduced the ratio from 23.3:1 down to 9.7:1 and effected a great improvement in the performance. After some trials in 1808, Trevithick calculated the fuel consumption and stated that allowing 'one third for friction . . . I think it will be about 110 million or 22 million for each bushel of coal burnt in the 104 minutes',[30] a result that was far in excess of ordinary engines of that day. However, Rennie tested the engine later in the month and found that 'the comparison was as 3 to 4 in favour of Watt's engine'.[30]

The description given in the 1804 patent and the type of valve gear described by Farey show that, to begin with, Woolf did not cut off the admission of the steam during the working strokes in either of the cylinders. The valves were all operated at the same time. In his patent, it was stated,

> When they [the pistons] have reached the bottom of their respective cylinders, the communications between the boiler and small cylinder, between the small and large cylinder, and between the latter and the condenser, must all be shut off.[31]

This would have largely negated the value of compounding because the high-pressure cylinder would have been filled with steam at the highest temperature for the whole of the stroke. On his double-acting rotative engines, in order to give an adequate exhaust to the high-pressure cylinder, its exhaust valve had to remain open all the time as well, and, as long as this valve acted as the inlet valve to the low-pressure cylinder, that cylinder too had to be worked without any cut-off. This lack of cut-off was necessary to clear the steam out of the pipe connecting the two cylinders because it was too small to act as a receiver to store the steam coming out of the high-pressure cylinder. Now the steam coming out of the high-pressure cylinder must have been at almost full boiler pressure. Because the connecting pipes were too small, there must have been some wire-drawing and loss of pressure between the two cylinders which would have given a difference in pressure. In addition, there was a further loss in pressure as the steam filled the transfer pipes and the spaces at the ends of the low-pressure cylinder. During the working stroke, the pressure and temperature would have fallen in both cylinders throughout the whole of the return stroke to the terminal pressure and temperature of the low-pressure cylinder. While this pressure was lower than in other contemporary designs and so Woolf's engine ought to have been more efficient,[32] it meant that the range of pressure, and so temperature, through which the steam was operating must have been nearly the same in both cylinders. This was wasteful because all the parts had to be reheated up to the incoming temperature of the steam at the commencement of the next stroke.

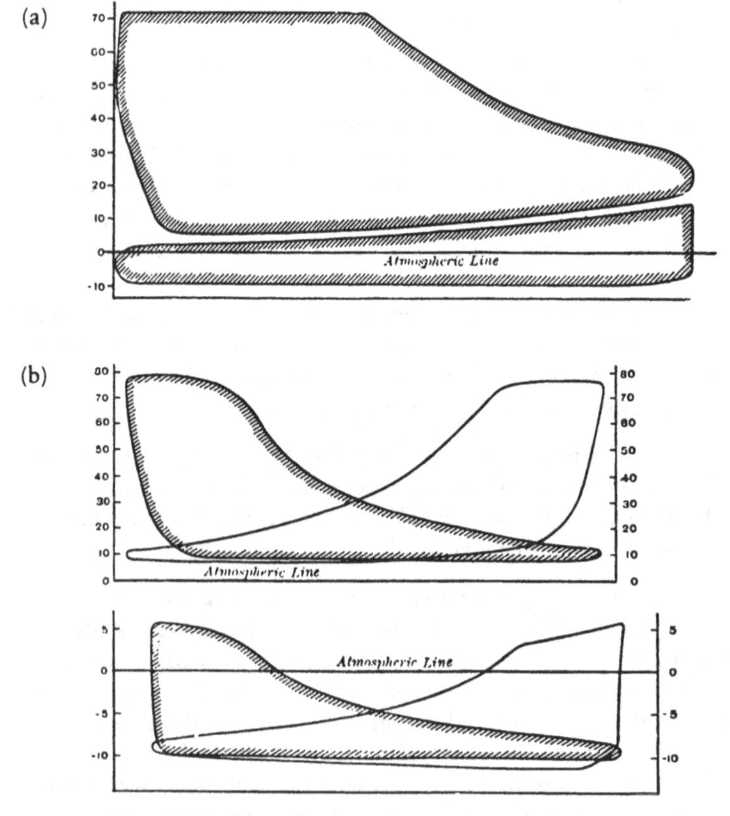

Figure 38 Indicator diagrams of engine compounded on (*a*) Woolf's principle and (*b*) the later receiver type. (Ewing, *Steam Engine*.)

The compound engine 'acts beneficially by diminishing the range through which the temperature of any part of the cylinder metal varies'.[33] To achieve efficient working as a compound engine, the steam must be admitted first to the smaller cylinder where it is cut off during part of the stroke. Then it expands, does work and cools. During the exhaust stroke, this steam passes to a receiver where it can be stored ready for use in the low-pressure cylinder. But this needs different valves on the high-pressure exhaust and low-pressure inlet. From the receiver the steam passes through the low-pressure inlet which again will cut it off during part of the stroke so it can be used expansively within that cylinder. This effectively separates the two cylinders and contains the pressure drops within each cylinder and not between both as was the case in the Woolf engine. From the low-pressure cylinder, the steam is exhausted through another valve to the condenser. In this way, useful work is done in each cylinder where the steam can be expanded separately to its appropriate terminal pressure through a suitable period of cut-off. However, it was to be many years before this was properly sorted out.

On one of his early rotative engines, Woolf used a five-way rotating cock as the distribution valve on the high-pressure cylinder and no separate inlet valve on the low-pressure. This did not allow expansive working in either cylinder. The steam passages were made too small, a feature Woolf continued on some other engines.[34] Later plug valves were introduced and finally the double-beat drop valves. Probably, it was not until after Woolf had returned to Cornwall in 1811 that any expansive working through cut-off was adopted, and probably then only on the pumping engines. At the later end of 1814, two of Woolf's new compound engines began working at Wheal Abraham copper mine and Wheal Vor tin mine. During the last six months of 1815, the average performance of Wheal Abraham was 48.63 million pounds and Wheal Vor 47.63 million. These engines gained a high reputation. On these engines, the steam could be cut off as it entered the high-pressure cylinder at about two-thirds of their stroke.[35] These were, of course, single-acting engines. In them, the steam passed out of the top of the high-pressure cylinder to the bottom of that cylinder through an equilibrium valve as the piston returned upwards. From the bottom of the high-pressure cylinder, the steam was admitted to the top of the low-pressure cylinder where it pushed down that piston before passing through another equilibrium valve to the bottom of that cylinder and so finally through another valve to the condenser on the final stroke. The bottom of the high-pressure cylinder would have acted as a sort of receiver and at least isolated the high-pressure end from direct communication with the low-pressure cylinder and given a more efficient engine.

However, the lack of any cut-off in the low-pressure cylinder and its connection to the high-pressure cylinder for the whole of its stroke still remained as inherent design defects in the Woolf rotative engine until the 1860s. Both cylinders worked in unison and the steam pipes connecting them were short so

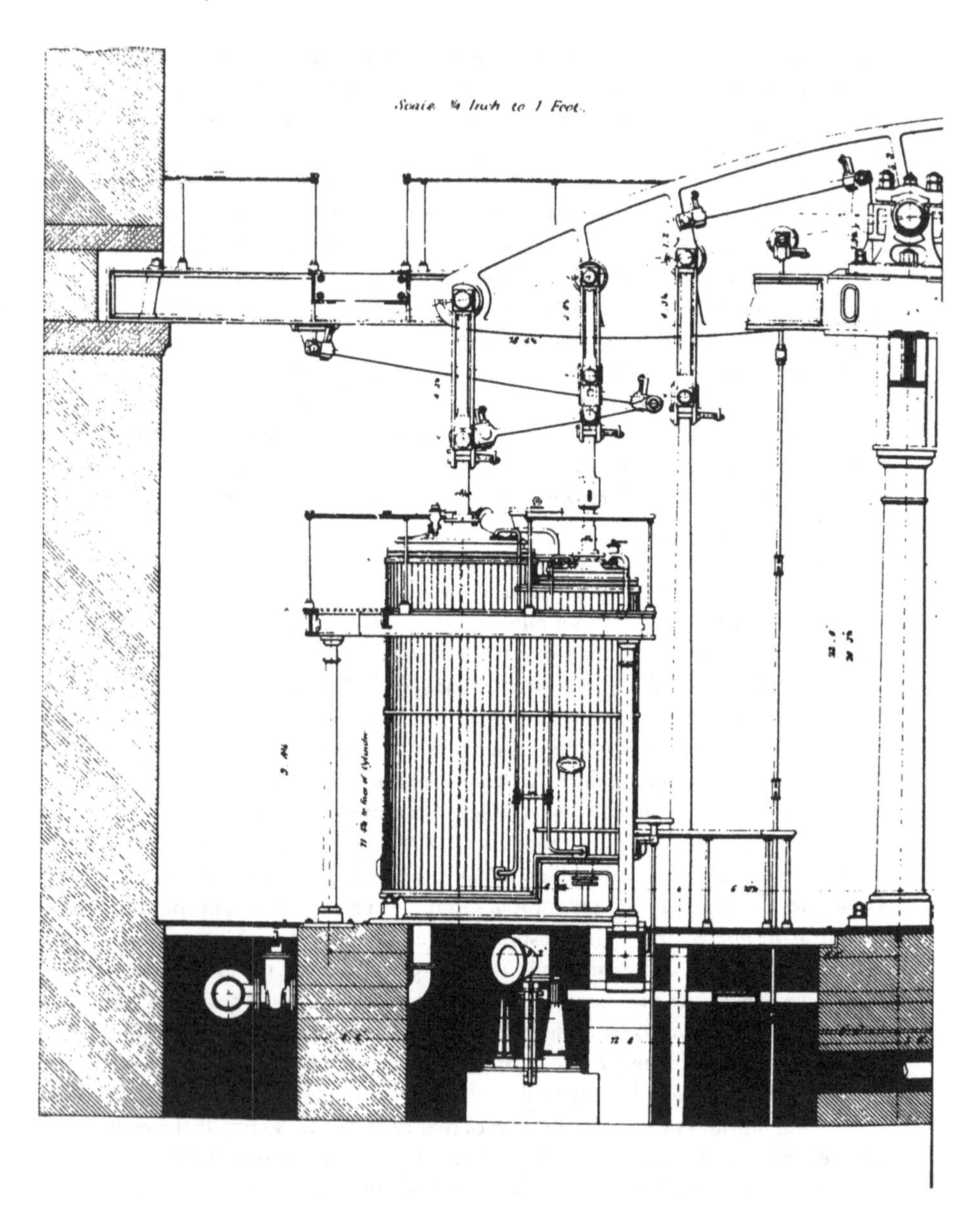
Scale ¾ Inch to 1 Foot.

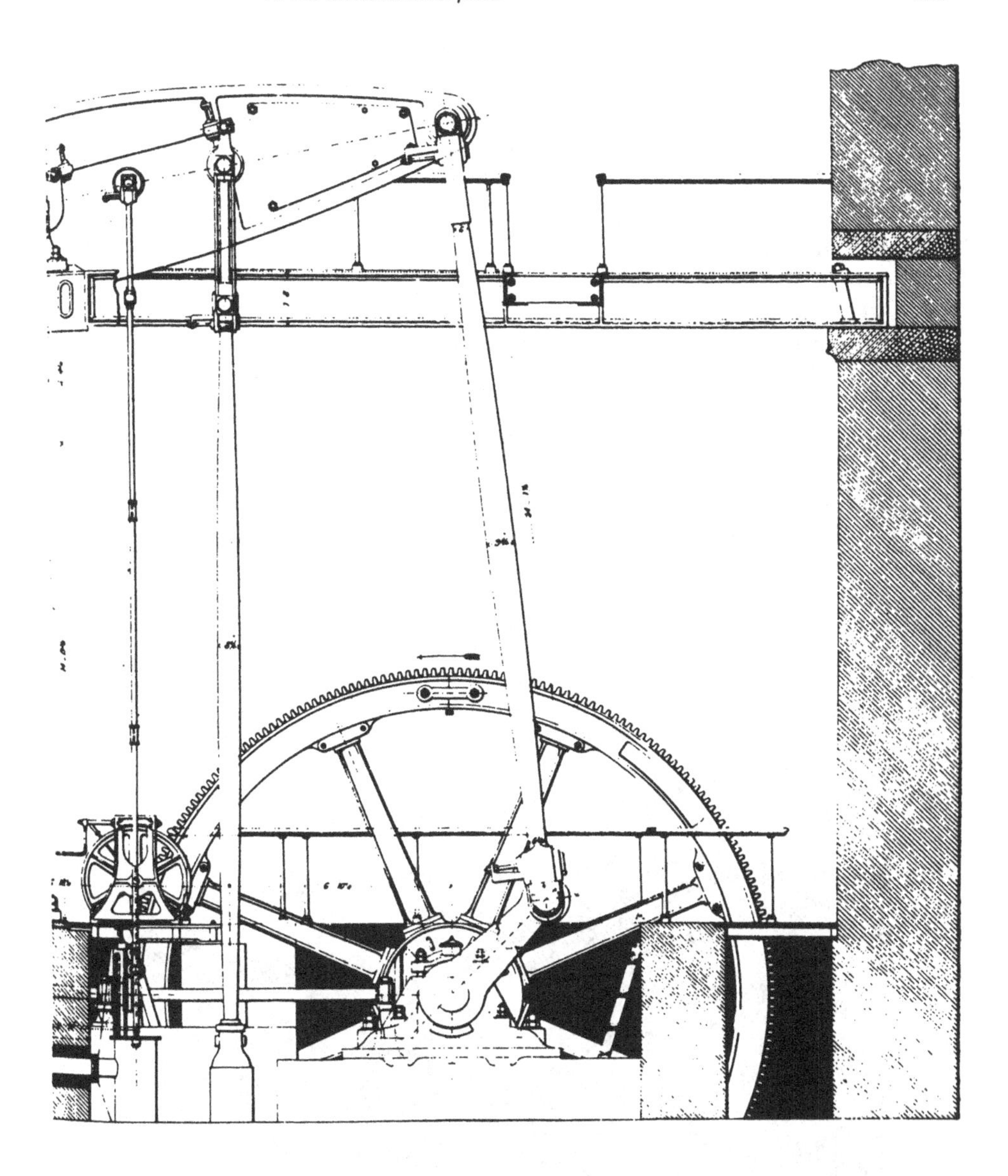

Figure 39 A rotative beam engine with compound cylinders on Woolf's principle which would have been the type driving textile mills. (Rigg, *Steam Engine*.)

could not act as intermediate receiving vessels. The inlet valve to the low-pressure cylinder had to be open for the whole of the stroke to clear the high-pressure cylinder of steam. Therefore there could be no early cut-off in the low-pressure cylinder to give advantages of expansive working in that cylinder too because there was nowhere to store the steam coming out of the high-pressure cylinder. Of course the steam did expand into the larger volume of the low-pressure cylinder almost isothermally, but that is different from closing the inlet early and allowing the steam to expand within the cylinder and do work there.

Here it is interesting to note a later development in the use of a type of Woolf compound by the Manchester firm of Galloway's. In 1873, they patented a side by side horizontal compound.[36] A form of straight link gear, with the position of the valve rod in the expansion link controlled by the governor, operated the inlet valves on one side of the high-pressure cylinder so that variable cut-off could be obtained. An eccentric worked the high-pressure exhaust valves which were also the low-pressure inlet valves. So the cylinders were placed close together with very short passages between them. The low-pressure piston moved in the opposite direction to the high-pressure but to give a little cut-off in the low-pressure cylinder, the cranks were set so that the low-pressure piston had a slight lead. More slide valves let the steam out of the low-pressure cylinder

Figure 40 Galloway's horizontal engine installed at the Oakenholt Paper Mill, North Wales. (Oakenholt Paper Mill.)

on the further side which must have given quite a good flow of steam through the engine from the thermodynamic angle. These engines were very compact and were often fitted with a horizontal condenser driven by an extension of the low-pressure piston rod. Many were supplied to cotton and paper mills for almost 30 years by which time the design was outmoded.[37]

As the advantages of using higher-pressure steam became recognised, so single-cylinder engines were adapted to work with shorter cut-offs and greater expansion. At the boiler pressures then in general use, this could have given them a greater advantage over the Woolf compound with its design fault and extra complexity of two cylinders. Certainly, the Woolf compound was abandoned in Cornwall after 1826 and also Fairbairn preferred single-cylinder engines at Saltaire Mill in 1853 because they had been improved to equal the compounds at the steam pressures then being used.

> The double cylinder or compound engine, in which high pressure steam was employed, expanded through three-fourths of the stroke, appeared to effect a considerable saving of fuel; but taking both engines worked alike, with steam of the same pressure similarly expanded, as is now the case in the best single cylinder engines, there appears to be no advantage in the compound over the simple single engine. On the contrary, there is a loss in the original cost of the engine, and the complexity of the one as compared with the other... I have therefore no hesitation in recommending the single engine worked expansively, as an efficient competitor of the compound engine.[38]

But Fairbairn was dealing with the Woolf engine and this problem was avoided in the next important type of compound engine, that patented by William McNaught of Glasgow in 1845.[39] On a beam engine of the ordinary Boulton & Watt design, he placed a high-pressure cylinder between the main column supporting the beam and the crank, where the cold water pump was normally fitted. This was a simple alteration which could be added to existing engines and became exceedingly popular to increase their power and efficiency. In his patent, McNaught stated that there would be

> the effect of increasing the power of the engine, of lessening the consumption of fuel in proportion to the power produced, and by working the steam expansively in the low-pressure cylinder a further saving of steam may be effected, and consequently a proportional saving of fuel.[40]

The situation of this extra cylinder, well away from the original one, meant that there had to be a long pipe connecting the two which could act as a receiver. Also there had to be two sets of valve gear so that the low-pressure cylinder could have its inlet valves adjusted to give an appropriate cut-off. The advantages of this were recognised by McNaught stating that the steam would be worked expansively in that cylinder. In this way, the McNaught design was an advance on Woolf's as is stated in the patent. Savings of fuel of up to 40 per cent were being claimed by 1854.[41]

There was still one disadvantage because, with both pistons still acting in unison, there were periods when the engine developed no power at the top and bottom of the stroke. Yet, in terms of smoothness, these engines gave a much more even power output than their single-cylinder predecessors and this point must be examined because this seems to have been the main reason for the popularity of McNaughting beam engines. McNaught claimed that one of his objects was to equalise the stress on the main working beam and all the parts connecting it with the framing and engine house. To give an example of one engine,

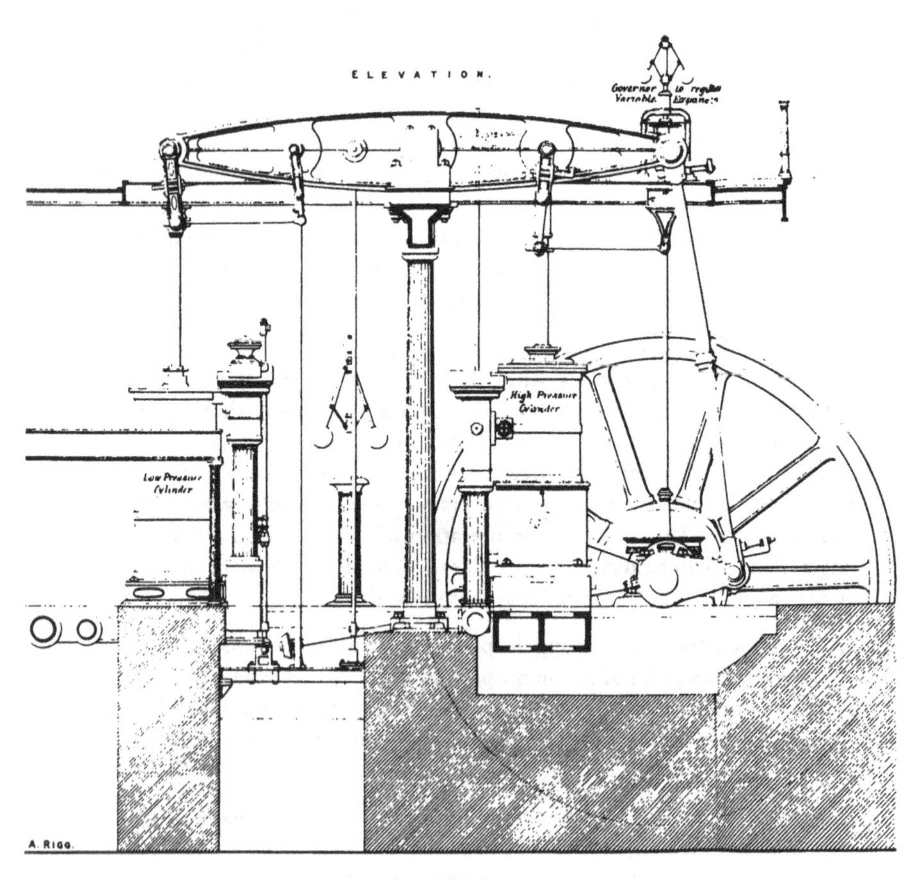

Figure 41 Layout of the compound engine patented by McNaught in 1845. (Rigg, *Steam Engine*.)

Before Compounding,	The pressure on the beam was 85,408 lbs.
	The pressure on the crank pin was 42,704 lbs.
After Compounding,	Pressure on the beam's centre reached only 656 lbs.
	Pressure on the crank pin reduced to 40,893 lbs.[42]

Thus, although the engine was actually doing more work, the pressures in both cases were reduced. In a patent issued five years later, McNaught proposed an arrangement for compensating the difference in the pressure of the steam on the piston at the beginning and end of the stroke when working expansively so that the stress on the parts of the engine would become more nearly a mean.[43] Again, in 1868, he patented mechanical devices with the momentum of weights on arms to equalise the power being developed during the stroke of an engine where steam was being used expansively.[44] Obviously he was still looking for a smooth running engine and he was not alone for these were ideas which others were promulgating at about the same time and will be examined later (see p. 201).[45]

Yet the nature of the compound engine did give an inherently smoother running engine and this was another point which helped the McNaught engine gain acceptance particularly in textile mills. The point has been made already with Hornblower's engine that the power developed in the two cylinders was smoother than in one, even when the cylinders were arranged to work in unison as in a beam engine or in the tandem compound layout. Just as in the Cornish pumping engines when high-pressure steam was admitted into a single-cylinder engine, there was a great shock to the pumping rods, so in a rotative engine there was a blow on the crankpin even though this was lessened through the rotation of the crank. When a high expansion took place in a single cylinder, there was a great difference between the initial and final stresses on the working parts and so in the turning force.

In a compound engine, the difference between the initial and final stresses in each cylinder was much reduced, and the two added together in the types of engines just mentioned were also much reduced when compared with a single cylinder. The ratio might be from 40 to 70 per cent of what it would be were the same total ratio of expansion employed in a single cylinder.[46] So the twisting moment of the crankshaft was more nearly uniform in the case of the compound engine and its parts could be designed lighter.[47] Another small but important point was that the pressure drop across the pistons in each cylinder was less and so leakage of steam was reduced. Therefore millowners had good practical and mechanical reasons for introducing compound engines. This became more true when the cross compound engine was introduced. In this type, there was one cylinder on either side of the flywheel with cranks set at right angles, thus combining the advantages of the smoothness of this layout with the greater smoothness of the compound. Clark commented on one running at the 1862 Exhibition that it was 'the nearest approach that it is possible to obtain to perfectly uniform rotative power'.[48] In both the cross compound and the tandem arrangement of the cylinders, each cylinder was fitted with separate valve gears and the transfer pipe connecting them could be made sufficiently large to act as a receiver.

The thermodynamic advantages of the compound began only with steam over 60 p.s.i. and really closer towards 100 p.s.i.[49] In the later compound designs, the temperature difference between the fresh high-pressure steam and that passing to the condenser was divided between the two cylinders. The steam entering each cylinder did not have to raise the temperature of the walls and the piston to such a great extent and therefore there was less condensation. The compound had the further advantage because the hottest steam first entered the smaller cylinder which was reduced in size compared with a single cylinder on an ordinary engine. Therefore the condensation in this cylinder was considerably less than was the case with a simple engine.[50] But in 1850, the thermodynamic advantages of using high-pressure steam were still much in question and many people preferred to play for safety by keeping pressures low. The mechanical benefits of compounding were recognised sooner than the thermodynamic advantages and indeed did much to bring it into favour before the pressures of steam had risen high enough to make compounding advantageous from the thermodynamic point of view.[51] Higher pressures without adequate

Figure 42 The cross compound horizontal engine became the most popular type in textile mills. This was one at Ace Mill, Chadderton. (Courtaulds Ltd.)

Table 8. *Types and operating pressures of steam engines in the Manchester area, 1859 (i.h.p.)*[53]

	Pressure in lbs per square inch					
Type of engine	−15	15–30	30–45	45–60	60+	Total
Condensing						
Low expansion	2,297	9,898	3,118	—	—	15,313
Intermediate	680	12,370	5,941	614	—	19,605
High expansion	—	4,477	3,731	—	—	8,208
Non-condensing	—	30	389	59	—	478
Compound	—	495	9,046	18,552	4,189	32,282
Total	2,977	27,270	22,225	19,225	4,189	75,886

methods of boiler construction, use and maintenance led to numerous boiler explosions so that by the 1850s, such occurrences were reaching the dimensions of a national scandal. The Manchester Steam Users' Association did not undertake the insurance of boilers working at over 60 p.s.i.[52] The introduction of stronger boilers and improved safety features in them such as fusible plugs caused the position to change quickly, as Table 8 shows. Another reason must be ascribed to a correct understanding of thermodynamics which emerged at this time.

9
The new theory of heat

> It is but just, however, to state that the new theory of *heat*, now being submitted to the test of experiment, will modify very much the theory of the steam-engine. Until the new views, however, have been conclusively affirmed, it would be premature to specify them.[1]

So wrote Robert Burn in 1854. For a long time people had been puzzling over what was the actual driving force behind the steam engine. It is interesting that Savery stated that his engine would work by the 'impellent force of fire' and that the early atmospheric engines were called 'fire engines'. Watt also was acutely conscious of the importance of heat in a steam engine which was shown by his use of the steam jacket and hence, in effect, lagging the cylinder. We have seen how this awareness was, in some degree, due to his understanding of the science of heat through his connections with the men of science of his day like Black who had just founded the new science of heat. Engineers after Watt, like Smeaton, Brindley, Ewart, and Fairbairn, had no such link with this science and they derived their theories from hydraulic engineering which thought in terms of water pressure or weight. The result was that often their concept of the operation of steam engines was deficient.

An example of these confused ideas was shown by the thinking of Farey. He could see that the role of heat was essential when he wrote in Rees's *Cyclopaedia*,

> Steam is a fluid so different from air, as to have no one property in common with it, except elasticity. This elasticity is wholly derived from the quantity of heat which it contains, and its force increases and diminishes with the quantity of heat; but by what law it increases or diminishes we are uncertain, because we have no measure of the actual quantity of heat which is contained in steam of any given elastic force. All we know with certainty is what is stated in our table of expansion, *viz*, that water, being converted into steam, and confined in a close vessel, when heated until the thermometer indicates a certain temperature, will have a certain pressure or elastic force ... It is the quantity of heat alone which determines the elastic force.[2]

Here Farey seems to consider that it is the heat which was the most important factor for the steam engine. Yet later, in *A Treatise on the Steam Engine*,

Historical, Practical and Descriptive, he could write as if it were the pressure which was vital.

> We must attend to the elastic force which steam exerts against the surfaces which form the boundaries of the space that it occupies; because the intensity of the elastic force indicates the manner in which the steam occupies the space. The elasticity is also dependent upon the temperature, and therefore we must take that circumstance into account; so that the three considerations of density, elasticity, and temperature, must be taken altogether, in relation to each other . . . The above tables show that the elastic force of confined steam increases more rapidly than its temperature, as measured by a thermometer.[3]

This fact, that the pressure increases more rapidly than the temperature, led some people to hope that the high-pressure steam engine would yield a cheap source of power and so looked on the pressure as the actual driving force. This view can be found in other parts of Farey's work, where he regards the pressure and not the heat as being the power behind the steam engine.

Many people were afraid of the dangers of using steam at higher pressures through the added responsibilities that this might entail, particularly when the lives of working people might be at stake. So it took a long time to convince manufacturers of the advantages of high-pressure steam particularly when there appeared to be no theoretical grounds advocating its use.[4] In some passages, even Farey seems to have regarded the use of high pressure as unnecessary.

> This circumstance has given rise to many fallacious notions respecting the mechanical power, which may be exerted by steam of a great elastic force, in preference to steam of ordinary elasticity; and because the elastic force which steam will exert to burst open any vessel in which it is confined, is found to be very quickly obtained by an increase of temperature, it has been supposed that a much greater mechanical power may be derived from the same heat when accumulated into steam of a higher temperature, than when applied in steam of a less elasticity. In reality, the quantity of heat required to produce a given quantity or bulk of steam, bears no direct proportion to the temperature of that steam, but is nearly according to its elasticity; so that highly elastic steam requires as much more heat for its motion, as it is more elastic; for it is in fact only a greater quantity of heat and water crowded into a smaller space; and hence, any greater power that it possesses, will be obtained at a proportionally greater expenditure of heat.[5]

As long as such views were held, there were no apparent advantages in using a little steam at high pressures or a greater volume of steam at lower pressures. People could still prefer to play safe and use a greater volume of low-pressure steam in larger engines to obtain the same power. G. de Pambour, writing in 1838, examined how the steam engine operated and showed how it was the pressure which overcame the various resistances opposed to it.[6] Fairbairn, writing around 1860, still regarded the steam engine as a pressure engine for he felt that the full advantages of expansive working at higher pressures had not yet been attained and that therefore they had not found the greatest economy in the use of steam as a source of power. Unfortunately he gave no reasons.[7]

While testing his high-pressure engines, Alban had found that they were much more economical in fuel and he was one person who thought that this could not be accounted for by pressure alone. The greater expansive force he attributed to the fact that in high-pressure steam there was more free caloric.

> This is an advantage, rather than the contrary, in as much as the greater or less elasticity of the vapour is not the effect of density alone, but also of its expansion by the free caloric it contains.[8]

It was the nature of this free caloric which brought these advantages through 'the fact of its exceedingly great subtlety and penetrating power, in which respect it is beyond comparison with any other highly compressed fluid'.[9] This led him to believe, correctly, that even more beneficial results would be obtained if the steam could be supercharged with this caloric.

Therefore the question has to be asked, what was this caloric which formed the basis of the theory of heat at that time? One problem is that, as more was discovered about heat, so the caloric theory altered over the years. John Dalton thought of caloric as an 'atmosphere' surrounding his billiard ball atoms. Caloric was supposed to be strongly attracted to ponderous matter but was also self-repellent; hence the expansion of heated gases. In 1834, Alderson tried to describe caloric.

> By its qualities only can it be examined, since it is a '*Proteus*' which has eluded the most vigilant search of all philosophers.
>
> Caloric is uniform in its nature; but there exist in all bodies two portions of caloric very different from each other. There is one called 'sensible heat', or free caloric, the other 'latent heat', or combined caloric . . . 'Latent caloric' is that portion of the matter of heat which makes no sensible addition to the temperature of the bodies in which it exists. Caloric, as it penetrates bodies, frequently forms a chemical composition with them, and becomes essential to their composition . . .
>
> Whenever caloric quits its latent state . . . it always resumes its proper qualities and character, and affects the thermometer and the sense of feeling as if it had never been latent . . .
>
> It insinuates itself among the particles of substances to which it is

> applied, and immediately separates them in some measure from each other . . .
>
> The large quantity of caloric latent in steam renders it an exceedingly convenient vehicle for concentrating and conveying heat.[10]

Some thought that caloric itself consisted of atoms much smaller than ordinary particles which had a strong repulsion for each other while being attracted to ordinary particles so that they adhered to them. Caloric was an indestructible, elastic, gaseous fluid which weighed nothing and could insinuate itself into other bodies causing them to expand or dilate as they changed from a solid to a liquid or a liquid to a gas.[11] Caloric was released from the substance in which it had been hidden when that substance was burnt. The caloric could penetrate other materials, even solids like copper and iron, so that it flowed through them into other substances, like water, in contact with them.

The nearest analogy to caloric was water and the steam engine was compared with hydraulic power. Hydraulic technology borrowed the ideas, designs and improvements of steam technology to advance the promising column-of-water engine. In England, men like Smeaton and Trevithick were equally at home with steam and water power. The analogy between steam and water power was well summed up by J. V. Poncelet who later wrote,

> We have discussed up to now developers of continuous motion whether by wind or water. These same receptors have, in the same way been applied to steam, but as their use seems difficult in this last case steam is usually applied to reciprocating engines. It is the same for great falls of water . . . Their pressures are applied to piston machines which are called column-of-water engines.[12]

The steam engine was also considered similar in some ways to a waterwheel. The pressure of steam was equated with the fall or head of water available to power the wheel. If that were correct, then provided the available head of caloric, or the pressure, were used in the most efficient way, we return to the point that it would make no difference whether the head were great or small as regards the efficiency. This seemed to be confirmed by 'Watt's law' which accepted the basic idea of caloric as a substance and stated that a given amount of saturated steam contained the same amount of caloric at all temperatures. In the normal course of expansion, no more caloric was added to the steam and so it would, at every stage of its passage through a steam engine, contain the same amount of caloric. If it were saturated initially, according to Watt's law, it would remain so throughout its expansion.

It was Sadi Carnot in France who pointed out that part of this analogy was wrong for it was not the pressure within the boiler which formed the basis for calculations but the true analogy had to start with the temperature of the fire in the furnace. He wrote,

> We can easily conceive a multitude of machines fitted to develop the motive power of heat through the use of elastic fluids; but in whatever way we look at it, we should not lose sight of the following principles:
> 1) The temperature of the fluid should be made as high as possible, in order to obtain a great fall of caloric, and consequently a large production of motive power.
> 2) For the same reason the cooling should be carried as far as possible.
> 3) It should be so arranged that the passage of the elastic fluid from the highest to the lowest temperature should be due to increase of volume; that is it should be so arranged that the cooling of the gas should occur spontaneously as the effect of rarefaction.
>
> The limits of the temperature to which it is possible to bring the fluid primarily, are simply the limits of the temperature obtainable by combustion; they are very high.
>
> The limits of cooling are found in the temperature of the coldest body of which we can easily and freely make use; this body is usually the water of the locality . . .
>
> We are obliged to limit ourselves to the use of a slight fall of caloric, while the combustion of the coal furnishes the means of procuring a very great one.[13]

In Carnot's book there were three vital principles. In the working of a steam engine, both heat and cold were necessary and for the most efficient one, it was necessary to use the greatest possible range. There had to be a flow of heat between these extremes and there had to be expansion of the working fluid, in this case steam. In the steam engine, the caloric had to be absorbed at the high temperature from one reservoir and it had to be passed to a second reservoir at a lower temperature. The caloric absorbed was equal to the caloric lost. As it passed from one to the other, a certain amount of work or 'motive power' was gained.

For Carnot, the water analogy was very appropriate in this argument. A given quantity of water will only generate power if there is a lower level to which it can flow. For Carnot, the same will apply to caloric where there is a temperature difference. For the generation of maximum power, there must be no useless flow of heat. In other words then, heat must flow smoothly and there must be no sudden falls in the pressure as the steam expands.[14] Carnot pointed out that in the ordinary steam engine there was a great deal of waste because, while combustion in a fire occurred at 1,000–2,000 °C, only a small part of this was used, for steam at 6 atmospheres pressure corresponded to 160 °C and condensation seldom took place much under 40 °C. Therefore we could use only 120 °C. It followed that the availability of heat for transformation into work depended essentially on the range of temperature between the hot and cold bodies. No mechanical effect could be produced from heat, however great the amount of heat present, if all the bodies were at the same temperature. Out of this would be formulated the Second Law of Thermodynamics which, in one of its different expressions, states that a self-acting machine cannot convey heat from one body to another at a higher temperature (see p. 7). It was such reasoning which caused Carnot to point out that it was easy to see the advantages of the high-pressure steam engine.

> *This superiority lies essentially in the power of utilizing a greater fall of caloric* . . . A good steam-engine, therefore, should not only employ steam under heavy pressure, but *under successive and very variable pressures, differing greatly from one another, and progressively decreasing.*[15]

From 1824 to 1834, Carnot's ideas remained forgotten until Emile Claperyon rediscovered them. He pointed out that it was solely through the use of caloric at high temperatures that improvements in the art of utilising the motive power of heat could be expected.[16]

Carnot had shown that, in the steam engine, there had to be a smooth transfer of heat from the hot to the cold sources, but he had not shown what caloric actually was. If it were some subtle fluid and had particles or atoms, then there must be limits to how much could be produced from a given source such as a burning lump of coal. However, in 1798, Count Rumford had communicated to the Royal Society the results of an experiment in which he forced a blunt boring tool at 10,000 pounds pressure against the 'head' of a casting of a six pounder brass cannon immersed in a trough of water. Much to the surprise of the spectators, the water boiled after two and a half hours, without any fire. Rumford was forced to ask,

> What is heat? Is there any such thing as an igneous fluid? Is there anything that, with propriety, can be called caloric? . . . It is hardly necessary to add that anything which an insulated body can continue

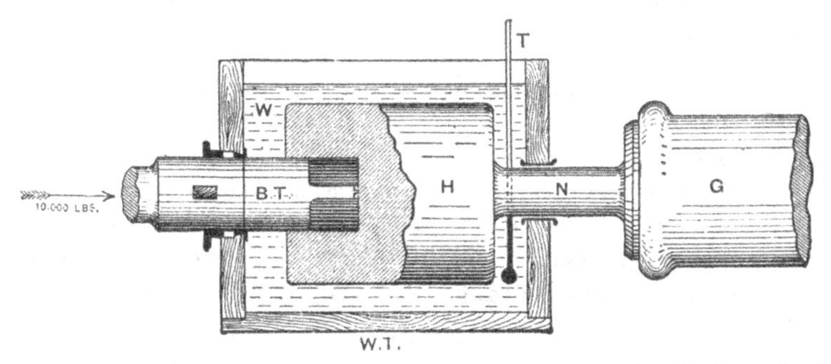

Figure 43 Count Rumford's experiment boring the 'head' of a casting for a cannon. (Jamieson, *Elementary Manual.*)

> to furnish without limitation, cannot possibly be a material substance; and it appears to me to be extremely difficult, if not impossible, to form any distinct idea of anything capable of being excited, and communicated in the manner heat was excited, and communicated in these experiments, except it be motion.[17]

In the following year, Davy carried out some experiments in which he was able to melt ice by rubbing two blocks together. Although it has been doubted subsequently whether he was actually able to do this because the melting point of ice is lowered under pressure, the importance of his result lay in his comment;

> The immediate cause of the phenomenon of heat is motion, and the laws of its communication are precisely the same as the communication of the laws of motion.[18]

The next important advances towards a correct understanding of heat did not occur for many years until the investigations of James Joule in Manchester. He had been led to consider the problems of energy conversion through an early interest in electro-magnetic machinery. In 1843, he published his important paper in the *Philosophical Magazine* 'On the Caloric effects of Magneto-Electricity, and on the Mechanical Value of Heat'. Here he demonstrated that the heat caused by the passage of an electric current was not transferred from another part of the circuit which was correspondingly cooled, but was actually generated. As the result of a number of experiments, Joule concluded that the mechanical value of a unit of heat was 838 ft lb of work expended to raise the temperature of one pound of water by one degree.[19]

Joule continued with a variety of experiments to determine the mechanical equivalent of heat and the principles connecting mechanical work and heat. In one of his most famous experiments in 1845, he rotated a paddlewheel to stir up a known quantity of water contained in a vessel. The movement was given by a weight falling under the influence of gravity. The work done as this weight fell was measured by the increase in temperature of the water through very accurate thermometers. After more experiments with different apparatus, Joule was able to state in 1849,

> 1) The quantity of heat produced by the friction of bodies, whether solid or liquid, is always proportional to the quantity of force expended.
> 2) The quantity of heat capable of increasing the temperature of 1 lb. of water (weighed *in vacuo*, and taken between 55 °F and 60 °F) by 1 °F requires for its evolution the expenditure of a mechanical force represented by the fall of 772 lbs. through the space of one foot.[20]

This has been termed 'Joule's Mechanical Equivalent of Heat'. From it, the first law of thermodynamics has been deduced which Rankine expressed as,

> Heat and mechanical energy are mutually convertible; and heat requires for its production, and produces by its disappearance, mechanical energy in the proportion of 772 foot-pounds for each British unit of heat.[21]

In other words, heat may be changed into mechanical work and conversely mechanical work may be transformed into heat.

This dynamical theory of heat startled men of science still accustomed to the caloric theory and was not immediately accepted. One problem was that the caloric theory stated that heat was always conserved as heat and could not be changed into mechanical energy. If it were some form of matter with particles or atoms, then it must behave like other matter which could not be destroyed. Even Joule himself stated,

> Believing that the power to destroy belongs to the Creator alone I affirm . . . that any theory which, when carried out, demands the annihilation of force, is necessarily erroneous.[22]

For a long time it was believed that the amount of heat put into a steam engine was the same as that which could be measured in the condensate. This seemed to have been confirmed by experiments carries out by Lee during his tests on his engines in Manchester shortly before the end of the eighteenth century. He

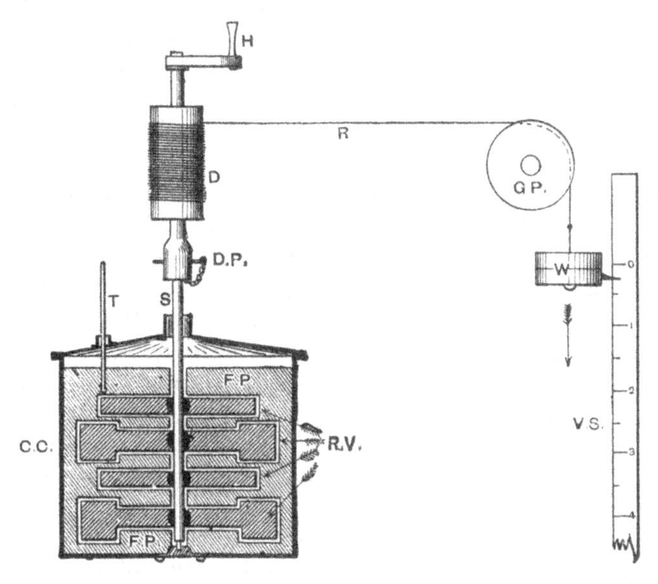

Figure 44 Joule's experiment with a paddlewheel stirring water. (Jamieson, *Elementary Manual.*)

compared the amount of sensible heat e.g. that derived from the temperature, entering a Watt engine with that rejected into the condenser and found that the two figures were approximately the same, as was to be expected with the caloric theory. His engine was never more than 2½ per cent efficient and his figures were obviously too inaccurate to distinguish between 97½ and 100 per cent of the heat reappearing in the condenser.[23] The thermodynamic efficiencies of early engines were so low that any transfer of heat into energy would have been difficult to detect. Having started with the assumption that 100 per cent of the heat would reappear, Lee would quite rightly feel that his assumption was justified by the results.

In some ways it was the development of the hot air engine which brought the problem of the conservation of caloric or heat to the fore. In 1833 Lt John Ericsson, a Swede, patented an air engine which incorporated a regenerator. He hoped to revolutionise sea transport because his engines would not need a heavy dangerous steam boiler and would recycle the caloric. The new feature was the regenerator or heat exchanger. Air in the cylinder was heated and did work by acting on a piston. When the stroke was completed, the caloric was absorbed by the regenerator at progressively lower temperatures until the piston was restored to its initial position and the air cooled down. The process was then reversed and the heat in the regenerator used to restore the temperature of the cold air which acted on the piston again. It was presumed that the only additional heat required to complete the next cycle would be that needed to make up for losses through radiation, conduction and convection. In short, it seemed that the air engine opened up possibilities of something approaching perpetual motion, or at least of substantial power at negligible cost. Ericsson built a ship with engines on this principle and claimed that people would soon be boasting that they had crossed the Atlantic by caloric. Unfortunately the engines with four cylinders 14 ft diam. proved to be complete failures on their test run in New York harbour in 1852.[24]

The descriptions of these hot air engines and the claims made on their behalf as presented to the Institution of Civil Engineers bewildered famous engineers like I. K. Brunel and Robert Stephenson. They were far too sensible to accept what amounted to claims that perpetual motion machines had been invented but they did not know how to refute these arguments.[25] It was left to an unknown engineer, Benjamin Cheverton, who read an interesting and revealing paper to that Institution in 1852. So provocative was his argument that the succeeding discussion lasted three evenings. Cheverton pointed out that the arguments in favour of these air engines amounted to nothing less than affirming the principle of perpetual motion, which was an absurdity. He saw clearly that if work is to be done, there must be a fall in temperature, the bigger the better. He pointed out that, even if all the heat is absorbed in the boiler, there is still a big gap between the temperature of the furnace and that of the boiler. This alone would account for the fallacy in the arguments for the hot air engine because that gap could never be bridged by regeneration. Cheverton accepted the dynamical theory of heat although he never mentioned Joule and recognised that heat and work are interchangeable. As a result of the discussion following Cheverton's paper, Charles Siemens was asked to present a paper to clear the matter up. He gave a straightforward account of Joule's work and of the dynamical theory of heat. He repeated Joule's assessment of the efficiency of contemporary steam engines and added, 'The comparatively small effect produced by the steam engines of the present day would seem to indicate that there is still much room for improvement'.[26]

In the meantime, from 1849 onwards the science of thermodynamics was developed with extraordinary rapidity in Britain by men like Rankine and Thomson (later Lord Kelvin). On the continent, there were Rudolf Clausius in Germany and G. A. Hirn in France. Their findings were applied especially by Rankine to practical problems in the design of engines.[27] Hirn had been convinced as early as 1845 of the correctness of the dynamical theory of heat and set out to prove it at Mulhouse where he lived. In the following decade, he carried out his series of experiments to examine all aspects of the performance of steam engines. It was a 120 h.p. Watt engine which he tested at Le Logelbach and wrote a report in a letter to the President of the Société Industrielle de Mulhouse, dated 21 October 1854.[28] He was able to use steam at nearly 200 °C, which was much higher than Lee, and found that it was impossible to account for some of the heat as it passed through the engine. He demonstrated that the amount of heat leaving a steam engine in the condenser was less than the amount entering it from the boiler. He succeeded in measuring the actual consumption of heat and showed that there could be a loss of between 10 and 20 per cent.[29] He went on to show that this loss was always equal to the magnitude of the work done divided by Joule's equivalent; that is the missing heat was actually converted into work.[30] Hirn's contribution was immediately recognised as a major advance to understanding the steam engine and helped to secure his election as a corresponding member of the Académie des Sciences. As a result of all these investigations, it was possible to show that what was happening was the conservation of energy rather than heat.

The work of Joule, Cheverton, Clausius and later Hirn showed that the steam engine was extremely inefficient and this puzzled Stephenson, by then President of the Institution of Civil Engineers. Where could such great losses occur? Thomson derived the key formula for the efficiency of a perfect heat engine and in the interpretation of this formula lies the answer to the problem that had puzzled so many engineers. The formula showed that the greater the temperature drop, the greater the efficiency of the engine, as Carnot had postulated. It also indicated that no engine, not even a perfect one, could have 100 per cent efficiency unless it had a condenser whose temperature was absolute zero, or −273 °C. So Joule's calculation was correct only in the case of an engine where the condenser was at absolute zero. A perfect engine was one which gave sufficient power to restore the original situation, so it would be possible for an engine to be perfect but, because it was not operating over the full temperature range to absolute zero, it would have a thermodynamic efficiency of less than 100 per cent.[31] This was the important point not clearly grasped at that time.

Finally, in 1850, Clausius succeeded in reconciling Carnot's arguments about the nature of heat in which there was a transfer of heat from a hot to a cold body with the dynamical theory in which heat was changed into energy by showing that both processes, flow and transformation, take place. This showed that it was the heat in a steam engine which was the vital driving force, and only secondarily the pressure. The ideas of Carnot and the work of Joule pointed the way for the acceptance of the high-pressure steam engine and gave the correct theoretical reasons for doing so. Then engineers had to discover the best methods of safely constructing engines which would work at pressures very much greater than any in use in 1850. In about 1884, T. M. Goodeve wrote,

> It is only within the last thirty years that a knowledge of the principles of the mechanical theory of heat has influenced the practice of those who are engaged in improving the construction of the steam-engine, and in seeking to obtain from it a larger amount of useful work with a given expenditure of fuel.[32]

How this was achieved will be examined in the following chapters.

10
The internal operation of the machine

The increasing boiler pressures, the increasing speed of engines, the need for more accurate control of the speed and the development of compounding all meant that the older forms of valve gear became increasingly inadequate. Once it was seen that it was necessary to allow the steam to expand in the cylinder as much as possible, and therefore utilise the maximum range of heat, there was a gradual change from controlling the speed and power of the engine with the throttle valve to devising ways of allowing the steam to enter the cylinder at maximum pressure and then cutting off the entry of the steam at a suitable point to allow it to expand before the exhaust valve opened. The ideal valve should open quickly to allow the maximum amount of steam to enter the cylinder at maximum available pressure, e.g. full boiler pressure, for the precise period of time necessary to generate sufficient power to drive the engine and then it should close quickly, leaving the steam to expand until the end of the stroke. The exhaust valve too had to open quickly to allow the steam to pass to the condenser as quickly as possible, ideally at the same pressure as the condenser. The exhaust valve should remain open for the duration of the exhaust stroke to keep any back pressure low until it closed before the piston actually reached the end of the cylinder to create some compression to act as a cushion. When the load on the engine varied, the closure of the inlet valve, the point of cut-off, should be altered to vary the amount of steam passing through. Therefore three phases of the timing of steam engine valves should remain the same while the fourth (actually the second in the sequence) varied.

The deficiencies of slide valves had been recognised for many years. Farey preferred the older engines of Boulton & Watt with their four plug valves to each cylinder.

> It is well known to engineers and manufacturers, who still have those old engines in use, and also engines of the modern construction employed in the same kind of work, that the old ones perform better than the new ones, supposing both to be in equally good order. The older engines are capable of exerting a greater power without becoming overloaded, and . . . also, when they are moderately loaded, they consume less fuel in proportion to the power they exert, than the modern engines, as they are usually constructed.

> This retrograding in the perfection of engine making, would have been avoided if the engineers of the present time had been as well acquainted with the internal operation of the machine as Mr. Watt was; but as they may easily acquire that knowledge, by the aid of the indicator, it is inexcusable that any engines now made, should be inferior to those which were made forty years ago.[1]

A start was made by improving the slide valve.

In 1842, James Morris patented an adjustable slide valve to alter the

> portion of the stroke during which the steam freely enters the cylinder by attaching the slide valves each to a right and a left handed screw by which they may be placed nearer to each other, or further apart. When by turning the screw the valves are brought nearer, the ingress opening is closed later and the steam enters during a greater portion of the movement of the piston and the expansion of the vapour is less.[2]

This sounds remarkably similar to the better known type patented in the same year by J. J. Meyer of Mulhouse.[3] The central box of an ordinary slide valve was lengthened at both ends with other boxes which were open at the top. These could be closed by a further pair of valves worked by a second eccentric. These cut-off valves could be adjusted by a rod screwed with a right and left hand thread to draw them closer together or to separate them. These valves worked on the inlet of the steam only and the exhaust passed through the central box in the normal way. Usually they were adjusted by hand through a wheel on the end of the screwed rod which passed through the valve chest to give a general setting while the governor moved the throttle valve for controlling fluctuations when the engine was running. Such an arrangement can be seen on the Musgrave engines illustrated in Evan Leigh's book on *Modern Cotton Spinning* which were installed in 1863.[4] These were pairs of horizontal engines powering the Gidlow spinning mill at Wigan through gear drive on the rim of the flywheel. One of the disadvantages of the Meyer valve gear was that it was difficult to work through a governor and was rather cumbersome. Also it suffered from the problems of all slide valves, that the steam pressure forced them against the

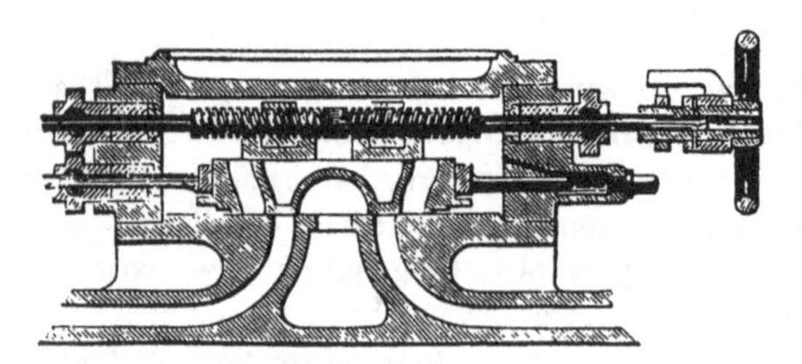

Figure 45 Meyer's slide valve with an expansion valve working on top of it. (Bourne, *Recent Improvements*.)

surface of the valve chest which caused excessive friction the more the pressure of the steam was increased.[5] On railway locomotives, it was estimated that they might expend 25–30 h.p. to work each valve, or about 1–2 per cent of the engine's power and that the savage cutting action of an unbalanced slide valve was not unlike that of a planing machine so that the surfaces wore quickly.[6]

Yet people persisted with slide valves or variations of them and many ideas for fitting them with adjustable cut-off secondary valves were patented in the next few years. In 1819, James Petrie had started to build mill engines in Rochdale and was joined in 1838 by William McNaught (not the McNaught of the compounding patent) who became his chief designer and superintendent for 20 years before starting his own works.[7] It is claimed that McNaught was responsible for designing the cut-off gear patented by Petrie in 1844.[8] In this case, the cut-off valves were circular and had sloping faces. They could be twisted so that the sloping face gave a greater or lesser opening. This motion could be connected quite easily to the governor which McNaught patented in 1850. Both Petrie and McNaught used this form of valve gear until well into the 1890s and other people developed variations of it.

Some other people in England who patented different forms of expansion valves at this time were William and John Galloway. In 1848, they had an expansion valve with a variable lift which could be altered by regulating the throw of a cam.[9] In the same year, J. L. Cole patented yet another form of 'moveable laps which cut off the steam' with rods and right and left hand threads.[10] In 1850, Charles Atherton tried varying the throw on an eccentric,[11] and something similar was patented by T. S. Holt three years later.[12] Interest in variable slide valves was to continue for many years and, to give but one example, the Rochdale engine builders, E. Earnshaw and T. Holt took out a patent in 1874 for 'A variable cut-off under the control of the governor . . . obtained by operating a cut-off valve on the back of the slide valve by a cam of peculiar form operated by screw gearing'.[13] Yet, while the slide valve was adequate at low pressures and continued to be fitted on low-pressure cylinders of compound engines for many years, as Burn commented in 1854,

> To obtain the full efficiency of the expansive method of working, it is considered best to have the cut-off instantly effected – this the slide valve cannot do.[14]

One of the main disadvantages of slide valves was their comparative slowness in opening and shutting. This caused the steam to be 'wire-drawn', that is to enter the cylinder at less than full boiler pressure, and so the pressure in the boiler had to be raised further than would have been necessary otherwise,[15] and the cut-off lengthened. A clearer understanding of the thermodynamic principles involved led people to realise that there was a loss in performance of engines through wire-drawing.[16]

The problem of the excessive friction of slide valves could be partially overcome by making balanced valves or by dividing them into two parts, one at each end of the cylinder. The thermal efficiency was improved by shortening the steam passages so that the hot inlet steam did not have to pass through such a long route which had just been cooled by the exhaust steam. This also helped to reduce the clearance space but the disadvantage of using the same passage for inlet and exhaust remained, for, if made large enough to give an adequate exhaust, it would have been far too large for the inlet. The idea of having separate slide valves for inlet and exhaust does not seem to have been tried until after the development of other types of valve gear which will be examined now.

Farey, following Watt's example, had realised the advantage of having four separate valves, an inlet and an exhaust at each end of the cylinder, because they would permit a better steam flow.

> The modern steam-engines are by their construction less capable of speedy exhaustion of the cylinder than the original construction, with four valves and handgear. Mr. Watt was very particular in attending to this circumstance in his engines, and hence his plan of inverted exhausting-valves, and it is from this cause that the old engines which he made and adjusted under his own superintendence, perform better than modern engines usually do.[17]

While it was usual to make both inlet and exhaust valves on the Watt engines the same size, Farey realised that the inlet could be made smaller because the fresh steam could be controlled by the throttle valve.

Separate inlet and exhaust valves were used by some engine makers. On a simple beam engine built for a cotton spinning mill in 1835 which rotated at 24 r.p.m. with a steam pressure of 30 p.s.i., Hick fitted 'Cornish' valves. He used them again in 1860 on a similar engine with the same boiler pressure but lower speed, 22½ r.p.m.[18] Presumably these were double-ported drop valves but such valves were difficult to open quickly at such slow speeds by mechanical linkages and, if the engine rotated faster, were difficult to close without knocking against their seatings. Farey described a system of plunger weights for opening them.

> Each plunger is fitted into a short hollow cylinder, like that of a pump barrel, which is fixed down in the condensing cistern, beneath the water. The lower end of each barrel is closed, but the bottom has a hole through the centre of it, which is covered by a leather clack-valve opening upwards. This clack will admit the water freely into the barrel, as the plunger weight is raised upwards in it, whilst the valves are closing; but when the plunger is left to fall suddenly by its own weight, in order to open the valves, the contained water must make its escape out of the barrel. The plunger does not fill the barrel very exactly, but a sufficient space is left, to allow the water to squeeze out

around the plunger on all sides as it descends; and so much resistance is thus opposed to the descent of the plunger, as will give it a suitable motion for opening the valves quickly, and yet without noise or concussion. This method allows the plunger weights to be made so heavy, that they can have no chance of sticking, or failing to open the valves; and yet they will act quietly, and without shaking and deranging the working gear.[19]

Fairbairn fitted drop valves on engines he built in the later 1840s. He wrote

The expansive principle upon which Steam Engines are now worked, and the economy which this system has introduced in the expenditure of fuel, has effected considerable changes in the working of the Valves, and has rendered the D and plate [slide] valves almost inadmissable for such a purpose.[20]

Fairbairn had found that there was little advantage to be gained with compounding at the pressures then employed, 15–30 p.s.i., and so for most of his engines he built paired beam engines with two high-pressure cylinders. The most famous were probably those at Saltaire. On these he fitted double-beat drop valves at the top and bottom of the cylinder. They were driven by a vertical rotating shaft with a disc close to the valves. On the surface of the disc three or four lifting cams were formed which opened and closed the valves through roller cam followers. The exhaust valves remained open for a fixed duration but the roller for the inlet could be moved across to a different cam which altered the cut-off in steps from ⅓ or ¼ to ½ of the stroke. The cams could be shaped to give a point of cut-off corresponding with the steam pressure and the load.[21] While they could be operated with steam at pressures ranging from 30 to 40 p.s.i., the mean pressure at Saltaire worked out at only 7 p.s.i. for two of the four engines and 13 p.s.i. for the other two. The speed was 25 r.p.m.[22] Similar engines were installed later in the 1860s by Yates of Blackburn at the India Mill, Over Darwen, but here the boiler pressure was 100 p.s.i.[23]

These engines at Saltaire lasted in their original form for only ten years before having the cylinders replaced by ones with Corliss valves.[24] There were two reasons for this. Such valve gear with all its separate parts for working four valves was difficult to manufacture accurately, maintain later and keep steam tight.

The greatest difficulty, however, is that the gearing must not only open the valve, but must also shut it again, in such a manner as to deposit it gently in its seat, that it may not receive an injurious shock from the action of the steam suddenly forcing it down. Moreover, the end of the motion must exactly correspond with the position of the valve in its seat, to avoid injurious strains.[25]

Fairbairn also failed to make the cut-off point on his valve gear controlled by the governor. The cam follower had to be set by hand on an appropriate cam before the engine was started to suit the average expected load. The governor varied the throttle valve to regulate the speed when the load changed, thus introducing wire-drawing into the system.

What was needed was an automatic cut-off which could be varied according to the load so that the amount of steam could be varied according to the load without any checks in the flow of steam from the boiler to the steam chest. Such a system could accommodate both fluctuations in load and boiler pressure and give the maximum amount of expansion under all conditions without wire-drawing.[26] While the Petrie system achieved this with his twist cut-off valve, it was fitted onto a slide valve which would have involved all the problems of lubrication, wear and slow opening and closing of the ports. A solution was pioneered in America but it took many years of development to turn it into one that worked satisfactorily.

The Corliss engine

Although he had started working out ideas for his engine in 1846, and had completed an engine in 1848 for the Providence Dyeing, Bleaching and Calendering Co.,[27] it was not until March 1849 that George H. Corliss obtained an American patent for 'certain new and useful improvements in Steam-Engines'.[28] The first improvement had little importance or significance for it outlined a scheme for adding tie rods and trussing to help strengthen the central column supporting beam engines and also the main beam. The other improvements concerned the valve gear for a beam engine and were the beginning of important developments because Corliss valves were to become the most popular type on the high-pressure cylinders of Lancashire mill engines.

Corliss used paired valves, two inlet and two exhaust. Eventually they were placed on opposite sides of the cylinder which gave good thermal properties because the hot incoming steam did not have to enter the cylinder close to the cold exhaust ports and the size of the exhaust port and valve could be larger than the inlet. Corliss must have realised the importance of short steam passages between the valves and the cylinder, for he sited his valves as close as possible to the ends of the cylinder. Therefore the mechanism operating the valves had to have long rods reaching across to them. He achieved this by having a single eccentric (a weak point) rocking a 'wrist plate', a disc, to which the rods were attached.

The movement of an eccentric will be slowest at either end of the stroke and quickest in the middle, but a steam valve should be opened quickly, i.e. the opposite to the motion actually given. Corliss made the eccentric rod oscillate the wrist plate. Four, or more, rods were attached to the circumference of the wrist plate, two operating the inlet valves and two the exhaust. The ends of the rods on the wrist plate will travel through an arc and it is possible to arrange the take-off point with such an angle that the rod moves quickly at first and then more and more slowly the further it leaves the tangent. The rod to the equivalent valve at the other end of the cylinder is arranged to have the opposite movement, so that one valve is virtually stationary, or closed, while the other is opening rapidly. Corliss claimed that,

> By this means I not only save much of the power due to the working of the valves when closed but at the same time I attain the important advantage of greatly accelerated motion of the valves whilst opening and closing the ports.[29]

The wrist plate was not a vital part of the Corliss valve gear for some later designers, such as George Saxon, did not use it. However it did give Corliss a means of opening his valves quickly without recourse to a weight or spring operated system such as the one described by Farey. The next part of the patent was vital and related

> to the method of regulating the cut off of the steam in the main slide valves, and consists in effecting this by means of the governor which operates cams, so that when the velocity of the engine is too great these cams shall be moved by the centrifugal action of the regulator

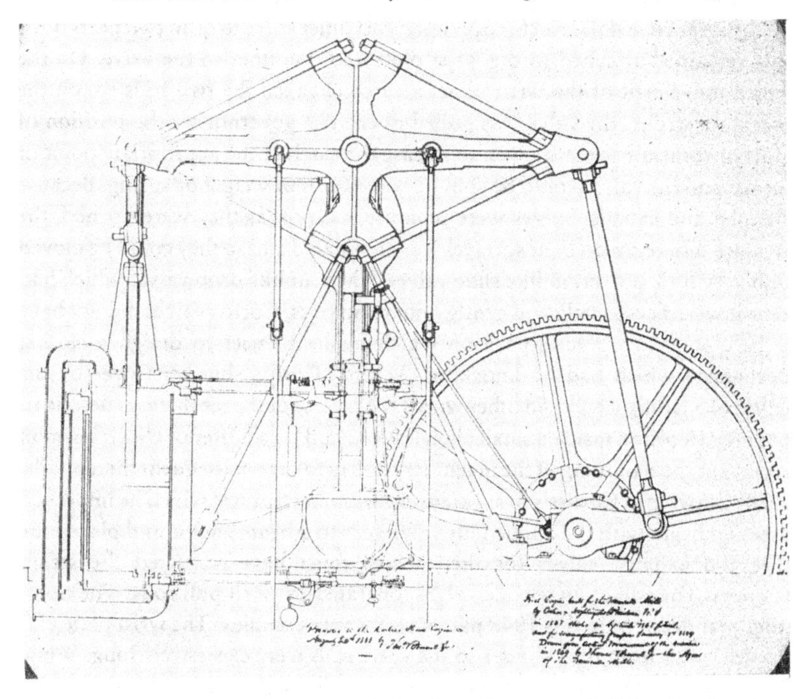

Figure 46 Corliss's drawing for his first valve gear with slide valves fitted to a beam engine. (Smithsonian Institution.)

> that a catch on the valve rods may the sooner come in contact with them to liberate the valves and admit of their being closed by the force of weights or springs, and thus cut off the steam in proportion to the velocity of the engine this being done sooner when the velocity of the engine is to be reduced and later when it is to be increased.

This mechanism was connected only to the inlet valves, for the exhaust valves had a separate cycle and remained open during the exhaust stroke for as long as possible. Steam at full boiler pressure could be admitted while the inlet valves were open and, as soon as the cut-off was released, the valve would close quickly so that the steam would expand in the cylinder without any wire-drawing. More or less steam could be admitted to the cylinder by altering the point of cut-off and making the inlet valves shut later or sooner. Corliss took the crucial step of making his governor control the point of cut-off so that he could regulate the speed of the engine by the amount of steam admitted at full boiler pressure and its subsequent expansion. The governor determined the position of the release of the catches which allowed the valves to snap shut. Very little power was needed to operate the release catches so that the governor worked more sensitively than in the older method of turning a butterfly throttle valve or some forms of cut-off valve such as Petrie and McNaught's.

Corliss arranged the rod for operating each inlet valve to be in two parts. One part remained attached to the wrist plate and the other to the valve. On the return movement of the wrist plate, a catch engaged the two rods so, on the operating stroke, the valve was pulled open. The governor set the position of cams or some other mechanism to release the catch at the appropriate point of cut-off and the valve would be shut by some form of weight or spring. Because the inlet and exhaust valves were separate, as soon as they were opened, the pressure difference on their surfaces was eliminated and so they could be moved with very little power, unlike slide valves. Then, unlike drop valves which had to be lowered accurately and gently onto their seats, Corliss valves did not have to be closed in a precise position which made it easier to design a closing mechanism which had no knocking. At first, Corliss shut his valves by the action of a falling weight, in other words the reverse of the mechanism described by Farey. A piston inside a small air cylinder acted 'as a buffer or elastic cushion to prevent the slamming of the machinery and breakage consequent thereon'.[30]

The drawing in Corliss's first patent shows many features which he improved later. To begin with, he applied his valve gear to a beam engine and placed the inlet and exhaust valves together, which must have restricted the steam passages. The valves were slide valves, operated by push-pull rods. The operating gear depicted in this 1849 patent seems very complex. The wrist plate was set well away from the cylinder so the valve rods were excessively long. While the exhaust valves were operating through bell cranks, the inlet valves were moved by rack and sector, surely a weakness. It is not possible to see how the catches were released or how any buffing mechanism worked, but the valves were closed by weights operating through bell-cranks. The weights fell only through the force of gravity, which was adequate for the slow speed of most beam engines of that period.

Some improvements were added quickly. A drawing of a cylinder for a horizontal engine supplied in 1849 to Phillip Allen & Sons, print works at Providence, Rhode Island, shows the valves placed at each corner of the cylinder.[31] The inlet valves were at the top and the exhaust valves at the bottom. Not only would this have given much better drainage of any condensate, but, by separating the inlet and exhaust passages, the thermodynamic layout was improved. Equally important was the possibility of increasing the size of the ports, which were soon enlarged to one-tenth of the area of the piston, allowing much freer movement of the steam.[32] This was the position of the valves which became the most common on later Corliss engines but the valves were still slide valves.

On this Phillip Allen engine, the wrist plate has been moved to a bracket on the middle of the cylinder, the usual place on later engines, but full details of the cut-off release mechanism have not been drawn in so we do not know how the valves were closed. However, it is clear that this new arrangement was much neater with shorter operating rods and another feature used on all later Corliss engines appeared. The operating rods alongside the cylinder moved levers fixed on the ends of more rods which passed across the barrel of the cylinder. As these second rods oscillated or twisted, they moved the valves inside the cylinder.

Some of these improvements were consolidated in Corliss's next patent of 1851.[33] The drawing shows a beam engine with the wrist plate in the centre of the cylinder and valves at each corner, the inlet on the left and the exhaust on the right. The patent is concerned mostly with a description of the disengaging gear, a subject later for many patents in many different countries because each engine builder seems to have developed his own type. Corliss fitted hardened steel plates, which could be replaced easily, to the faces of his trip gear. He retained slide valves which were closed by weights. In this instance the falling weights were turned into pistons inside cylinders where in their fall they compressed the air which acted as a cushion to prevent concussion. In order to allow a longer travel of the weight, there was a hole in the side of the dashpot cylinder through which air could escape until it was covered by the falling weight and then the rest of the air was compressed.[34] It would seem that these weights constituted one of the limitations to the development of the engine because, working by gravity, they did not close the valves quickly enough to allow the speed to rise above 40 r.p.m.[35]

Although Corliss had envisaged in his 1849 patent closing the inlet valves by springs, he did not patent a system until 1859.[36] Springs could be set with an initial compression or tension that would close the valves more quickly than gravity working on weights. One of the problems he faced was that a coil spring varied in the amount of force it would give up again the further it was compressed which would alter the closing of the valve. Therefore he developed a leaf spring with a moving fulcrum. An interesting example of this type of valve gear is preserved on an engine built in France in the 1870s, now displayed in the British Engineerium at Brighton. It was an unnecessary complication and was quickly replaced by coil springs. The importance of this development was that closure of the inlet valve by springs enabled engine speeds to rise to 60 r.p.m. quite safely.[37] Because the Corliss valve did not close against a seat, the dashpot did not arrest the motion of the spring or weight until after the valve had closed which could be at full speed, whereas on the drop valve the dashpot had to be used to slow down the valve to prevent it hitting the seat, so introducing an element of wire-drawing.[38]

The patent drawing of 1859 shows another important development which had appeared by 1850.[39] This was the cylindrical oscillating valve which is the type always later associated with Corliss engines. The centre of the oscillating valve rod passing through the cylinders became the centre of the cylindrical valve too so that the valve could be turned by it. Such valves could be moved with less friction than earlier ones. At last the Corliss engine had developed far enough to start challenging other types. The first one set to work in Britain was purchased from the Corliss Steam Engine Company in America by Alexander Pirie and erected at his Stoneywood Paper Mill, near Aberdeen, in 1861 or 1862.[40] This was a horizontal type, 20 in bore by 4 ft stroke and ran usually at 60 r.p.m. but sometimes at 68 r.p.m.

Two continental manufacturers exhibited horizontal Corliss engines at the International Exhibition of 1862. One was built by the Wilhelmshütte Machine-works near Sprottau with a cylinder 17½ in bore, 2 ft 10 in stroke running at 30 r.p.m. with a Watt governor turning at 85 r.p.m. The other was built by the United Hamburg–Magdeburg Steam-Navigation Company of Buchau near Magdeburg which had a Porter governor.[41] Both engines had weight operated valve gear and failed to impress the British. D. K. Clark commented,

> Another great objection is to be found in the irregularity of the cut-off, owing to the want of precision of the valve gear, – varying occasionally from an admission of steam throughout the stroke, to an admission of nothing at all, – and involving a very heavy fly-wheel to neutralize those violent fluctuations of force.[42]

The fault may have lain with the weights for, while the Watt governor may not have been sensitive enough, the new Porter type ought to have been for these governors provided a solution to governing at the higher speeds being reached and will be examined later. Porter governors were regularly fitted to Corliss engines by the end of the 1860s.[43]

The Corliss engine had to overcome the prejudices of people in Britain probably because it was an American invention. John Bourne reflected the attitude of many others when he wrote,

> This form of engine . . . is of American design, and in some quarters its advantages have been loudly vaunted. But it is not pretended that it is able to work with greater economy than ordinary engines, while the complications of its valve gear are manifest and enormous. There are four separate valves or cocks for the admission and emission of the steam, and the gear is governed by the aid of air cushions, springs, and other rattle traps which by their complexity throw the old hand gear into the shade. Apart from these disfigurements, however, the general plan of the engine is not so good as that of some other engines now in common use. But with these additions the engine may be pronounced to be as bad a one as perverted ingenuity could easily have constructed. The American example of this engine shown at Paris in 1867 was of admirable workmanship, and was radiant with the silver in which the cylinder was enveloped. But these aids to acceptation were incapable of concealing from competent observers the inherent vices of the design, which appears to reckon complication as a merit, and which seeks to achieve no advantage in economy by a great multiplication of parts.[44]

Nevertheless prejudice was overcome gradually, particularly by William Inglis, who had designed several Corliss marine engines for use in Canada. It seems that John Frederick Spencer persuaded Inglis that the Corliss engine had a future in Britain, so that in 1863 Inglis first went to Edinburgh and a couple of years later to Manchester as a Consulting Engineer.[45] Robert Douglas, one of the leading mill engine builders in the east of Scotland, built an engine to Inglis's designs in 1863 for the papermill of David Chalmers near Edinburgh. By the middle of 1867, Douglas had built, or had in hand, a total of 17 Corliss engines, many of them to the designs of Inglis. After he moved to Manchester, Inglis linked up with Hick Hargreaves & Co., Bolton, and became their engineering manager in 1868.

Spencer, who came from Newcastle upon Tyne, received an order for a vertical single-cylinder engine in 1863 and then designed several engines which were built by Losh, Wilson & Bell of Newcastle. Both he and Inglis patented forms of releasing gear for the trip mechanism and improved the springs for the dashpots. Inglis patented the use of rubber or steel coil springs in 1863.[46] A single spring sufficed for the dashpots for both ends of the cylinder, so the dashpots had a common spindle running through them. As the rubber acted under tension and the coil spring under compression, the layout for each had to be slightly different.

Inglis also described two forms of releasing mechanism. In both, a foot on the end of a lever, linked either to the governor or a hand setting wheel, raised a catch to release the valve rod which slid along part of the rod from the wrist plate. In the first form, there was a curved lever which was lifted while in the second there was a long flat spring with the catch near one end. This was the basis of later development. It is probable that these mechanisms tended to stick, particularly at higher speeds. Neither was supported in any way between the wrist plate and the valve operating rod. The releasing lever pushed down against the parts that had to slide apart and worked on only one side of the mechanism so it was unbalanced. Also neither type was suitable for application to vertical cylinders.

In his patent of 1865,[47] Spencer improved both the release mechanism and the dashpots. He specified only one release mechanism which he derived from the earlier one by Inglis. It had a lever with a double foot to open a pair of flat spring catches, one on either side of the rod from the wrist plate. In this way it was better balanced and could be made more sensitive as the two clips did not need to be so deep in mesh as a single one. It would also operate in any position, and a final point was a 'reduction of risk, as in the event of one clip giving way, the other one would act until a convenient opportunity for repair'.[48] It was an elegant gear to watch at work because the catches on their long flat springs looked rather like fingers. After being forced apart by the double-footed cam to release the valve, they snapped shut again as they grabbed the rod to pull the valve open once more. It was a highly successful type of release mechanism and Inglis used it for many years on the engines he built at Hick Hargreaves. Some of their engines were running with it in the late 1960s, 100 years after its invention.

Spencer also patented the double-ported slide or Corliss valve to allow more steam into the cylinder without excessive movement of the valve itself. 'The less the motion of the valve spindles or levers, the more certain is the action of the escapement gear, and the engines can be worked at an increased number of revolutions'.[49] But the engines could not be speeded up without improvements to the dashpots. While Inglis had developed a neater layout than Corliss, it still needed quite a lot of space and, as the length of the dashpot was very short, the travel of the valve may have been restricted.

Spencer's drawing shows that, in his spring-driven dashpots, proper pistons were fitted in place of the earlier weights. For the first part of their travel, the air could escape through holes or vents while a screw valve controlled the cushioning effect on the last part of the distance. He put separate springs for the valves at each end of the cylinder and mounted the dashpots back to back in the middle of the cylinder, a neat layout. To help the coil spring give a quicker action (or even sometimes replacing the spring), he had a piston in a cylinder worked by the pressure of the steam, or by air, gas or water. With the improvement in steel for springs, this idea was rarely used although it was reintroduced in America later. The Spencer–Corliss valve gear was neatly laid out with a minimum of parts. It was certainly simpler than the type patented by Corliss in 1859 and was probably a great improvement which may have helped the introduction of the

Corliss engine to Britain. In 1868, Inglis stated that more than 60 Corliss land engines were at work and in 1870 Spencer claimed that 'several hundred' engines had been made in Britain.[50]

One of Inglis's early designs was built in 1866 for driving a rolling mill at the Crewe locomotive works of the London & North Western railway.[51] The governor was based on the Porter pattern and the valve gear on Spencer's patent. Two eccentrics on the crankshaft drove separate wrist plates for the inlet

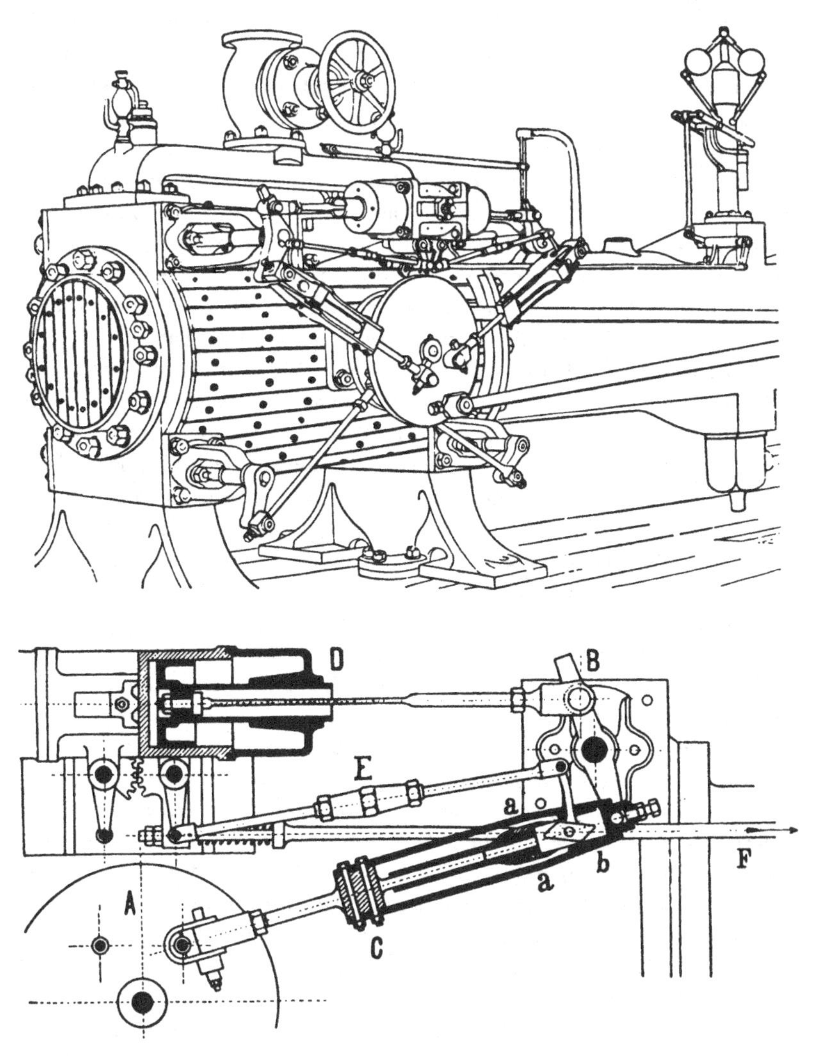

Figure 47 Corliss engine cylinder fitted with the later type of Spencer–Inglis valve gear. (Ewing, *Steam Engine*.)

and exhaust valves so their timing could be varied independently. The lead on the exhaust valves could be adjusted without interfering with the inlet valves. Separate eccentrics were to become common on later engines. The Corliss valve gear gradually gained popularity particularly for high-pressure cylinders of compound engines.[52] When beam engines were McNaughted, the new high-pressure cylinders frequently were fitted with Corliss valves because this valve gear helped to give close control of the speed which was particularly important for textile mills.

The large Corliss engine exhibited at the Centennial Exhibition in Philadelphia in 1876 caused a great sensation in America but it is difficult to establish when mill engine builders in Britain adopted Corliss valves owing to the fragmentary records that have survived. J. & E. Wood had patented double-ported Corliss valves in 1867 and their archives show that they fitted Corliss valves on all their engines from 1875.[53] Daniel Adamson patented oscillating cut-off valves within Corliss valves in 1869. These valve gears worked without the wristplate which necessitated two eccentrics, one for the inlet valves and the other for the exhaust which could be timed separately. Musgrave took out two patents in 1876 for actuating and releasing cut-off valves but it is not until 1883 that the Corliss valve is actually mentioned in one of their patents.[54] Goodfellow built their first Corliss valve engine in 1877 but it was not until after 1881 that Corliss valves became more popular than slide in their designs.[55] On the other hand, McNaught persisted with their slide and twist cut-off valves until 1892 when they installed the triple expansion engine at Ellen Road Mill with Corliss valves on the high-pressure cylinder.[56] This engine is now preserved but the high-pressure and intermediate cylinders were replaced early in the 1920s. Galloway drawings do not feature a Corliss valve engine until 1896 but they may be very incomplete.[57]

Most British engine builders followed the same general layout with the Corliss valves at each corner of the cylinder. On the cross compound engine at the Wiseman Street mill, Burnley, the Burnley Iron Works fitted a high-pressure cylinder with the inlet and exhaust valves side by side at the top. Not only must this have restricted the steam passages but any condensate would have had to be let out through separate drain pipes in the bottom of the cylinder. Some manufacturers followed the arrangement where both valves were placed underneath the cylinder. Daniel Adamson, Goodfellow, Musgrave and Wood all designed engines like this, perhaps using the American Jerome Wheelock's patents. His first British patent was taken in 1878 when he also exhibited an engine at the Paris Exhibition.[58] This was followed by further patents in 1885 and 1889.[59] A steam pipe came up underneath the cylinder and divided to the inlet valves at each end. The steam was admitted by the inlet valves past the exhaust valves and into the cylinder. While the steam pipe acted as a partial steam jacket, the exhaust pipe passed below it and so must have cooled the incoming steam. Also the fresh steam passed over the exhaust valve, losing more heat. Wheelock's later patents concerned grid valves for the exhaust.

The engines of J. & E. Wood were beautifully designed and looked very neat. The blue planished steel cladding swept across the whole of the top of the cylinder without any interruption except for a brass oiler. The valve gear was placed neatly at the bottom with plenty of bright wrought iron. However, the steam passages must have been tortuous, restricted and so not very efficient. This, in addition to the thermodynamic disadvantages of the common inlet and exhaust ports may have been one of the reasons why J. & E. Wood ceased building engines before the beginning of the First World War. Combe Barbour, Belfast, placed their Corliss valves across the ends of the cylinders. This resulted in shorter steam passages, but on horizontal engines could increase the problem of drainage.[60] Also, if the cylinder cover had to be removed, the valve gear had to be dismantled as well.

As the size of engines increased, double-ported designs were introduced to pass the greater volumes of steam more quickly, particularly through the exhaust valves. The steam passages were also shaped more smoothly to help improve the flow of steam. Instead of passing up the sides of the cylinders and through sharp bends into the cylinders, the steam pipes swept from above into the cylinder tops. Each maker added his own variations to the valve layout and trip gear.[61] J. & W. McNaught put the springs and dashpots on the floor where they collect the dirt. The Dobson design used by Saxon dispensed with the

Figure 48 The Dobson layout of Corliss valve gear fitted to the engine at Wye No. 2 Mill, Shaw, No. 2 engine, 1925. (Author.)

wristplate. The catchplate, springs, dashpots and valve operating rod were all in the same line and were operated by a rod from one eccentric, while another eccentric worked the exhaust valves. This dispensed with most of the rods used by other manufacturers and the whole layout was exceptionally neat.

A few manufacturers developed other forms of trip mechanisms without Corliss valves. One was John Turnbull of Glasgow who patented sliding grid expansion valves in 1869.[62] The grid valve could open a large port area very quickly with very short steam passages but of course it could be the cause of great friction. Turnbull fitted his grid valves on the backs of slide valves but he arranged a trip mechanism and dashpots to close them quickly like Corliss valves. Two years earlier, the Sulzer brothers displayed an engine with an entirely new valve gear at the Paris Exhibition. This was the horizontal drop valve engine with dashpots to control the closing of the valves which had a far-reaching effect upon Continental practice.

> On the Continent, an Engineering mania may be said to have sprung up since the Sulzer valve-gear came into public prominence, so resulting in an affluence of patterns both surprising and perplexing.[63]

The drop valve was eventually to replace the Corliss but higher standards of engineering were necessary to make it operate satisfactorily.

The Corliss valve gear became obsolete through increasing boiler pressures and increasing speed. The rise in boiler pressure increased the friction as the valve rotated in its seat. Stronger springs had to be used to close the valves, which of course wasted power. The increase in speed caused difficulties with the trip gear because the catches failed to catch. In 1900, it was generally accepted that different valve gear had to be used at speeds much in excess of 100 r.p.m. Designers tried to overcome these problems. For example, the Harrisburg Foundry & Machine Works of Pennsylvania reverted to Spencer's idea and replaced the springs in the dashpots with pistons operated by steam at full boiler pressure, which they called the 'steam accelerating valve gear'.[64]

As late as 1925, Allis-Chalmers of Milwaukee were still issuing a catalogue for their heavy duty Corliss engines.[65] They stated that, as the oldest and largest manufacturers of Corliss engines in the world, they had built in the last 45 years engines developing over 5½ million horsepower. They claimed that their design of valve gear could cope with high speeds, up to 150 r.p.m., high pressures and the high temperatures of superheated steam. Their design reduced the clearance volumes to a minimum, and had ample steam passages to lessen wire-drawing. Yet by this date, they must have been among the last few manufacturers to be offering steam engines with Corliss valve gear.

The horizontal engine

During the period of the introduction of Corliss valves, the beam engine was gradually replaced by the horizontal. When Trevithick moved to

London, the first high-pressure engines he set up were horizontal ones with the cylinder inside the boiler.[66] However, the first true horizontal engines are usually ascribed to Taylor & Martineau, of City Road, London, who built such engines for factory purposes in 1825.[67] On the Leicester & Swannington Railway opened in 1833, a horizontal winding engine was installed to haul wagons up the Swannington incline. This engine ran until 1947 when it was preserved at the York Railway Museum where it can still be seen. Possibly the railway locomotive was another influence which helped their introduction. For example, a horizontal engine was built at John Dickinson's Apsley paper mills in 1856 by Leonard Stephenson who had worked at Crewe locomotive works.

> It was of the horizontal type with central plain and expansion valves, two 12 in. diameter cylinders by 24 in. stroke. The main cast iron oblong bedplate and frame was of the open box-girder and bracket design. Mounted on a cross bracket was the belt driven governor of the open two round ball type. The speed of this and subsequent engines varied from 60 to 120 revolutions per minute according to the work and arrangement of driving.[68]

By this time, the horizontal engine was being developed quickly for marine use. The screw propeller had to be driven by a shaft lying close to the keel. I. K. Brunel fitted double diagonal engines in the *Great Britain* which drove a crank-

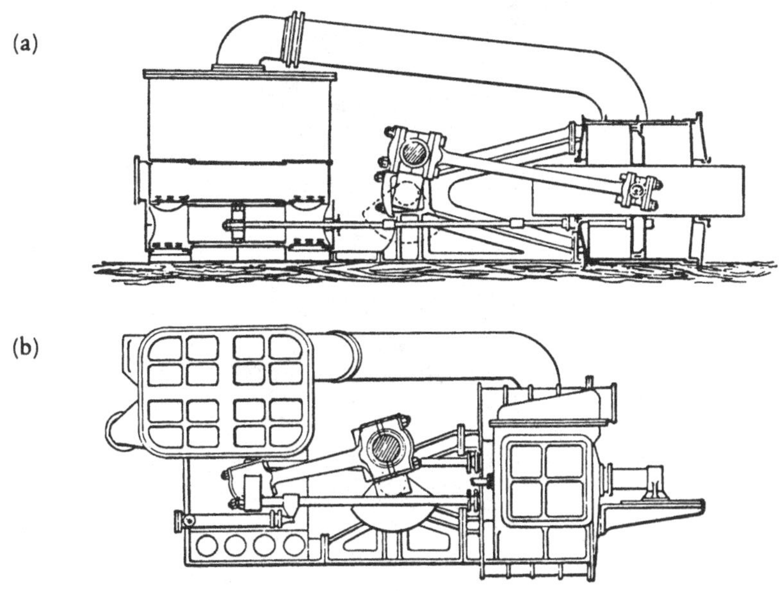

Figure 49 Arrangement of horizontal marine engines: (*a*) Penn's horizontal trunk engine and (*b*) return connecting rod engine. (Seaton, *Marine Engineering.*)

Figure 50 Horizontal tandem compound engine with trunk guides for the crosshead slides. (Nasmith, Cotton Mill Construction.)

shaft just below the upper decking and were connected to the screw shaft through a chain drive. Although this enabled the screw shaft to be geared up as it turned more quickly than the engines, it was a cumbersome arrangement and the Admiralty was to insist that the mechanism of steam engines in naval boats be kept below the waterline. This resulted during the late 1840s and early 1850s in a spate of designs for horizontal engines which had to be fitted into the confines of the hull and drive the propeller shaft directly. Because low-pressure steam was all that was available, in order to develop sufficient power within the space available, these engines had to have short strokes and large diameter pistons. In 1848, John Penn invented the trunk engine which was widely adopted by the Admiralty but this type proved to be unsuitable for higher pressures.[69] Elder patented another compact type of layout in 1854[70] and many others were to follow. Reporting on the type of engine most suitable for use in the Royal Navy, in 1858 an Admiralty committee said,

> That of all the variety of engines that have been purchased by Government for our screw ships of war, the following are so far superior to all others, that no engines of an older make should ever again be put on board. The engines to which they now refer are:–
> 1) The single piston-rod engine, with the connecting-rod attached direct to the crankshaft, and with a single flat guide.
> 2) The engine commonly known as the trunk engine, and patented by Messrs Penn and Sons.
> 3) The double piston-rod engine.[71]

These naval engines proved that there need not be excessive cylinder wear from the weight of their enormous pistons and must have influenced the designers of mill engines for it is after 1860 that we find a rapid introduction of the horizontal engine into textile mills. Earnshaw & Holt had built the single-cylinder slide valve engine for Durn Mill, Littleborough, in 1863, which has been preserved in the Manchester science museum. The horizontal engines shown by Leigh were built in 1863 for Gidlow Mill, Wigan, and were described as being the largest simple expansion engines in a mill for many years. By 1870, the horizontal engine could be described as standard.[72]

While the horizontal engine did not have the mass of reciprocating parts to help even out the force in the cylinder when using steam expansively, it was more compact and could be built to higher standards of engineering which meant higher speeds. Because it did not take up as much space, engine houses were smaller and so cost less. Fewer parts with no beam and all its supporting ironwork also reduced prices. But more important was the way the engine itself was built. All the moving parts, the crankshaft, crosshead and the piston in the cylinder, could be aligned on a single bedplate, or sections bolted together. Not only would this give much greater accuracy but, because the bedplate could be placed on solid foundations, the engines were much more rigid. While at first the beds were treated as plates to which the cylinders, guides and bearings could be bolted with the beds adding stiffness, soon they began to be designed to take the stresses with the metal in the correct places. One example of this was the change from open crosshead slides to the trunk pattern to which the whole face of the cylinder could be bolted and not just the bottom.[73] Although a few beam engines continued to be built until 1900, the horizontal engine had really replaced them by 1880. One of the last beam engines was that designed by J. H. Tattersall for the Nile Mill in Hollinwood. It was built in 1898 by Buckley & Taylor as a pair of triple expansion engines at a cost of over £10,000. With steam at 160 p.s.i., 2,400 h.p. was developed and it ran until the closure of the mill for cotton spinning in 1960.[74]

11
Such absolute smoothness

The Allen engine

Visitors to the 1862 International Exhibition in London were surprised to see some cotton and woollen looms being driven by a single-cylinder horizontal steam engine at 150 r.p.m., or about three times the speed customary for that day. Their credulity was further strained when they discovered that this engine was the result of inventions which had originated in the United States of America. One of these was a new type of steam inlet and exhaust valve recently patented by John F. Allen of New York. Charles T. Porter contributed a 'loaded' governor which he had patented in 1858.[1] Porter had asked his associate, Charles B. Richards, to design the third invention, a high-speed indicator which he did by the summer of 1861.[2] Its importance lay in its ability to produce accurate diagrams at high speeds. Without it, valve-setting on this engine would have been exceedingly difficult, for it enabled the engineer as it

Figure 51 The governor of Magnet Mill, Chadderton, showing the crossed arms to the weights, the Porter type central weight and the Lumb compensator mechanism. (Author.)

were to see inside the cylinder. The valves themselves were driven by the valve gear designed by Pius Fink.

Allen had been trying out improvements to valve design for some years, probably through his experience with railway locomotives. In his patent of 1856, he was concerned at first with getting the steam out of the cylinder

> to obtain a free exhaust of the steam to the termination of the stroke of the piston or, till the lead of the valve for the induction of steam commences, thereby obviating the resistance which is caused by what is known to engineers as the 'cushioning' of the steam in the cylinder.[3]

Allen achieved this by placing an intermediate valve between the main slide valve and the cylinder ports. This patent had nothing to do with the 'Allen' engine and it is almost ironical that criticism was voiced later about the valve events on that engine giving excessive cushioning. There was, at that time, little understanding of the thermodynamic importance of the compression at the end of the exhaust stroke.

Allen took out two patents in 1862 which were both concerned with valves to let more steam more quickly into the cylinder. The earlier one[4] was primarily for railway locomotives. The faces round the steam ports into the cylinder were raised so that the valve faces of an ordinary slide valve could travel beyond the port faces and open an extra passage inside the valve itself. This passage communicated with the inlet port opening at the opposite end of the cylinder, so that the steam entered the cylinder port through two routes, the usual one past the end of the valve and the second one through this extra passage from the far end of the valve.

Allen's second patent in 1862 was the crucial one.[5] The first point to be noticed is that separate valves were to be used for the inlet and exhaust, so their timing could be adjusted independently. The next point is that there were four valves, paired at opposite ends of the cylinder. In this way, the steam passages could be made short. Allen's real intention was to improve the steam flow into, or out of, the cylinder. He used the idea from his previous patent of raising the port faces and had 'cup' valves slide across them. At one end of the valve, the steam entered the port in the usual way, while at the other, a second opening was exposed through which the steam passed under the valve face, over the port face, into the 'cup' of the valve, then through the port and down the passage into the cylinder. Single or double versions of these valves could be employed.

It was Porter who persuaded Allen to patent this type of valve after Allen had shown him his ideas chalked out on the floor of his engineering shop one winter's day in 1860–1.[6] Porter realised that here was the key to the high-speed engine because the valves could be driven positively and yet the travel of the inlet valves could be varied to give the maximum expansion at different cut-offs. Porter had received a classical education and became a practising lawyer. He moved to New York and suddenly, when a family man of almost 30, abandoned the profession of the law and became an engineer.[7] His first venture was a stone-dressing machine which he had to redesign to make work properly. He found it was vastly improved when it ran faster. While trying to get the steam engine that drove the stone-dressing machine to run more smoothly, he made an invention of incomparably greater importance, for he improved the original Watt type of centrifugal governor.[8]

The Watt governor was still the accepted means of controlling steam engines, and Porter tried to find a way of increasing its sensitivity. Such governors usually rotated at the same speed as their engines, e.g. 30–50 r.p.m., and had to have heavy flyball weights to overcome the friction in the joints of the rods linking the governor to the butterfly valve and the friction in that valve as well. According to the laws of the centrifugal pendulum, the displacement of the weights or flyballs is a function of speed only, and is independent of the mass of the balls. The function is non-linear, for the ratio between displacement and speed, that is the governing sensitivity, decreases with rising speeds as the balls move higher. A simple calculation shows that the optimum range of operation lies well below 100 r.p.m.[9] Therefore the speed of the Watt governor could not be increased and, if the weight of the balls were increased to develop more power, the inertia was increased which slowed down the rate of response. The Watt governor could require variations in speed of from 15 to 40 per cent to make it work.

However, on some engines, a weight was added just above the governor itself on the rodding that connected the arms of the governor to the throttle valve. This weight would have helped to close the valve, particularly in a case of emergency, and it would also have helped to make the governor respond more quickly to a decrease in engine speed by forcing the flyball arms down and overcoming any friction in the operating rods. The weight would also have prevented the flyballs moving outwards until their speed had risen sufficiently for the centrifugal force to overcome the mass of this weight. This would have made the governor more sensitive at particular speeds. When this was introduced is not known. One of the engines built in 1824 at Pode Hole near Spalding for pumping water out of the Fens with a scoop-wheel had such a weight but this seems early and may have been a later addition. This feature appears more regularly on engines built from around 1840 onwards.[10] Scott Russell described a slightly different arrangement in the seventh edition of the *Encyclopaedia Britannica* just after 1840. It consisted of a long horizontal bar with heavy weights on it which would dampen the oscillations of the flyballs. These weights would also make the governor react in the same way as the weights placed on the rodding and so they would seem to pre-date part of Porter's work.

What Porter did was to reduce the size of the balls and increase the speed by a factor of about ten. In this way he gained a motion that was powerful. To make the device sufficiently responsive at the right speed, he balanced the centrifugal forces by a counterweight. In his patent model,[11] this counterweight was placed on a lever, but soon it was moved to the central shaft of the governor itself. He made this weight hollow so it could be filled with lead till it weighed from 60 to 175 lb. The balls were only about 2½ in in diameter and lifted the weight at about 300 r.p.m.[12] Balls weighing 1 lb, rotating at 300 r.p.m., exerted the same centrifugal force as balls of 36 lb at 50 r.p.m. The counterpoise weight prevented the balls flying out until the optimum speed was reached, so the steam valves remained fully open until that point. Then the balls would react more quickly to slight variations in speed. Porter claimed that a speed variation of between 3 and 5 per cent could be achieved from the regulating valve being closed to being held wide open.

The importance of this governor was recognised quickly because it was so sensitive and prompt in action.[13] The *Scientific American* wrote the following eulogy in 1858:

> There are three requisites for a perfect governor. It must be extremely sensitive, so as to begin to open or to close the regulating valve instantly on the slightest variation in the speed of the engine; it must effect the whole movement necessary to open or close the valve very rapidly; and a force must be developed by almost inappreciable variations in speed sufficient to overcome all hindrances to its action. These requirements seem at last to have been met. The improvement in centrifugal governors illustrated in the accompanying engraving, if not absolutely perfect in its action is so nearly so, as to leave in our opinion nothing further to be desired.[14]

At first, Porter manufactured and sold his governors himself. He claimed that, to get a perfect action from them, a degree of accuracy in their manufacture was necessary which could be obtained only in a specially equipped factory.

> By the use of a variety of tools, devised expressly for performing each operation, and by making the governors in large numbers, I am able to afford them at prices at which a small number could not be sold; at the same time the highest excellence in workmanship is uniformly attained.[15]

Porter's experience with speeding up his stone-dressing machine and his high-speed governor convinced him that a high-speed steam engine should be equally successful, and this became the goal of his engineering career. In the cases of his stone-dressing machine and his governor, Porter had found that higher speed had brought the advantage of smoother running and he expected the same benefits from his steam engine, but, at first, this was not the case.[16] It was to take extensive design modifications over several years before his vision was realised.

In fact the engines displayed successively at the London International Exhibition of 1862, the Paris Exposition Universelle of 1867 and the Exhibition of the American Institute, New York, in 1870, all show fresh improvements.

Any new engine Porter designed had to compete with the Corliss engine which was beginning to find favour even though its valve gear at that time limited its speed to around 40 r.p.m. Porter realised that if the valves were linked all the time to the mechanism of the engine, then they would open and shut in unison with the motion of the engine. The problem Porter had to solve was how to vary the cut-off while retaining the maximum expansion of the steam at all settings without wire-drawing the steam. The answer lay in a new valve gear patented by Pius Fink in 1857 (this patent does not appear in the English list) which found one of its first applications in the Allen engine. All the valves were operated by a single eccentric. The eccentric strap was extended into a slot for an expansion link motion. A pivoted arm prevented the strap from rotating but allowed the link to oscillate. The top of the link drove the

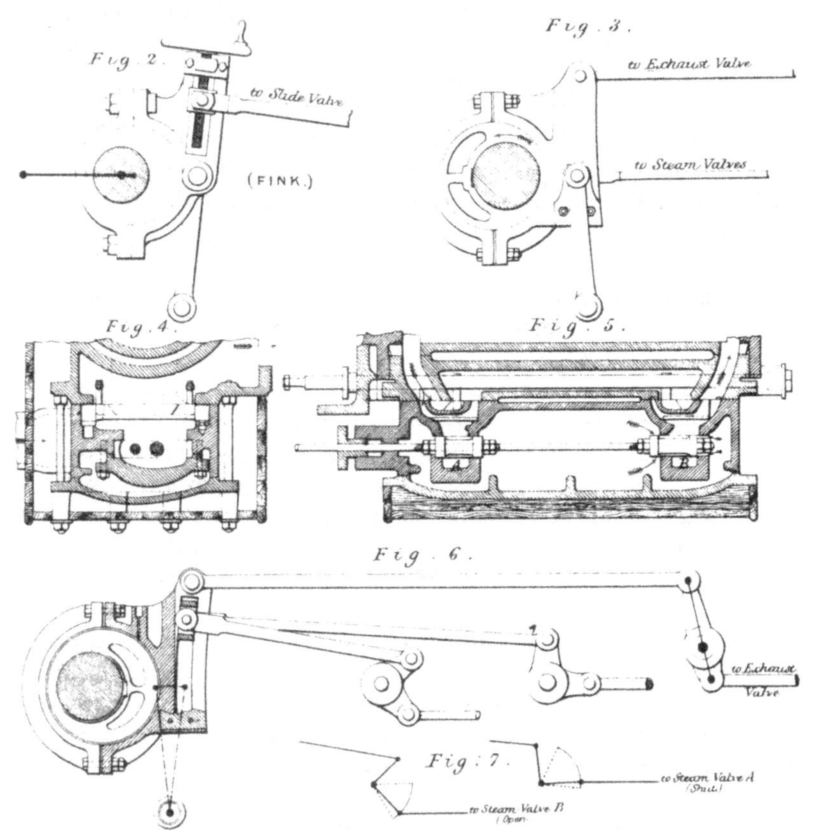

Figure 52 The Fink valve gear and Allen's valves on Porter's high-speed engine. (Rigg, *Steam Engine*.)

exhaust valve rod which was connected to a fixed pin to give the maximum travel which was never altered. On the other hand, the travel of the inlet valves could be varied by sliding a die-block up or down the slot in the eccentric strap. Porter arranged his governor so it would lift or lower the die-block in the expansion link. In this way, the governor acted directly on the distance the inlet valves travelled and so controlled the cut-off which determined the admission of steam to the cylinder. Much to the surprise of his contemporaries, Porter was able to dispense with the throttle valve and the engine was entirely self-regulating.

From the die-block, separate rods worked the inlet valves at either end of the cylinder. By driving through bell cranks set at different angles, it was possible to arrange the motion so that one valve was stationary and the other moving. While the Fink valve gear was the simplest of the various link motions, it performed best only when set to give short periods of admission and high expansion.[17] However, it had a defect because the 'lead', or the point of entry of the steam, varied as the cut-off was changed which made it almost impossible to obtain quiet running with the earliest engines. The answer was to re-set the valve gear so that greater compression was obtained at the end of the exhaust stroke.[18]

Indicator diagrams taken on the engine at the 1862 Exhibition[19] show a good expansion curve but also an early, and quite high, terminal compression. Until the thermodynamic advantages of this were fully recognised, it was considered a fault on which Arthur Rigg commented adversely in 1878,

> Such a link motion, when set at high rates of expansion, produces an unsatisfactory compression of exhaust steam . . . A comparison of the steam port closing and range of expansion may now well be made between this system and others, without its being very clear why so much has been done to get from one eccentric a series of movements that can be easily obtained from two eccentrics by a judicious construction of double valves. Such valves, whether moving on fixed seats or within other valves, will produce a closing of steam ports nearly as rapid as given by the Allen gear, and of equally convenient range; while the exhaust ports are regulated by an invariable law.[20]

Porter visited many engineering workshops in America in order to find the best way to construct his engine. He lagged his cylinders better than most engine builders of the period. First there was a layer of felt and then a wooden covering over the cylinder barrel. The cylinder heads, or covers, were hollow to prevent external cooling.[21] Porter considered that steam-jacketing was advantageous only where there was a steam supply separate from that being used in the cylinder and which both had a higher temperature and could drain freely back into its own boiler.[22] He also realised the problems of the expansion of the cylinder through heat so, on his later engines, he designed it to hang beyond the main bed where it was free to expand and contract.

In order to obtain a smooth running engine, Porter used a disc crank with a counterbalance weight. Although balance weights had been used on the driving wheels of railway locomotives since Roberts introduced them in 1837,[23] this was an early example on a stationary steam engine. The part of the disc opposite to the crank was cast hollow and filled with lead to balance not only the crank itself but the reciprocating parts as well.[24] In this way, during the summer of 1861, Porter had found himself in possession of all the necessary inventions which enabled him to assemble a realistic solution to the problem of the high-speed steam engine.

His next step was a common one among American inventors of the day. He took his engine to England where he realised that the London International Exhibition of 1862 would provide the golden opportunity to show its merits. The engine he displayed there ran at 140–150 r.p.m. with a piston speed of 600–750 ft/min. It was tried later at 225 r.p.m. The inlet and exhaust valves were placed on the same side of the cylinder. It is difficult to make out the precise details from the drawings, but enough can be seen to show that the hot incoming steam passed across the backs of the exhaust valves, a poor thermodynamic arrangement. The routes of the steam pipes cannot be determined either, but both inlet and exhaust pipes were situated underneath the cylinder, again a poor layout. Porter realised the disadvantage of this arrangement and at first tried to lessen the effects by 'a belt . . . interposed to protect the cylinder from the cooling blast of the exhaust vapour'.[25]

Although the engine for the International Exhibition had been made by the London firm of Easton Amos & Sons, Porter formed an association with Joseph Whitworth in Manchester for building subsequent ones. High speed demanded high precision in construction which Whitworth's works ought to have been able to provide but relationships between the two men broke down and Porter returned to America in 1868.[26] In the meantime, Porter carried out many improvements to his engine. In America, he had not met with either condensing or compound engines and it quickly became evident that a non-condensing engine was unmarketable in England and so he had to design an airpump that would operate at high speeds. He placed a horizontal condenser behind the cylinder and used a ram or plunger for the airpump piston which was connected directly to an extension of the piston rod through the cylinder.[27] The plunger thus added to the reciprocating masses of the engine. The first condensing engine was completed late in 1866 and ran at 150 r.p.m. with unprecedented quietness and smoothness. Porter was puzzled: how would an increase in the reciprocating masses improve high-speed performance? He had no time to investigate this phenomenon before the next important world exhibition in Paris the following year.[28]

At the Paris Exposition Universelle in 1867, Porter and Whitworth exhibited

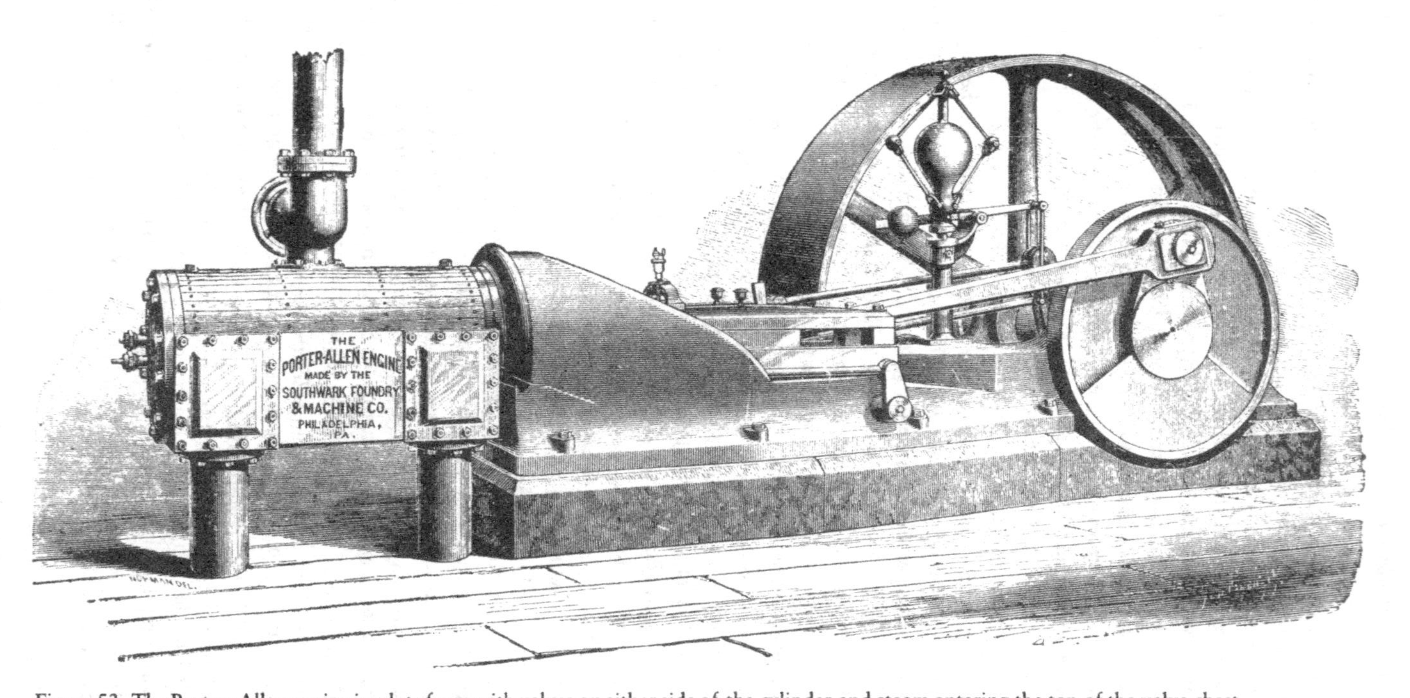

Figure 53 The Porter–Allen engine in a late form with valves on either side of the cylinder and steam entering the top of the valve chest. (Smithsonian Institution.)

at least four engines. Two drove electric generators at 300 r.p.m., a task for which these engines were to be used extensively in later years. The largest, with a cylinder 12 in by 24 in was a condensing engine which drove the English machinery at a speed of 200 r.p.m.[29] A smaller one, 6 in by 12 in, ran at a speed of 400 r.p.m. but could be speeded up for demonstrations to 2,000 r.p.m. or more.[30] These engines performed flawlessly. However, a drawing of the condensing engine shows that the valves were still set together at one side of the cylinder and that the steam entered below the cylinder and was exhausted out of the top to go to the condenser. This was a poor thermodynamic route and not good for draining away any condensate.[31] John Hick, then probably the leading builder of stationary steam engines in England, watched one of these engines in Paris and said, 'No amount of testimony would have made me believe that a steam engine could be made to run at that speed, with such absolute smoothness'.[32] His reaction was probably typical but the Porter engine seems to have had little direct influence on the design of mill engines as far as speed was concerned. It was more important in other industries such as generating electricity.

After the Paris Exposition, three changes occurred which must have improved the thermodynamic performance and lessened steam consumption considerably. Manufacture of the engine ceased between 1873 and 1876 when Porter had a chance to revise some of his ideas.[33] The inlet valves were linked together and placed on one side of the cylinder. The exhaust valves were moved across to the other side so that not only were the tortuous passages eliminated but the thermodynamic layout was improved too.[34] Then the exhaust valves were set low down so that the bottom of their ports was below the level of the bottom of the horizontal cylinder. This allowed any condensate to run away naturally through the valves and obviated the need for drain cocks.[35] Finally the steam pipe was changed so it entered from the top of the cylinder and a pair of exhaust pipes left below. In this form, the engine continued to be built by the Southwark Foundry in the United States until well after 1900. The use of slide valves by the Manchester firm of Galloway has been mentioned earlier (see p. 156).

After the Paris Exposition, Porter had time to investigate why his condensing engine ran so smoothly. What he found surprised him, for it was the reverse of engineering practice at that time[36] which was to make the reciprocating parts as light as possible. Porter claimed that,

> The reciprocating parts of an engine are properly considered as a projectile, which is put in motion by the force of the steam . . . But in this engine, the reciprocating parts are made as heavy as possible, without giving to them a clumsy appearance, in direct violation of the received maxim, that these parts must be made as light as possible.[37]

The length of the piston was made equal to half the stroke, a feature probably introduced after 1867. Porter claimed that, with a heavy piston, the forces acting on the crank were more even because part of the initial force of the steam was absorbed in giving motion to the piston and other reciprocating parts. Then, during the second half of the stroke, these parts would be providing energy as they were slowing down at the same time as the force of the expanding steam was diminishing.

Porter aimed to adjust the weight of the reciprocating parts so that their initial acceleration required about half of the initial pressure of the steam. Porter wrote in 1867,

> The reciprocating parts thus act as a reservoir of force by which, precisely as the flywheel, they become not the mere means by which, but rather the medium *through* which the force of the steam is transmitted to the crank. They remedy the great defect of the crank-motion, which is, that at the beginning of the stroke, the force of the steam is exerted more directly to punish the engine than to rotate the crank. They remove the serious objection to working steam at a high grade of expansion, that the engine is driven by a succession of blows. They save the crank and shaft and frame from sudden shocks and cause the engine to run with a gliding motion that is really surprising to witness.[38]

Porter deliberately kept the stroke of the engine as short as possible, compared with other engines of the period, to reduce the piston speed. Later, a typical engine with a 20 in by 36 in cylinder would have been running at about 100 r.p.m. with a piston speed around 600 ft/min.

Porter's claims aroused a great deal of interest and a diagram was produced showing how the steam and reciprocating forces were balanced out. Edwin F. Williams developed a sort of indicator which produced a diagram on a piece of paper graphically illustrating the acceleration and retardation of the reciprocating parts.[39] The Southwark Foundry was making the same claims for their engines 30 years later and even said,

> This engine having no features which impose a limit to its rotative speed, we are at liberty to employ those speeds which are, on the whole, found to be most desirable.[40]

Yet other engineers were quick to realise the defects of the system which worked only at one set of conditions for load, speed and steam pressure. Except at these optimum settings, there would be considerable and unnecessary transfer of the work from the first to the second half of the stroke and

> an amount of irregular action on the crank nearly as pernicious as if the whole initial pressure came upon it at the beginning of the stroke, like the case of an ordinary slow-going engine working with the same expansion.[41]

Rigg commented that 'This remarkable engine carries sound theoretical principles beyond the prudent restraints of a more refined investigation'.[42]

In tests at the Exhibition of the American Institute, New York, 1870, the steam supplied to the Allen engine was superheated to some extent, with a pressure of about 70 p.s.i. and working non-condensing. The superheat averaged 35.5 °F (or 343.9 °F) but it was admitted that 50 °F superheat would have been desirable to reduce the initial condensation in the cylinder. Running at 125 r.p.m., it indicated 125 h.p. with a consumption of 2.87 lb of coal per indicated horsepower per hour.[43] These tests showed the benefit of superheat in checking condensation of the steam during admission. It was later claimed that, by running these engines at high speed, a great weight of steam was worked through the cylinders and so the percentage of loss through condensation was reduced. Also, by 1890 at least, the advantages of the compression of the residual steam at the end of the stroke were beginning to be understood. The Southwark Foundry catalogue stated,

> The speed at which our engine is operated enables us to secure in the construction of the valvegear, the compression at the end of each stroke, which completely fills the end clearances in the cylinder. This compression is so adjusted by the peculiarities of our valvegear that it is constant at any power which the engine is called upon to develop. By this means we obtain a higher temperature in the cylinder itself, and at the same time avoid the use of live steam for filling the end clearances.[44]

This principle was to be taken up in the last type of reciprocating engine, the Uniflow, but is very different from what Allen was trying to do in his first patent of 1856.

Later types of governors

The Porter governor seemed to unleash a great variety of different types of governors over the next few years, far too many to cover in detail here. However, we may trace the broad development along several lines. These were first to improve the sensitivity through changing the geometrical layout. This could be further improved by making it possible to alter the load weights while the engine was running or by replacing them with springs. Then there were devices to compensate for variations in the load and finally ways of lessening the power the governor had to exert by introducing servo-mechanisms which would make the alterations in the setting of the valve gear or throttle valves.

As with so many other parts of mill engines, each manufacturer developed his own variations of governors and remained faithful to that type.[45] By altering the positions of the arms of the flyballs, a Watt governor could be made more sensitive; in fact it could become too sensitive so that it would over react and cause the speed of the engine to fluctuate wildly. Much the same thing happened when the balls were caused to move out in a parabolic arc because there would be one position of equilibrium at one particular speed of rotation but the balls would fly up to the top of the arc if the speed were very slightly increased and would sink to the bottom if it were reduced.[46]

Ironically on such governors, it was necessary to introduce damping mechanisms. The simplest of these consisted of a piston in a cylinder. The ends of the cylinder were linked with a passage through which the flow of either air or oil could be restricted as the piston moved up or down.[47] A much more sophisticated device was Higginson's regulator. This consisted of two vessels partially filled with mercury. They were attached to the ends of arms mounted upon the lever mechanism of the governor and connected by tubes. At the correct speed, both vessels would be level but when the speed, and therefore the position of the lever operating the control rods, altered, the level of the vessels would change. Mercury would flow along one pipe to the lower vessel (the other pipe allowed air to pass in the opposite direction) and its weight exercised a considerable steadying action upon the governor, preventing oscillations of a very short period.[48]

One of the disadvantages of the Watt pendulum governor was that, the further the flyballs rose, the less force they exerted as the arms approached a horizontal position. Also the force developed was dependent upon gravity working upon the flyballs until Porter introduced the loaded governor in which the force was created by gravity working on the load weight. It was possible to arrange the flyball linkage so that the balls moved in a horizontal path which brought them under the control of centrifugal action only. Wilson Hartnell devised one such governor which of course still needed some form of load or counterweight to counteract the centrifugal force.[49]

Hartnell also invented one form of shaft governor. Such governors became very popular in America where they were fitted inside the flywheels. On later drop valve engines in Britain they were operated by shafts rotating at the side of the cylinders. Here, because the governors were placed horizontally, there could be no use of gravity and springs had to be used to give the restraining force on the weights. The important feature of shaft governors was the use of the inertia of the weights as well as the centrifugal force. The weights were pivoted so that, if the speed of the engine suddenly slowed down, the inertia of the weights would cause them to fly outwards, and vice versa when the engine speeded up. At the correct speed, the position of the weights (no longer shaped like balls) would be balanced by the centrifugal force acting against the springs but a change in speed caused the inertia of the weights to alter the steam inlet.[50] Such governors could exert great power and became popular on high-speed engines running at over 150 r.p.m.[51] They reacted very quickly to changes in speed.

The problem remained how to compensate for changes in load. The first method was to alter the effective mass of the central weight of the Porter governor. To begin with, Porter had placed his load weight on an arm which projected to one side of the governor. By sliding this weight along the arm, the effective mass could be varied. Porter-type governors on some much later mill engines were fitted with such an auxiliary weight which could be easily adjusted. A more sophisticated variation was fitted to a McNaughted beam engine built in Glasgow in 1880 by Turnbull, Grant & Jack for Power's Distillery in Dublin. In this case, the weight was moved along the arm by a screw so it could be adjusted by the engineman while the engine was running.

Weights on governors suffered from the same objections as weights on Corliss valve gears because the force of gravity was slow in operation. Therefore weights were sometimes replaced by springs which had the additional advantage of being easily adjustable to give different tensions. Springs had no inertia to overcome and so their reaction was generally swifter.[52] The most important of the spring-loaded governors for mill engines was the type patented in 1894 by H. F. C. Whitehead who worked for Scott & Hodgson. It consisted of a

> centrifugal governor in which, when the centrifugal forces alter owing to movements of the governor in response to fluctuations in the speed of the motor, the amount of resistance opposed to the centrifugal forces is automatically varied.[53]

The essential feature was a pair of compressed springs, one inside the other, which exerted their pressure on opposite sides of a plate. There was also a cylinder filled with oil and a piston inside it to reduce the tendency of hunting. At the correct speed, the force of the springs exactly balanced the centrifugal forces on the flyballs. When the speed rose, the flyballs moved out and compressed the outer spring. This in turn moved the piston in the cylinder and

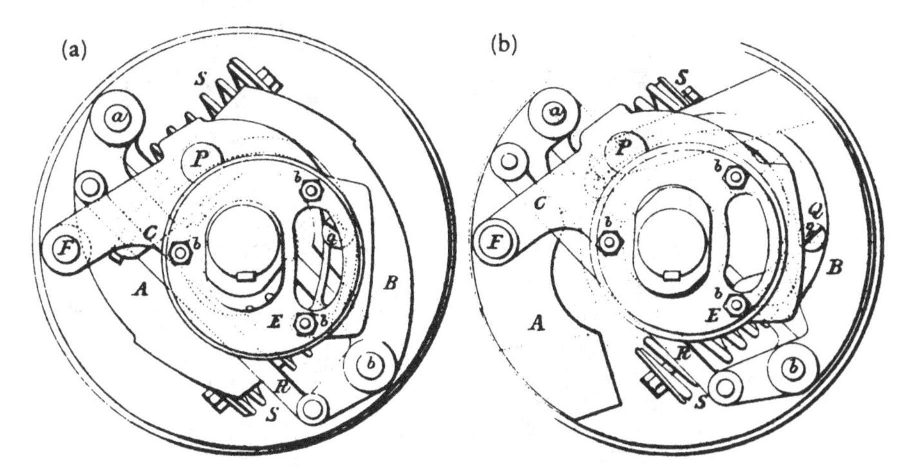

Figure 54 Hartnell's shaft governor which worked with both centrifugal force as well as inertia shown as (*a*) stationary and (*b*) full speed. (Rigg, *Steam Engine*.)

transferred some of the load to the other spring, relieving the outer spring so that the speed would return to normal again.[54] Scott & Hodgson fitted these governors to most of their later engines and this type was used by many other manufacturers too.

The problem with all the governors mentioned so far was that, when there was a direct connection between the governor and the steam adjusting gear, whether a throttle valve or a trip gear mechanism, there was only one setting which corresponded to one particular speed and load. Full power would be developed at one speed and lower powers gave correspondingly increased speeds at the same throttle or cut-off setting. To maintain the same speed, adjustments had to be made in the lengths of the rods connecting the governor to the steam controlling mechanism which at first was done manually through turning the knuckle joining rods threaded with right and left hand screws.[55] Ways of doing this automatically had to introduce another intermittent driving force into the system.

One of the earliest mechanisms to give an automatic variable adjusting motion was developed by William and John Yates of Blackburn. They supplied an engine to Russia before 1878 on which the governor shaft was fitted with a pair of friction cones. These could drive a third cone covered with leather, which remained stationary when the engine was running at its proper speed with the ordinary load. When the speed varied through a change in load, the

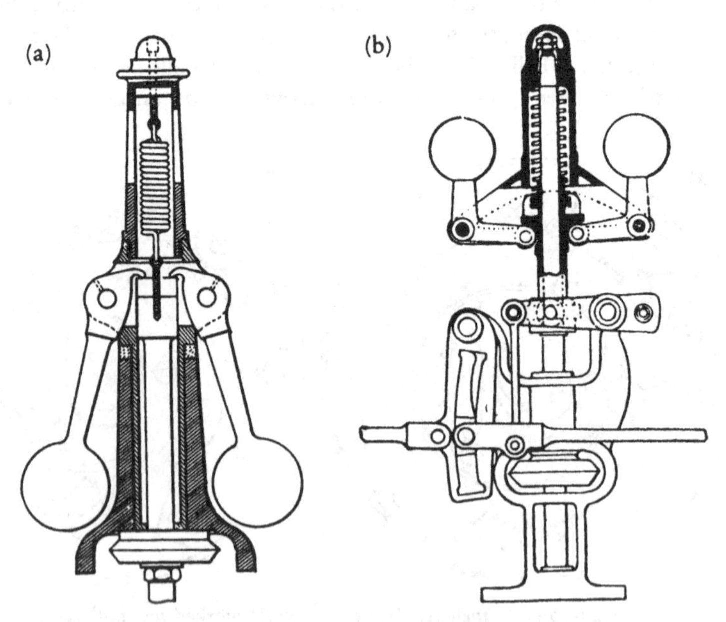

Figure 55 Two types of spring-loaded governors. (*a*) Tangye's and (*b*) Hartnell's. (Ewing, *Steam Engine*.)

cones on the governor shaft moved up or down into engagement with the friction cone and turned it. The friction cone rotated the knuckle on the governor rodding and altered the length to vary the steam inlet according to the new load. When the speed had returned to its proper setting, whichever cone on the governor shaft had been in contact with the friction cone disengaged and the adjustments ceased.[56] In a similar system, W. Knowles used a supplementary governor to provide the power to rotate the knuckle on the main governor.[57]

Probably the most sophisticated and successful of these devices was the Lumb regulator. The original patents were taken out by R. Wilby in 1886 and 1887[58] but the device does not seem to have become popular until after it had been improved by James Lumb in 1900.[59] A drive was taken from an oscillating shaft, such as the rod to the Corliss valves, which could turn either one of a pair of ratchet wheels. The teeth on the ratchet wheels faced in opposite directions and were fitted onto a shaft which turned the knuckle on the governor rods through bevel wheels. The arms which rotated the ratchet wheels oscillated up and down all the time but were brought into engagement only when the speed of the governor varied. Another linkage connected to the flyballs of the governor could be finely balanced by a small weight to bring the ratchet operating arms into and out of engagement. While the Lumb regulator still depended for its operation on a change in speed, it was reckoned that variations in speed could be kept to less than 3 per cent.[60] Most of the mill engines still running in the 1960s were fitted with these regulators because they could give

Figure 56 Whitehead's spring-loaded governor, the most successful of such governors on mill engines, on the engine at Saxon Mill, Droylsden. (Author.)

such a close control of the speed and even out the minor fluctuations of machines being stopped or started.

The final phase of governor development was to reduce the force which was needed to operate the rodding and linkages. This would leave the governor free to react to changes in speed more quickly. When the force had to be supplied by the governor, the variation in speed had to be sufficiently extensive to produce a corresponding change in the centrifugal force and so the speed variation became much greater than was desirable.[61] The answer was to introduce a servo-system so that the governor controlled only the valves which worked the servo-cylinder and the piston in that cylinder drove the rodding. In a patent of 1888, S. Buckley and J. Taylor specified an idea for

> a mechanism through which the fluctuations of any suitable ordinary governor are transmitted so as to control on the relay principle a variable cut-off gear.[62]

While steam could be used, such systems really had to wait until the introduction of pressure lubrication in which the oil could provide the power in the servo-cylinder.[63] This did not happen until after the turn of the century and so few mill engines were ever fitted with servo-mechanisms. Around 1900 too, various devices were fitted on governors to trip out and stop the engine should the speed become too fast or too slow. The most popular were Tate's which could be linked to emergency buttons throughout the mill. When the emergency button was pushed or the glass in the trip panels broken, an electric circuit closed the main steam valve and opened a valve in the condenser to admit air and break the vacuum so that the engine was brought to a halt quickly.

Rope driving

The traditional way of distributing the power from the steam engine to the rest of the mill was through gearing and line-shafting. A heavy vertical shaft had to be employed to take the drive from the engine to the various floors of the mill. At each level, bevel gears linked the vertical shaft to the horizontal shafts running along each floor. This system was noisy with the rumbling of the gear teeth. It needed a great deal of maintenance and constant oiling was necessary. Should a spinner or weaver let the drive into one machine too quickly, there was a danger that teeth might be stripped. This caused stoppages for repair. There was an additional problem that the speed of line-shafting rose more quickly than engine speeds and so it was necessary to install a higher ratio of gearing to overcome the difference. While there was an enormous inertia in all this rotating metal, there was also an enormous loss in power through friction in the gears and bearings.

While individual machines had been driven by belts from the line-shafting for a long time and it is possible that a few small engines drove machinery through belts, it was in 1828 at the newly built Appleton Mills in Lowell, New England, that belting took the place of gear drive to connect the waterwheels to the rest of the machinery. Paul Moody, the millwright in charge of construction,

> did away with the heavy English-type gearing which had creaked and groaned in every American mill from 1790 on ... [and] touched off a new American style which soon came to constitute an important distinction between English and American mills.[64]

From a drum driven by the waterwheel, a broad flat leather belt would be taken around guide pullies to all the other floors of the mill where it would pass round pullies on the line-shafting. The belt drive ran far more smoothly and quietly, cost less, and was more readily installed and repaired. Breakdowns were fewer and shutdown times shorter. The disadvantage was that, if the belt broke, the entire mill had to be stopped for it to be replaced and the immense length, weight and the difficulty in handling it were strong objections to this arrangement. Fairbairn, who had, of course, introduced the system of lighter, faster line-shafting, continued to favour gear drive because experience had convinced him that there was less loss of power with it. Later it was also claimed that the system of ropes absorbed 5–10 per cent more power than direct drives.[65]

During the 1870s, some belt drives were installed in Britain. Once again, there probably always had been a few on smaller engines but by 1877, both Galloway and J. & E. Wood were building larger engines with flat belts. If the mill were extensive, then more than one belt might be necessary. Each floor would be driven by its own belt so that, if a belt broke, only that floor would be stopped until it could be repaired during a convenient time. In 1877, J. & E. Wood built a pair of tandem compound engines developing 800 i.h.p. at 46 r.p.m. for Armitage & Rigby of Warrington. This engine had a flywheel 28 ft diam. and 7 ft broad round which were three belts, each 26 in wide.[66] Similarly Derker Mills in Oldham were equipped with a pair of Galloway side by side slide valve horizontal engines, one either side of a 22 ft diam. flywheel with probably four flat belts round it.[67] A Galloway side by side horizontal engine drove the English machinery at the Paris Exhibition in 1878. In this case there was a single flat belt, 3 ft wide, round an 18 ft diam. flywheel on the engine crankshaft which rotated at 60 r.p.m.[68] This firm also supplied a similar engine to drive machinery for the Woollen Exhibition at the Crystal Palace in 1881 but in this case the flywheel was only 15 ft diam.[69]

Flat belt drives survived for a few more years and from about 1910 a few steel belts were used which ran on flat rims faced with cork. Generally in Britain belts were superseded by another form, multiple rope drives. While there were a few previous attempts at using ropes, the history of the textile mill rope drive can be said to date from 1863 when one was installed at Falls Foundry, Belfast, with 200 i.h.p. capacity.[70] It was James Combe of Belfast who made a number of experiments with pullies in which grooves of different angles were turned. Ropes were passed over these grooves and it was found that an angle of 45° gave the best results. The first ropes used by Combe were made of leather, cut from the hide in the longest possible strips and made up into a circular rope. But their cross-section varied and they were liable to become untwisted. Then manilla hemp was tried and this proved to be successful. Later cotton ropes were introduced which proved to be longer lasting but more expensive.[71]

The flywheels on the mill engines were grooved to take the ropes. The amount of power required to drive each floor of the mill would be calculated to find the

Figure 57 Rope drive at Mons Mill, Todmorden. This flywheel was one of the widest in any mill. (Cotton Board.)

number of ropes which would be necessary. One of the widest flywheels installed in Britain was at Pear Mill, Bredbury, 13 ft 8 in across, with grooves for 73 ropes. A pulley with the appropriate number of grooves would be fitted on the line-shafting on each floor and aligned so that the ropes would fit across the width of the engine flywheel. It was a magnificent sight to see the ropes reaching up from the massive flywheel to the top floors of the mill. One of the problems which had to be overcome was to splice the ends of the rope so that the diameter remained uniform and did not stick in the grooves. Also the length had to be determined accurately to give the tension necessary for driving without slipping or being too tight and causing friction. Starting up on a Monday morning might prove hazardous if the ropes had become damp over the weekend and had shrunk.

Rope drives began to be installed in the Lancashire mills during the later 1870s. In 1877 Galloway built a complex arrangement for a mill in Hadfield with a shaft drive supported on columns across a yard to the lower floors while two higher floors were driven by ropes.[72] More significant was an order the same year for a mill in Eccles where the steam engine had a 25 ft diam. flywheel grooved for 17 ropes which took the drive to three different levels. The drawings of both Galloway and Goodfellow show that during the 1880s, rope races took over from gear driving. J. & W. McNaught supplied its first engine with rope drive in 1881 and really had switched to rope drive by 1890 although a few spur drives were fitted later.[73] With J. & E. Wood, although they probably installed more engines with flat belt drives in the early 1880s, the change to ropes was even quicker for by 1885 most of their engines had this form. Hick Hargreaves too must have made the change at the same time for in 1883 they patented flywheels for rope drives which could be made of up various rings bolted together to give the width necessary to take the large number of ropes which would be needed in a big mill.[74] A rope with a diameter of 1½ in at a velocity of 4,700 ft/min. would transmit about 40½ h.p. At much greater speeds, the effect of the centrifugal force would tend to throw the ropes out of the grooves so they slipped.[75]

Rope drive affected the design of textile mills and engine houses. The greater breadth of the engine meant that it would take up more space and so engines began to be built in separate houses on the outsides of mills. Then the rope drive also required more space than the previous vertical drive. Sometimes the rope race would be placed towards one end of a mill where its walls would separate the opening processes with the attendant risk of fire from the rest of the mill. With cotton ropes stretching the full height of the mill, the rope races themselves were a fire hazard. Later, the rope races were built at one end of the mill, leaving the whole of the floor space on each floor unobstructed. These flywheels with their ropes became too large to be turned by hand so the engine builders began to fit small steam barring engines to turn the main ones round for maintenance or to a starting position. Musgrave progressed from a hand barring gear in

1881 to a steam one in 1883.[76] Hick Hargreaves were taking out barring engine patents in 1883 and 1884[77] while J. & W. McNaught had one in 1890.[78] In fact this was another instance where every mill engine builder developed their own style of equipment.

Ring spinning

Another important development that came from America which was to make smoother running in textile mills was the ring frame. Various attempts had been made to improve the water frame and the throstle but it has been claimed that Jencks was the true inventor of the ring frame in 1832.[79] The flyer was replaced by a metal loop which travelled around a ring to give the twist to the cotton fibres during spinning. This was a continuous action and so needed a steady source of power. It produced yarn quickly but at first only the coarse counts could be spun. Although it was tried in Britain soon after its invention, it was not considered satisfactory but in America it came to replace the throstle. Leigh, writing around 1870 said, 'Notwithstanding that about forty years have elapsed since it was brought into England, not a single spinner has yet ventured to try it on a large scale'.[80] Gidlow Mill at Wigan was equipped with throstles and mules in 1863. Some ring frames began to be introduced in the 1880s and more after 1900. However, despite its growing popularity in other countries, most British spinners continued to prefer the spinning mule until well into the present century. Their confidence in the mule was shown in a massive investment in them for, between 1905 and 1914, the number of cotton spinning spindles increased by 15 million to a total of almost 60 million. Of these more than 80 per cent were mule spindles. Yet the owner of Elk Mill, Chadderton, was thought to be out of his mind in 1926 when he filled it with medium count spinning mules which were considered to be obsolete by that time. Some of these were still running in 1973 when they were the last of their type to be operated in the world.[81] Where ring frames were installed, the engines needed to be more powerful because not only was the spinning continuous but ring frames occupied less floor space and so many more could be fitted into a mill and their spindles ran faster. This led to the demand for bigger, more powerful mill engines.

12
Twinkle twinkle little arc

They had no vision amazing
Of the goodly house they are raising,
They had no divine foreshadowing
Of the land to which they are going;
But on one man's soul it hath broken,
A light that doth not depart.
A. W. E. O'Shaughnessy.

The development by Joseph Swan and Thomas Edison in 1879 of the incandescent electric light bulb was to have far reaching effects because it launched the electricity supply industry. Up to that time, generators were used in plating works which did not need much power. There were arc lights but while these illuminated large spaces such as railway stations, they were too bright and unsuitable for small rooms in offices or domestic houses. The incandescent light bulb did not give off the fumes or heat of gas lights or candles and the energy needed to power them could be transmitted comparatively easily through wiring. The electric motor followed and created new demands for electrical power, particularly with the development of transport systems on both railways and tramways. Magnus Volk opened his small railway on the front of Brighton on 4 August 1883,[1] but development was slow elsewhere for some years. Electricity had many advantages over older forms of power distribution. Compared with line-shafting, it could be distributed over longer distances and to less accessible places. As long as line-shafting was running, there were frictional losses in the bearings but, although electricity had to be constantly available, distribution losses were not apparent to the domestic consumer. Probably most important of all was that electricity could be usefully consumed in small amounts such as in light bulbs and this would eventually lead to an unprecedented demand for power.

What is considered as the first public electricity supply in Britain was started at Godalming in 1881 with a water-powered generator but it was discontinued in 1884 as unprofitable. Here not only were the streets lit but current was offered to private customers too. If we confine ourselves to undertakings which have had a continuous existence to the present time, there seems no doubt that the honour of priority should be given to that of Brighton, which was offering to supply current to anyone who desired it as early as February 1882.[2] This first station was powered by a 12 n.h.p. Robey engine driving at 900 r.p.m. a Brush arc-lighting dynamo with an output of 10.5 A at 800 V. From such small beginnings, it could hardly be thought that the electricity supply industry soon would be a forcing ground for the development of steam power and indeed that one day it would make the reciprocating steam engine itself redundant in textile mills and industry everywhere.

The operating conditions of engines driving generators varied enormously depending upon the type of installation. In a textile mill, once the mill had been started up first thing in the morning or after the mid-day break, the load would be fairly constant because most of the machines would be operating all the time. There would be minor variations as one machine was stopped and another started. With engines driving dynamos for electro-plating, the demand for current was fairly constant but this was not so with electric lighting. To begin with, demand was mostly for a short period of the day, or rather night, and, while probably there would not be sharp fluctuations because not everybody would switch on their lights at the same moment, the engines would have to cope with a wide range of loads without allowing the lights to flicker, grow dim or burn too brightly. To try and even out the demand, particularly during the day so that the plant would be utilised more economically, was one of the reasons for some municipal supply systems installing tramways. However on traction systems, the demand for current was constantly varying as one tram, or more, might start up and then suddenly cut off power. This imposed severe conditions on the steam engines driving the generators to ensure a steady supply.

This chapter will investigate three areas where the electricity supply industry affected the design of steam engines. The first is in the area of increasing the speed of rotation. The second is in the growth of the size and power output of engines to meet the ever increasing demand for current and the third is the types of engine which were evolved to meet both these requirements. Some of the technological innovations which first appeared in engines for supplying electricity were applied to those in other industries.

The early bipolar generators had to be rotated at considerable speeds to generate any useful current. The 900 r.p.m. of the Brush set at Brighton has been mentioned already. The station at Colchester established in 1884 by the South Eastern Brush Electric Light Co. Ltd had dynamos running at 750 r.p.m.[3] In America at Lincoln, Nebraska, a Westinghouse alternating machine running at about 1,650 r.p.m. was installed in 1887.[4] No steam engines could run at such speeds and so generators had to be driven through some form of belt or rope and pulley system. There were many disadvantages to this for a great deal of space was required necessitating a large building as well as special separate foundations for the generator. There was the cost of belting to be considered but the real disadvantage was the loss of power absorbed in the friction of the extra bearings on the shafting and in the belting. This has been estimated as being between 5 and 10 per cent of the power output of the engine.[5]

To avoid this loss, engineers began to increase the speed of steam engines and drive the generators directly. The advantages of higher-speed engines were that: they could be smaller to produce the same power; because there were more power impulses per minute, they would be smoother and for the same reason the governing systems would act more quickly. During the later 1880s, electrical engineers contributed by producing generators which did not need to be rotated so fast. Multi-polar types appeared, originating in Germany, which could be coupled up to the engines themselves. For small units, speeds might be 150–300 r.p.m. and, for larger ones up to 1,500 kW, around 70–150 r.p.m.[6] The first direct coupled sets in the United States of America were built in March 1892 when Reynolds of Allis-Chalmers contracted to build two Corliss cross compound engines for the Narragansett Light Co. of Providence, Rhode Island,[7] but the principle was well established in Britain by that date.

The Willans engine

Peter Willans was one of the pioneers of the high-speed engine for he had patented a design for a launch engine in 1874.[8] In his first layout, there was a line of three single-acting vertical cylinders so arranged that an extension on the top of one piston acted as the valve controlling steam admission to the next and, in compound or triple expansion working where the cylinders were placed on top of each other, the piston itself uncovered ports which exhausted the steam to the next cylinder. The design was improved in 1880 when a special valve was fitted which allowed the engines to work as simple when starting or compound to give better economy when running.[9] To further enhance the smooth running of these engines, the reciprocating parts were fitted with air-cushions in a patent of 1882,[10] although the wisdom of this has been doubted as it would absorb power. However, it did prevent knocking which was one of the disadvantages of the early double-acting high-speed types. Lubrication was by the cranks dipping into the oil in the bottom of the crankcase which was totally enclosed.

It was probably this type of Willans engine which was installed in 1886 at the Westbourne Park power station of the Great Western Railway which provided the electricity to light the Great Western Hotel as well as Paddington station. Three Willans engines of the three crank tandem compound type drove directly Crompton 25 kW dynamos at 400–500 r.p.m. to provide 110–130 V d.c. as the excitation current for the main alternators.[11] Steam pressure was 160 p.s.i. However, in 1884 Willans had patented an engine, which had valves running up the centres of the pistons themselves, called his central valve engine.[12] The pistons were fitted with a pair of connecting rods and this central valve was worked off an eccentric fitted between the rods. The valve gave better timing for the steam events than on his earlier engines. The single-acting vertical layout and the air-cushioning buffer were retained so the engines ran very smoothly.

The Willans engine was quickly adopted in power stations, particularly those generating d.c. where the sets could be run in parallel. The Kensington Court Electric Light Co. started operations in early 1887 with a Willans engine of the old launch pattern, pending the completion of a central valve type which was late in delivery. This was directly coupled to a 35 kW Crompton dynamo which it drove at 500 r.p.m. with steam at 160 p.s.i. This marked the beginning of the supremacy of the Willans engine in central power station work.[13] Other early stations supplied with these engines were Whitehall with three 170 i.h.p.

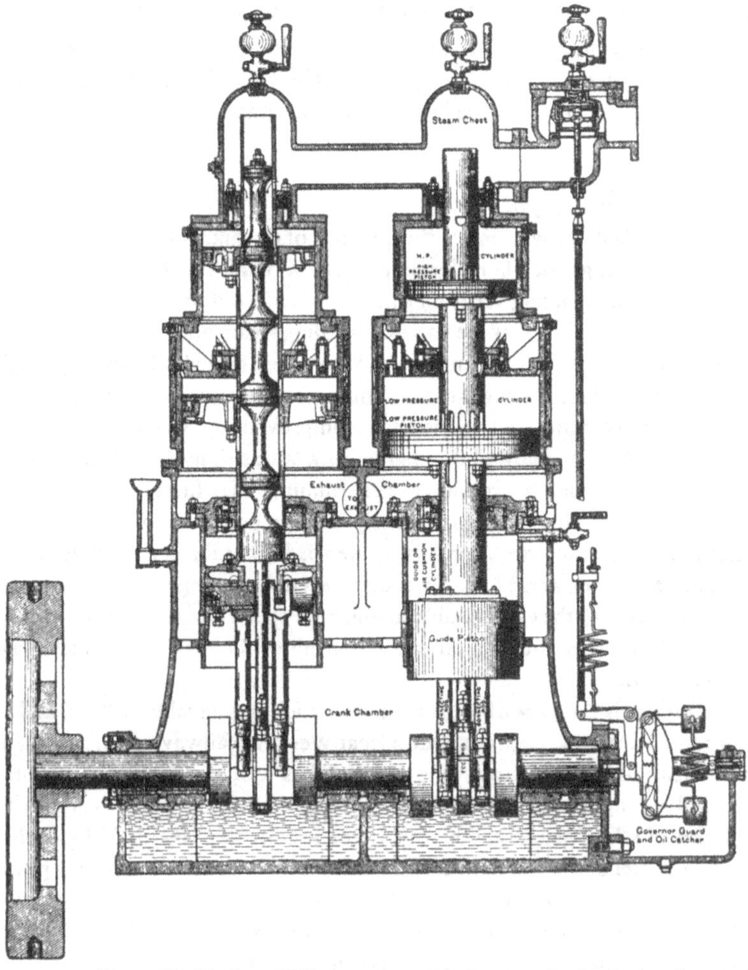

Figure 58 The later Willans engine with the central valve in the piston rod. (Dickinson, *Short History*.)

running at 350 r.p.m., and Charing Cross with four, both in 1888. In 1889, the St James' and Pall Mall Electric Lighting Co. had a pair of 80 i.h.p. and four 210 i.h.p. Willans engines, with 150 p.s.i. boiler pressure, all driving dynamos. When this station was completed in 1891, a further six of the larger engines had been added.[14] The popularity of the Willans engine is shown by the fact that, in 1892, the aggregate capacity of the engines at work in British central power stations possessing 300 h.p. and over was just under 33,000 h.p. Of this, 22,300 h.p., or 68 per cent, came from Willans direct coupled engines. By 1895, of some 101,390 h.p., 53,340 h.p. was generated by Willans engines. In 1903, the Westminster Electric Supply Corporation had altogether 49 Willans engines of an aggregate capacity of 9,330 kW in its three generating stations.[15]

Some larger Willans engines were installed in textile mills. For example Baerlein & Co., Manchester, had a 575 i.h.p. engine in 1894. A couple of years later the fine spinners, McConnell had two 760 h.p. engines at their Great Ancoats Mill where steadiness with the thin yarns was vital. They also installed one of 800 h.p. at their Droylsden mill so they must have been satisfied with the performance. The largest size seems to have been 825 i.h.p., one of which replaced a beam engine in 1903–4 at Langworthy Bros., Greengate Mill, Salford.[16] Such engines would have rotated at around 300 r.p.m., much faster than ordinary mill engines. They were coupled to a small diameter pulley grooved for driving ropes which distributed power to the different floors of the mills in the usual way.

These engines were Willans's normal vertical, single-acting designs, generally with three sets of cylinders. They were smaller and more compact than ordinary mill engines so presented a low initial capital outlay. The parts of cylinders, pistons, valves,. etc. were standardised and a stock maintained at Rugby so repairs, when necessary, could be effected quickly. More important was the smoothness in running. Each crank had the same cylinder arrangement so they balanced each other and also developed the same power. The high speed gave steadier control and running while the lubrication did not need the attention of the engineman every few minutes. Engines could be built as simple with only one cylinder on each crank and later have additional parts added to convert them into compound or triple expansion without having to scrap the whole engine. They were also well suited to being supplied with superheated steam and so were very economical as well as steady engines. It was claimed that, in a mill where a Willans engine replaced an ordinary horizontal cross compound, this steadiness resulted in an increase of 10 per cent in spinning due to the saving in broken ends.[17] What the Willans literature does not state is that the single-acting design was more bulky than a double-acting one.

The small Willans engines and generating sets would have been suitable for generating the current for electric lighting in mills. The Willans Co. advertised them for this purpose and no doubt many were installed but the author never came across any. Separate lighting sets were necessary because the mill had to

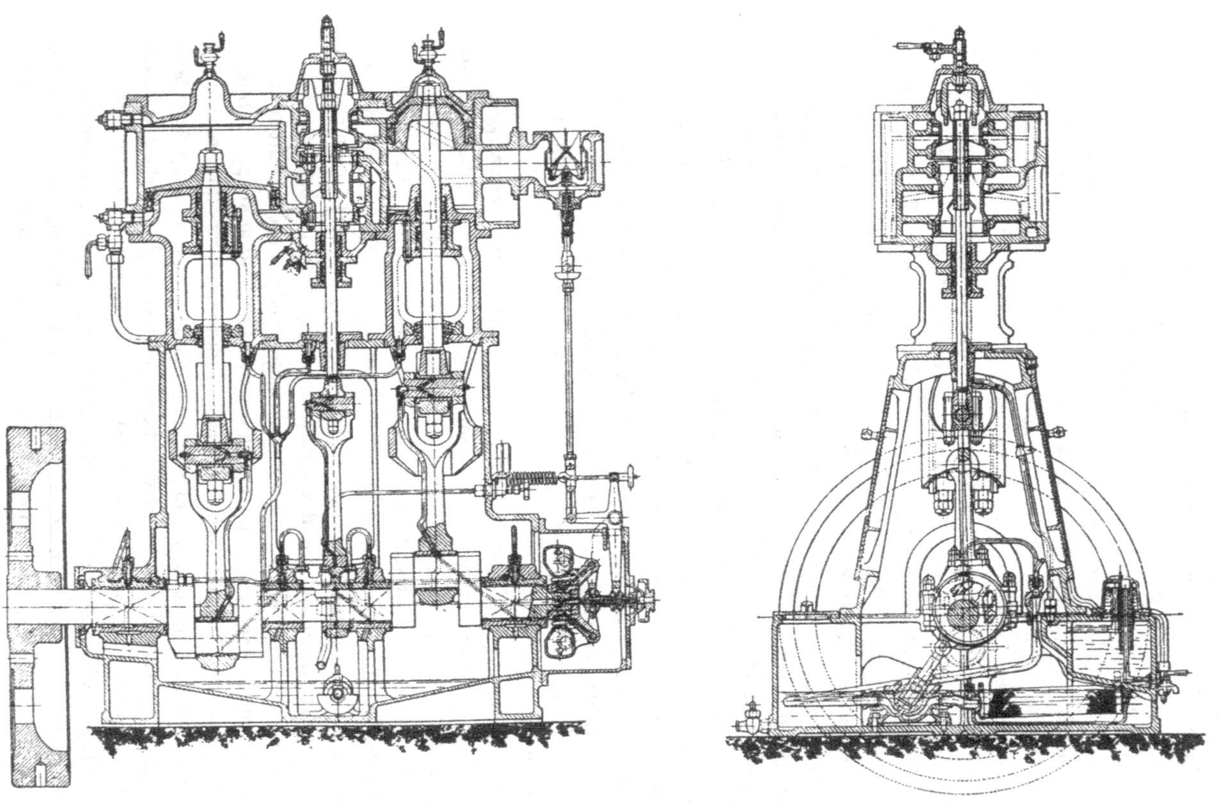

Figure 59 Belliss's self-lubricating high-speed compound engine. (Dickinson, *Short History*.)

be lit to allow the operatives to come and go, or maintenance to be carried out, while the main engine was not running. Although some other companies built single-acting engines on the lines of the Willans, for example the Alley MacLellan which had a central valve common to a pair of cylinders, the development of the high-speed totally enclosed engine followed double-acting designs. The disadvantage of the double-acting engine, that is the knocking of the reciprocating parts through the push and pull of the connecting rods on the crank pins, was solved by G. E. Belliss of Birmingham. The success of his engine has been attributed to the system of forced lubrication to all parts which was patented in 1890 and 1892 by A. C. Pain, a draughtsman in Belliss's works.[18] An oscillating pump worked from the valve eccentric supplied oil at a pressure between 10 and 30 p.s.i. The crankshaft was drilled to take the oil to the big ends and through them up to the little end bearings. Pipes distributed oil to other parts. These engines were vertical, with compound cylinders in line and totally enclosed to prevent the oil splashing everywhere.

Quickly many other manufacturers of similar engines appeared. Allen of Bedford, Galloway of Manchester, Browett & Lindley of Patricroft were some. W. M. Musgrave took out a patent in 1898. In 1901, W. A. Ashworth and W. S. Parker established a new works at Bury specifically for constructing these engines.[19] Accessibility to working parts was improved by designing larger removable side plates. The problem of water from the steam cylinders running down the piston rods and into the oil in the crank case was tackled with better glands and easier access to the packings. Belliss continued to build these engines until the 1950s because they were so popular for driving generator sets in hospitals, paper machines and the like where process steam was available. Such engines coupled to generators made by Mather & Platt, the Electric Construction Company and others were the most common type of lighting sets employed in cotton spinning mills. Ashworth & Parker made engines in sizes from 5 to 1,200 b.h.p., and in compound or triple expansion versions with three or four cylinders. One four crank triple expansion 500 h.p. superheated engine was installed in a woollen mill at Helmshore coupled to an alternator and a dynamo for d.c. current. Another drove a cotton mill at Todmorden through ropes but such installations were never common.[20]

The Ferranti engines

When Sebastian de Ferranti was appointed Chief Engineer of the Grosvenor Gallery Station on 13 January 1886,[21] he found an assortment of generating equipment. This station had gone into service towards the end of 1885 with a pair of Siemens single-phase alternators, the largest that company had built to date, driven by belts from a countershaft to which any or all of three horizontal Marshall engines could be clutched according to load. The two smaller engines developed 225 i.h.p. at 80 r.p.m. and the larger one 450 i.h.p. Ferranti reorganised this and installed two of his own alternators, designed to rotate at 160 r.p.m. One was driven through flat belts by the existing engines and the other by an additional 700 i.h.p. Corliss valve horizontal single-cylinder engine built by Hick Hargreaves with rope drive from the flywheel.[22] Because at this date it was very difficult to run two or more alternators in parallel due to the unsteadiness of the steam engines and the difficulty in governing them, the method of running the Grosvenor Gallery was this. One of the main alternators was driven by one of the small Marshall engines at times of light load. As the load increased and this engine reached the limit of its capacity, another small engine was run up to speed and clutched in so they shared the load. Then the third engine would be added. When the load increased beyond the capacity of these engines, the Corliss engine and its alternator were run up to speed and some of the circuits transferred to them.

It was obvious that the Grosvenor Gallery site was too small to cope with the increasing load and in August 1887 a new company, the London Electric Supply Corporation Ltd was formed to take over the Grosvenor Gallery Station and replace it with a vast new generating station eight miles away at Deptford from where the current for lighting 2 million lamps would be distributed to the whole of London. It was envisaged that engines developing 120,000 h.p. would be needed for supplying mains at a pressure of 10,000 V, which would have created a concentration of power unrivalled in the world.[23] The new station at Deptford was opened to members of the press in October 1888 who

> came away with a vivid impression, not only of the magnitude of the undertaking but also of the sheer audacity . . . of the promoters. We venture to assert that in the whole range of engineering enterprise there is nothing which has demanded more skill, ingenuity and daring than the . . . lighting of central London from Deptford Station.[24]

Initially there were to be four 10,000 h.p. engines, later reduced to two, and a couple of 1,250/1,500 h.p. engines for periods of light load. These 10,000 h.p. engines, had they been finished, would have been larger than any running on land at that time and had to be such a size to cope with the maximum load envisaged for the circuits on a single alternator. The designs of Ferranti's engines both at Deptford and those he installed in other stations later show the problems encountered in developing reciprocating steam engines for faster speeds and higher boiler pressures.

We have seen already that the boiler pressures used in some electricity generating stations were much higher than in contemporary textile mills. In 1888, it was stated that

> the boilers of the common marine design can be made to work satisfactorily at a pressure of 150 lbs. per square inch, and even higher . . . Until science and skill have discovered new materials, or other applications of old ones, there will not be much practical advantage in

> using steam at higher pressures than now obtained and 200 lbs. absolute pressure seems about the limit at which theoretical economy is swallowed up by practical losses.[25]

However at this time, a new application of boiler design had arrived in Britain from America in the form of the Babcock boiler. George Babcock and Stephen Wilcox took out their first American patent on 26 May 1867 and one for an improved design in Britain in 1874.[26] In 1881, an office was opened in Glasgow and Job No. 7 was an order received for a pair of boilers for the Edison Electric Light Co. Ltd power station on Holborn Viaduct in London. At Grosvenor Gallery, Babcock boilers were also installed with a pressure of 130 p.s.i. Cast-iron parts on the boilers were replaced by wrought iron in 1887 which enabled them to withstand higher pressures. The advantages of the Babcock boiler were that it occupied a small floor area and had good circulation of water. Because it was made from so many smaller parts, the thickness of the metal was less, with the result that the heat transferred more quickly and it could withstand higher pressures. The Deptford station was designed to house 80 boilers on two levels and 24 were actually ordered in March and October 1888, each rated at 250 h.p. to work at just under 200 p.s.i.[27]

It could well be that the pioneering work of the electrical engineers helped to introduce higher pressures in textile mills. The Babcock boiler was used extensively in generating stations but rarely in mills. Yet Lancashire boilers could be built to withstand these pressures. In Hampstead, a municipal electricity supply was started in October 1894. The steam-raising plant consisted of four Lancashire boilers with a working pressure of 150 p.s.i.[28] The 1890s was the era of triple expansion engines in mills which needed these sorts of pressures. What is clear is that Ferranti's pressure of 200 p.s.i. was exceptionally high in 1888.

Even the two smaller 1,500 h.p. engines Ferranti commissioned for Deptford were stretching the limits of engineering practice at that time. The larger textile mill engines in that day would have been horizontal compounds with probably two high- and two low-pressure cylinders in tandem. Their speed would have been 45–50 r.p.m., producing around 1,000 h.p. The engines built by Hick Hargreaves were vertical cross compounds running at 80 r.p.m. Cylinder dimensions were 28 and 56 in with 4 ft stroke which was short for the period but designed to keep the piston speed low. The 22 ft diam. flywheel weighed 60 tons and was grooved for 40 ropes which drove the alternator. It is doubtful if any textile mill engine of that day had as many ropes. Rope drive was necessary because the alternators rotated at 120 r.p.m.[29] to give 83⅓ cycles per minute. Corliss valves, almost certainly with Inglis and Spencer trip gear, admitted the steam. These ought to have operated satisfactorily at such a speed but the governing system is unknown. These engines ran without much mechanical trouble for a number of years until at least 1901.[30]

The vertical layout was unusual for textile mill engines. Galloway had constructed a small vertical blowing engine in 1878 but apparently not another until 20 years later. Goodfellow had built four vertical engines between 1871 and 1885. J. & W. McNaught supplied their first vertical type to Grove Mills in 1905, a side by side compound with Corliss valves developing 1,100 h.p. at 80 r.p.m.[31] Vertical engines had been accepted in screw-driven ships since the

Figure 60 One of the 1,500 h.p. Hick Hargreaves vertical cross compound engines being erected at the Deptford power station. (Ferranti Ltd.)

1860s because the propeller shaft lay near the bottom of the vessel in a position where it could be coupled easily to a vertical engine. Although in some ships the dimensions of the cylinders approached those which Ferranti planned for his large engines, because the boiler pressures were lower, it is doubtful whether many developed as much power as he proposed. His use of the vertical engine may have originated with marine practice for in a note book dated 1883 we find Ferranti making enquiries for the Grosvenor Gallery from the following manufacturers:

> Hick Hargreaves & Co., Bolton,
> Denny Brothers, Dumbarton,
> van de Rerkoux [*sic*], Gent,
> Galloway, Manchester,
> Babcox [*sic*] & Wilcox Co., New York & Glasgow,
> John Blair, Sunderland,
> Triple Expansion Engines by R. Napier & Sons, Glasgow.[32]

Denny's had built a design of engine for the *City of Rome* in 1881 with six cylinders developing 10,000 h.p. and constructed others subsequently. For his 10,000 h.p. engines, Ferranti started building a tandem compound vertical design with a pair of cylinders either side of an alternator. The layout with one cylinder on top of another was extremely rare and the height would have been nearly 50 ft from the engine room floor with another 17 ft below that.[33] They would have weighed 500 tons. The tandem compound layout would have given him an engine with a more even turning moment and, with a pressure of 200 p.s.i., compounding was justified, as in fact a triple expansion engine would have been. The reason for a vertical layout was that some of the reciprocating forces would have been absorbed directly into the foundations which may have been vital because there was no balancing of the moving parts. The high-pressure cylinders were 44 in diam. and the low-pressure 88 in. The stroke has been given variously as 6 ft 3 in and 6 ft 6 in. While the valves were the Corliss oscillating type, the photograph of the cylinders being assembled in Hick Hargreaves works, where erection had to be carried out with the cylinders lying horizontally because they were too tall to stand vertically, shows that the operating mechanism was not their usual Inglis and Spencer arrangement, but what it was cannot be identified.

To save the power lost in the drive to a separate alternator, Ferranti turned his engine flywheel into the rotor by placing windings round its circumference and making it revolve between the field coils. Now the problem he faced was that the alternating current had to be generated at 83⅓ cycles, so the rotor had to pass between the correct number of magnetised coils to give this frequency. The slower the rotation of the engine, the greater had to be the diameter of the alternator to fit in the requisite number of coils. In the middle 1880s, the larger cotton mill horizontal engines rotated at 50 r.p.m., so not only were Ferranti's smaller engines at 80 r.p.m. fast for their day but the proposed speed of 60 r.p.m. for his larger engines would have been fast as well. Even at this speed, the flywheel alternators had to be massive, weighing 225 tons. The 36 in diam. crankshaft was turned from the largest casting ever made in Scotland weighing 70 tons in the rough. The diameter of the outside of the moving coils was 46 ft (other sources give 40 or 42 ft) and that of the flywheel ring on which they were mounted 35 ft (or 34 ft).[34]

One of the advantages of the vertical design with cylinders on either side of the flywheel was that there were only two main bearings to align. Had he chosen an in-line type, there would have been up to five bearings including those on the alternator to line up. Ferranti took many precautions to avoid bearing trouble. In 1889, he patented a system of pumping water through the backs of the bearing shells in main and crank pin bearings to keep them cool.[35] In this patent, he also curved the bottoms of the bearing shells so they could take up any misalignment, a feature to which we will return on Edwin Reynolds's design of the Manhattan engines. Originally the smaller engines at Deptford had plain bearings which gave trouble with overheating. These were changed for swivelling ones on both main and crank bearings and the trouble ceased.[36] To deliver adequate oiling, Ferranti had installed a circulating system at Grosvenor Gallery with oil flowing to the bearings from an overhead tank. Pictures of the

Figure 61 A pair of cylinders of the 10,000 h.p. engines for Deptford under construction at Hick Hargreaves's works at Bolton. (Ferranti Ltd.)

original alternators at Deptford show that there was a similar system of positive lubrication through the pressure head from the tank. Again this was in advance of mill engine practice where each main bearing had only a limited drip-feed arrangement and other bearings had oil cups.

The London Electric Supply Corporation was not granted a concession for supplying such a large area as had been envisaged and so the 10,000 h.p. engines were never completed. Ferranti continued to develop the vertical design of reciprocating engine with alternators between cross compounded cylinders when he established his own manufactory at Hollinwood near Manchester in 1895. By this time, it was possible to run alternators in parallel so that power stations could be equipped with a number of engines linked together electrically. Ferranti's vertical engines took up little floor space, the foundations were simple and cylinder wear was reduced. In a compound design there were fewer moving parts than in a triple and the governor could act more rapidly to change the volume of steam passing through two cylinders rather than three, regulating the speed more smoothly and quickly.

If the steam engine could be made to rotate more quickly, not only would it run more smoothly through the greater number of power strokes per minute but it would be governed better for the same reason. Then it would produce more power size for size with a reduction in capital expenditure, and the same applied to the electrical equipment. Engineers in England, and Ferranti in particular, took the lead in pioneering high-speed engines from their American and Continental counterparts.[37] Ferranti dramatically increased the speed of the engines he built at Hollinwood. An advertisement of 1895[38] mentions medium speed engines of 150 revolutions and in 1896 there are notes on a 2,500 i.h.p. engine with the same speed. A 600 kW engine supplied to Hanley in 1897 reached 170 r.p.m. In 1901, in a paper presented by Charles Day there is mention of speeds at 214 and 240 r.p.m.[39] A report written early in 1902 about the performance of the Ferranti engines states, 'The only serious problem left on engines is that of the 1,200 [kW] revolution engines'.[40] An engine installed at Wakefield in 1901 had a speed of 258 r.p.m.[41]

Such speeds were not attained without problems. Forced lubrication was essential and this meant that the engines were totally enclosed. The increase in speed accentuated another problem, that of balancing, and the Ferranti engines seem to have been early examples of stationary engines being fitted with balance weights. At these high speeds, both the valve gear and the governors had to be redesigned. The limit of speed with ordinary Corliss valves was generally accepted as 100 r.p.m. although improvements to the trip gear and return springs had raised this to 150 r.p.m. on the American Allis-Chalmers engines by 1906.[42]

Ferranti developed a most ingenious grid slide valve which he patented in 1898.[43] With it, the ports could be made very short and the clearances reduced to a minimum. He arranged it so it would shut in the middle position of its travel and opened when moved to either side. In this way its speed could be reduced to half that of the engine itself and gave quick opening and closing with a very short movement. On a 1,000 h.p. engine, the gross movement of the high-pressure valve in one revolution was 1¼ in, which at 214 r.p.m. gave the low valve surface speed of 22 ft/min.[44] Ferranti also arranged to drive his valves positively by eccentrics or cams. The inlet valves were opened by one cam which had a fixed angle of advance to give a constant lead. To close the valves, the

Figure 62 One of the later Ferranti high-speed cross compound vertical engines driving a 300 kW alternator supplied to Huddersfield in 1895. (Ferranti Ltd.)

angle of the cut-off cam could be controlled by a spiral or scroll operated by the governor to set the cut-off point sooner or later. In this way the valves matched the speed of the engine the whole time and there was no period when they were free to move by springs. The whole of the valve operating mechanism was contained in a cast-iron box situated between the cylinders and flooded with oil to give a long operating life.[45]

One of Ferranti's engineers commented,

> Some of the old jobs have been hung up for a long time whilst an entirely satisfactory governor was made. Several designs were tried without a sufficient degree of success, but now we have got a governor which seems entirely satisfactory.[46]

The governing system had to be redesigned to permit high speeds and yet put no load on the governor. On the later Ferranti engines, the lever from the governor was connected to a small piston control valve. Movement of this valve would open a port and allow oil under pressure to operate a relay piston which operated the cut-off linkage. To prevent hunting through the relay piston moving too far, a dashpot and spring were fitted to steady the system and also allow adjustment of the valve settings to compensate for the load changing. Emergency governors set to trip in at slightly higher than the normal speeds were fitted on some engines too.[47]

A list of 'Work done on Old Engines' prepared by Ferranti's at the beginning of 1902[48] seems to suggest that the designers were going over the limits of the available technology because most engines had to be modified. The list of repairs to the 1,500 kW engines supplied to the City Co., Bankside, in 1896 is astonishing.

> *Work done on Engines at City Co.*
> Trip gear of HP and LP cylinders replaced by cam gear.
> New throttle valves fixed and governor connections altered.
> Several new valve spindles and sets of packing rings for valve spindles supplied.
> Cylinders replaced and new piston and piston rings.
> Arrangement fitted for cooling oil.
> Splash covers and glands fitted on top of columns.
> Dash pots supplied for governors.
> Stronger bolts put in balance weights.

Most of the other engines that Ferranti supplied had to have similar repairs. The 300 kW set sent to Tunbridge Wells in 1897 must have been a complete disaster, for it 'was returned to shops for general overhaul, beds, crankshaft and dynamo being only parts left at Tunbridge Wells'. The performance of the City Co.'s engine is described thus:

Table 9. *Engines built by Ferranti 1896–1901*[53]

Period	Engines	Total kW	Price £
Jan. 96–Jun.	3	2,100	12,641
Jun. 96–Dec.	4	2,700	15,983
Jan. 97–Jun.	12	2,450	18,561
Jun. 97–Dec.	1	600	3,658
Jan. 98–Jun.	5	2,500	15,255
Jun. 98–Dec.	None		
Jan. 99–Jun.	3	2,200	15,020
Jun. 99–Dec.	1	800	5,500
Jan. 00–Jun.	5	3,500	21,700
Jun. 00–Dec.	5	2,200	12,044
Jan. 01–Jun.	9	3,300	17,765
Jun. 01–Dec.	12	10,200	49,144

> The Ferranti engines, put down in 1898 and 1899, had to have the ingenious valve gear of their inventor much simplified before they could be got to work satisfactorily. They also swayed to such an extent that it was said that a man had to have 'sea-legs' to remain on their platforms, while the vibration they caused resulted in the disintegration and ultimate fall of the statue of St George and the Dragon which adorned the frontage of the station. They ran, nevertheless, until 1911, being the last of the reciprocating engines to be taken out of the main engine-room when turbines came into use.[49]

Yet by 1900, Ferranti had overcome these problems and the orders he received in 1901 were 80 per cent higher than the year before.

In July 1905, all the Ferranti engine building plant at Hollinwood was offered for sale by auction because the construction of reciprocating engines had ceased through competition from the turbine.[50] In the six years during which he had been constructing engines, the time taken to execute an order fell from an average of two years from the date of receipt to the completion of erection to 8¾ months by 1901. There was a loss on engine orders up to 1899, which was turned into profit for 1900 and 1901 even though the price calculated on a per kilowatt generated basis fell. The first engine on which any profit was made was one of 1,000 kW sold to Lambeth in 1899 at £4,922 which cost £3,928.[51] In 1896, the cost was £6 per kW spread over the first three engines which rose to £7.6 for 11 engines in the first half of 1897. By the second half of 1901, this had been reduced to £4.8 which included commissioning.[52] Production figures from January 1896 to December 1901 are shown in Table 9. Sizes ranged from 100 kW up to 1,500 kW. In October 1896, a 2,000 h.p. Ferranti alternator was supplied to Deptford which was said to be the largest in Europe and one of 2,500 h.p. was built for the City of London Electric Lighting Co. in the following year.[54] As these alternators could run in parallel, the growth of generating stations was even quicker with multiple sets being installed.

Generating engines of other manufacturers

Before Ferranti developed his own high-speed engines, he turned to other manufacturers for adaptations of the more conventional mill engines to power his alternators. The Portsmouth power station was officially opened on 6 June 1894 where a Parsons turbine with a capacity of 150 kW at 3,000 r.p.m. was installed alongside a pair of slow-speed reciprocating sets, each rated at 212 kW at 96 r.p.m. The alternators were built by Ferranti to generate 2,000 V single phase current at 50 cycles. Yates & Thom of Blackburn built the horizontal side by side compound condensing engines with Corliss valves which gained a reputation for economy and reliability.[55] For the Southport station which opened in 1895 with another Ferranti alternator, this time the cross compound Corliss valve engine was manufactured by Hick Hargreaves.

Musgrave of Bolton were very active in supplying engines for power stations. In 1901, the Manchester Corporation Bloom Street station opened with four 1,800 kW Musgrave vertical engines driving Westinghouse generators and the Stuart Street station, which came into operation the next year, had six 1,500 kW Yates & Thom vertical cross compound engines. In 1905 at Stuart Street were added a pair of engines from Wallsend Slipway,

Figure 63 High Hargreaves's horizontal cross compound engine driving a Ferranti alternator for the Southport Electricity Supply Co.'s station, 1895. (Ferranti Ltd.)

Newcastle upon Tyne, which were said to be the largest reciprocating engines in any power station in Europe, delivering 6,000 h.p. or 4,000 kW at 75 r.p.m.[56] Musgrave also supplied six 2,000 kW engines to the new power house for the Bankside station, London, four arriving in 1903 followed by another couple later.[57] Glasgow Corporation Tramway Power Station received from them 5,000 i.h.p. compounds with two low-pressure cylinders.

By that time, there was competition from overseas. At Bankside, the first engines to be installed in a new power house were three Allis-Chalmers engines of 1,000 kW capacity and at about that time the same company had supplied the Glasgow United Tramways with a 4,000 h.p. three cylinder in-line compound engine driving the generator directly.[58] Sulzer Bros., Winterthur, Switzerland, had in 1901–2 sent engines to the stations at Central Bow and Willesden in London. These were three cylinder in-line vertical direct coupled engines developing 5,000 h.p. at 75 r.p.m. At Bow a vertical cross compound design of 3,500 h.p. was installed by Sulzers as well.[59] These are just a few examples of engines supplied to electricity generating stations for there were many others of increasing size.

It is to America that we must turn for the largest reciprocating engines ever built and for a layout that would influence mill engine design. Reynolds, of Allis-Chalmers, was asked to go to New York in 1898 for consultation about some high-horsepower engines for the Manhattan transport system. Here 225 railway locomotives which used 226,925 tons of coal and 535,000 gallons of water per annum were to be replaced by trains to be hauled electrically powered from one generating station producing 60,000 h.p.[60] Reynolds sketched out his ideas for 10,000 h.p. reciprocating engines while on the train. Because space was limited, Reynolds planned a compact layout which became known as the 'Manhattan'. There were four cylinders, two high and two low pressure. Their bores were identical with Ferranti's Deptford design ten years earlier, namely 44 and 88 in, but the stroke was reduced to 5 ft so the speed rose to 75 r.p.m. The originality of Reynolds's design lay in the position of the cylinders for the high-pressure ones were horizontal and the low-pressure ones vertical. One high- and one low-pressure double-acting cylinder were placed on each side of a flywheel alternator. The pistons on each side drove a common crank pin but the cranks were set at 135° so that there were eight strokes on each revolution instead of the usual four.

Vertical and horizontal pistons connected in this way had no dead position and this, together with the eight power impulses, gave an exceptionally smooth running engine. The valve gear on all cylinders was the Corliss double-ported type. Boiler pressure was 175 p.s.i. and the steam was generated by Babcock boilers. The total weight of engine and alternator was 720 tons of which the field coils weighed 185 tons at 32 ft diam. The weight of the complete alternator was 445½ tons.[61] The weight of the armature was so great that the crankshaft was deflected even though it was 37 in diam. This forced the central portion of the shaft down so that the crank pins revolved in a plane slightly out of the perpendicular. Therefore the vertical cylinders were built inclined inwards, so that, in a height of about 30 ft, they were closer together at the top by 7⁄32in.[62] This shows the importance of Ferranti's spherical bearings on engines with massive alternator flywheels. Steam consumption was not to be more than 12¼ lb per hour when indicating 7,500 h.p. with a vacuum of 26 in. The engines could indicate up to 11,000 h.p. Eight engines were supplied.

These engines were so successful that when, in 1902, the New York Interborough Rapid Transit Co. wanted to operate about 800 electric trains on 13 miles of three and four line track, similar engines were ordered except that they were superheated and were fitted with drop valves on the high-pressure cylinders.[63] The generators were the largest ever built by the Westinghouse Electric and Manufacturing Co. and weighed, together with crankshaft and flywheel, over 165 tons. They were one of the earliest sets to generate 10,000 V 3 phase. The power house was designed for 12 engines but only 9 were installed at first. As the rush hour traffic built up, one engine had to be started every 20 minutes to help share the increasing load.[64] The last one of these units was scrapped in 1958.

The contract for the Interborough Rapid Transit engines was awarded in September 1901 after careful consideration had been given to using turbines. At that time some 3,500 kW Parsons turbines were being built at the works of Brown, Boveri in Switzerland, but there were no designs for the capacity needed for the Interborough Rapid Transit system. The results obtained from the use of smaller turbines did not justify the purchase of such equipment to give successful operation on such a scale. Yet by October 1904, when these enormous reciprocating engines had been commissioned, steam turbine design had advanced so quickly that turbines would be considered for any future increase in generating capacity.[65]

The last reciprocating steam engines in British generating stations

The 'Manhattan' type of engine was copied in Britain for both textile mills and electricity generating stations. Saxon, for example, built two cylinder types for Hall Lane Spinning Mill in Leigh and the Pear Mill in Stockport. The London County Council opened its Greenwich station in May 1906 which was planned to have eight units but only four were ever built by Musgrave. These were rated at 3,500 kW each at 94 r.p.m., with steam at 180 p.s.i. The valves were Corliss with cylinders H.P. 33½ in, L.P. 66 in diam. and stroke 4 ft.[66] In this case the two high-pressure cylinders were placed in the vertical position to give better drainage. These sets were magnificent examples of British engineering but were obsolete before they left the drawing board. In 1910, the Greenwich station had been augmented with four 5,000 kW turbo-generators. The first of these Manhattan engines was scrapped in 1914[67] and the last by 1922.

The decision to install reciprocating sets at Greenwich did not reflect much credit on the consultants because, in 1902, a Parsons steam turbine had been tested at the Neptune Bank station of the Newcastle Electricity Supply Co. against the slow-speed marine-type engines installed there and had shown its superior performance. 'Indeed it was there that the test between our "Slipway"

Figure 64 The 'Manhattan' engine for Hall Lane Spinning Mill, Leigh, built by George Saxon in 1909. (Cotton Board.)

engine and the Parsons turbo influenced the Government in subsidising the S.S. Lusitania and Mauretania'.[68] By 1904, Parsons turbines were running at Carville station on Tyneside rated at 4,500 kW and an overload capacity of 6,000 kW with a steam consumption of only 15.4 lb per kW per hour. In 1905 contracts were closed for a 10,000 kW Brown-Boveri-Parsons turbine for the Westphalian Electricity Works, Essen, Germany,[69] so the turbine had overtaken the reciprocating engine. Already Ferranti had seen where the future lay for, in 1902, he had taken out his first patent for a steam turbine, followed by four in 1903 and another four in 1904. This accounts for the sale of his reciprocating engine works at Hollinwood in 1905. The firm of Willans & Robinson too by 1906 had started to make turbines based on the designs of Parsons,[70] so laying the foundations of the present G.E.C. works at Rugby.

13
The drive for efficiency

The textile industry continued to expand in spite of various slumps up to the beginning of the First World War. Not only were more engines needed for the growing number of mills but the mills increased in size too. In Oldham, the average size of a spinning mill in 1873 was around 50,000 spindles. In 1883–4 75,000 spindles and in 1890 90,000.[1] Gidlow Mill at Wigan must have been one of the biggest in its time for Musgrave built two horizontal engines, each with two cylinders, 40 in bore by 6 ft stroke, which have been described as 'undoubtedly the largest simple expansion horizontal engines in a mill for many years'.[2] They drove 154 throstle frames with 200 spindles on each, 2 pairs of mules with 600 spindles on each and 13 pairs with 1,076 spindles each, making a total of 61,376 spindles together with the preparation machinery.[3] The increase in mill size may be seen by the Lion Mill, Royton, which was built in 1890 with 109,000 spindles, then the largest single mill. The engine here developed 2,000 h.p. When it was planned in 1898, the Nile Mill at Hollinwood with 104,000 spindles was the largest ring spinning mill in the world. In 1900, the Eclipse Mill, Rochdale, was designed to contain 118,000 spindles.[4]

The increase in size of mills continued during the boom of the opening years of the twentieth century. The Oldham area was prominent in the large size of its mills which specialised in spinning and it was in these mills that some of the biggest engines were installed. Monarch was built in 1903 to contain 132,744 spindles. In 1905, there were Durban & Laurel with 120,000 spindles and Grape with 126,324. In 1906, there was Roy with 140,000 and in 1907 Hertford with 120,000. Production started at Pear Mill, Stockport, in 1913 on 137,312 spindles. In 1915, the Sun Mill was recorded as having 157,000 spindles and was questionably the largest mill to be built in the district but it did not all date from the same time as did these others.[5] The mills such as Ace and Argyll near Oldham and Wye No. 1 in Shaw, which were built at the time of the First World War, contained fewer spindles. So did Wye No. 2, which was the last mill to be completed in Shaw in 1925–6 with only 96,712 spindles. This engine developed 2,500 h.p. but it must be remembered that spinning machines would have increased in speed during these years and so consumed more power.

This increase in size of mills had to be matched by an increase in the size of the engines that drove them. While the horizontal tandem compound and the cross compound types sufficed for smaller mills, something larger was needed

Table 10. *Wind, water and steam power, 1760–1907 (h.p.)*

	Wind		Water		Steam		Total
	h.p.	%	h.p.	%	h.p.	%	h.p.
1760	10,000	11.8	70,000	82.3	5,000	5.9	85,000
1800	15,000	8.8	120,000	70.6	35,000	20.6	170,000
1830	20,000	5.7	165,000	47.1	165,000	47.1	350,000
1870	10,000	0.4	230,000	10.0	2,060,000	89.6	2,300,000
1907	5,000	—	178,000	1,8	9,659,000	98.1	9,842,000

From Kanefsky, 'Power Technology', p. 338

for the bigger ones. At first, and indeed right through to around 1905, the horizontal four-cylinder engine arranged with pairs of tandem compounded cylinders, having one set either side of the flywheel remained popular. Quite an early example which has been removed for preservation was one with gear drive built by Buckley & Taylor for Fern Mill, Shaw, in 1884. At 42½ r.p.m., this developed 1,900 h.p. with latterly a boiler pressure of 160 p.s.i. in H.P. cylinders 22 in diam., L.P. cylinders 48 in diam. and stroke 6 ft. Although Buckley & Taylor preferred slide valves and indeed the low-pressure cylinders were so fitted, Corliss valves were fitted on the high-pressure cylinders. Quite a late engine was the 1,500 h.p. engine built by Scott & Hodgson for Dee Mill in 1907. While having Corliss valves on the high-pressure cylinder, this firm, as did McNaught, preferred piston valves on the low-pressure cylinders. By this time, the speed had risen to 60 r.p.m. but the stroke had to be reduced to 5 ft in order that the piston speed should be kept low. In order that the peripheral speed of the flywheel should not exceed the safe limits of cast iron, the diameter of the flywheel was 26 ft and it weighed a mere 58 tons. This engine showed one of the very few design faults seen on a mill engine for cracks began to appear in the bosses where the flywheel segments were bolted to their spokes and these had to be stitched. Saxon preferred Corliss valves on all the cylinders of his engines. The engines at Dawn and Magnet Mills were so fitted. These engines were nearly identical and the same patterns may well have been used for both. Dawn, built in 1902, developed 1,400 h.p. and Magnet 1,700; the greater power being accounted for by boiler pressures of 140 and 160 p.s.i. respectively. At 26 ft diam., their flywheels weighed 80–90 tons. Speed was the standard 60 r.p.m. for this period and the stroke the standard 5 ft.

At the same time, there had to be an improvement in the way the steam was used in these engines because the amount of coal they consumed was enormous and it all had to be transported and fed into the boilers. Boiler pressures were raised, speeds became faster and so the engineering techniques had to be refined. There was great debate about whether there was any gain to be had by

Figure 65 Scott & Hodgson's pair of tandem compound engines built for the Dee Mill, Shaw, in 1907. (Courtaulds Ltd.)

jacketing the cylinders with steam and the best way of doing this. With higher boiler pressures, it became worthwhile expanding the steam in three or four stages in triple or quadruple expansion engines. All this time, the theoretical advantages of superheating had been recognised but it was not until lubricating oils were improved in the 1880s and steel superheater elements were introduced around 1900 that superheating really became practical. This affected the design of engines with a reversion to the simpler layout of the large cross compound engine.

Steam jacketing

The idea of lagging the outside of the cylinder, and indeed the steam pipes as well, to prevent loss of heat does not seem to have occurred to people until well into the nineteenth century, although Farey did point out that, if the cylinder could be covered in some perfect non-conducting material, there would be no need for a steam jacket to prevent heat losses. It was not until 1828 that Grose discovered that, by 'clothing' the cylinder, steam pipes and valves of the pumping engine at Wheal Towan, he could increase the duty by about 50 per cent. But people outside Cornwall were slow to learn the lesson. The shape of the casting of the cylinder for the Haydock engine, which probably dates from about 1830, suggests that it was never lagged. The Hick beam engine now preserved in Leeds has never been lagged either and that was installed around 1848 at the warehouse of the Manchester, Sheffield & Lincolnshire Railway in

Manchester. Alban found it necessary to comment on the advantages of lagging when he was writing in 1843.[6] Later mill engines were of course lagged to prevent heat loss from the main surfaces of the cylinder castings. Polished strips of wood, often mahogany, with brass banding, were used as lagging at first but after the introduction of asbestos during the 1880s, this was replaced by sheets of blue planished steel, again with brass strips to secure it. Sometimes on later engines, the edges were finished with shining steel strips. Heat losses from the outer surface of cylinders caused condensation of the steam inside which it was necessary to avoid to gain the maximum efficiency.

It would seem that at first people did not realise the importance of avoiding external radiation of heat but were more concerned with lessening the condensation inside the cylinder which was caused by the expansion of the steam during the working stroke. In an ordinary engine, the fresh steam entered a cylinder which had been cooled by the steam expanding and doing work during the previous stroke. The colder exhaust went out from the same end of the cylinder as the hotter incoming steam entered. Therefore any new steam had to heat up the residual steam and all the cold metal parts before it could begin to do any useful work itself and this was one major cause of condensation in the cylinder. Then as the steam expanded, it cooled and there was further condensation. In fact Rankine concluded that

> the greater part of the liquid which collects in unjacketed cylinders and which was once supposed to be wholly carried over in the liquid state from the boiler (a phenomenon called "priming") is produced by liquefaction of part of the steam during its expansion.[7]

As soon as Watt had realised the importance of keeping the steam cylinder hot, he tried to maintain its temperature by putting the working cylinder inside a container and passing the incoming steam through the space between them to keep the main cylinder warm, in other words, he placed a steam jacket round the cylinder. He fitted steam jackets on even his first experimental engines. As Farey pointed out,

> As the condensing apparatus is immersed in water, to be kept cold, so the cylinder should, if possible, be immersed in steam to be kept hot; for which purpose, Mr. Watt from the first used a casing or jacket round the cylinder, and also at the top and bottom: this was attended with very beneficial effects, although it enlarged the steam surface, and exposed the external jacket to a more rapid condensation than would have taken place from the surface of the cylinder itself. But to have the vacuum as perfect as possible, it is necessary that the cylinder be kept up to such a temperature, as to prevent the least condensation of the steam upon the internal surface . . . because, if the sides of the

> cylinder were to be wet, as in the common atmospheric engine, the vacuum would be vitiated.[8]

On a jacketed engine, because heat from the steam in the jacket passed through the cylinder walls, it raised the temperature of the cylinder and the working parts of the engine as well as the steam inside and helped to prevent condensation.

The advantages of lessening this condensation were, as Farey pointed out, that a better vacuum could be achieved. Water on the cylinder walls and probably also in the form of water droplets suspended as mist and spray not only acted as a distributor of heat and equaliser of temperature from the hot part of the cylinder to the cold but, as soon as the pressure was lowered, would flash back into steam, absorbing heat to do so and also tending to destroy the vacuum. This was the problem Watt had faced when he tried to increase the vacuum in the model atmospheric engine before he thought of his separate condenser. With a steam jacket, heat passed from the steam in the jacket to that inside the cylinder and kept the cylinder walls hot so condensation did not form there. The steam was cooled instead in the jacket where the presence of liquid water produced no ill effects as it could be drained back to the boiler.[9]

In 1824, Woolf conducted a series of careful experiments on the Wheal Alfred engines with steam jackets and concluded that the advantage of using jackets was in the region of 10 per cent. But interest in them waned with arguments against them following two lines. First, was there any point in taking heat out of the steam before it entered the cylinder? Alban recognised this.

> It is natural that there must be advantage in supplying free caloric to the expanding steam in the cylinder where it is deficient; but whether it is advantageous to abstract such caloric from the steam about to enter the cylinder, is quite another question (Note; The Cornish engines are not so arranged. The jacket is supplied by a separate communication from the boilers).[10]

The answer to this problem was to jacket the cylinder with heat from another source, whether from a secondary boiler at a higher pressure than the main one,[11] or, as was tried occasionally, with the heat from gases coming from the boiler flues.[12]

The other argument against the use of steam jackets was that, in a counterflow cylinder, the heat from the jacket continually warmed up the exhaust steam before it left through the ports. Therefore

> this heat from the jacket is thrown away, and would be entirely wasted were it not true that the walls of the cylinder are not chilled to the same degree as when the jacket is not used, and that the initial condensation is thereby reduced.[13]

The reason for this waste lay in the method of jacketing. Watt had tried to totally encompass the whole of the cylinder with a jacket but of course this raises the question of how there could be a gradual change in temperature of the steam from that of the full boiler pressure to that of the condenser. Later, it was customary to jacket the cylinder walls but not the end covers. This again heated the wrong parts because the mean temperature of the steam in the centre of the cylinder ought to be less than that at the inlet ports. Therefore with a cylinder jacketed in the usual way, there must have been heat losses through reheating the exhaust steam. This point will be examined when discussing Professor Johann Stumpf's work in developing the Uniflow steam engine where he jacketed only the cylinder covers. This would probably have been best on counterflow cylinders too, which indeed Elder did recommend for marine engines. While some of the exhaust steam would have been reheated, the parts closest to the fresh steam would have been kept hot and the initial condensation reduced as well as some of the condensation during the working stroke.

One reason for the lack of interest in jacketing cylinders up to the middle of the nineteenth century was the deficiencies in the theories of heat. The caloric theory postulated that, as steam expanded, it would become superheated. Therefore the presence of condensation was attributed to priming and the real source of this water was masked until the new theories of heat were established. One of the earliest conclusions drawn from the new principles of thermodynamics was that the specific heat of saturated steam was negative. This meant that, as the steam expanded adiabatically, it would condense, and this accounted for some of the condensate. But the major cause of condensation in a steam engine cylinder was the inevitable cooling of the cylinder walls as the steam expanded, the effect becoming much more marked near the ports. The cooled metal absorbed heat from the incomg steam which then condensed. Thus a greater quantity of steam had to be admitted than was theoretically necessary to perform the duty required.

Until 1850, practising engineers reasoned correctly from the existing caloric theory that the steam jacket was an unnecessary encumbrance and abandoned its use, thereby saving themselves some very difficult moulding and casting problems. It was not until 1854 that Hirn began a series of experiments on the engines at Logelbach and found that, although steam in the jacket was condensed, so losing a certain amount of heat, the increase in power compensated for this loss. He gained 23 per cent in power by using a jacket. Steam jacketing was used in a paddle steamer around 1838 but it did not become popular at sea until much later. Brunel proposed that the engines in his *Great Eastern* should be jacketed but she was launched in 1857 without them. Steam jacketing was popular in textile mills during the 1880s and for the next 20 years. It would seem that first of all steam jacketing worked best where the steam was supplied from a separate boiler but no case of this has been discovered in a textile mill. Then jacketing gave greater advantage when it was applied to engines working with a high degree of cut-off and expansion. In this case, there would normally be a wide range of temperature difference between the fresh and the exhaust steam and so jacketing reduced the initial condensation. It was claimed that there could be a saving of 8–10 per cent in fuel.[14] Also slow speed, long stroke engines (e.g. 5–6 ft stroke and 50–60 r.p.m.) gained more from jacketing because more condensation could take place against non-jacketed walls than when speeds were higher.

By the beginning of the twentieth century, it was claimed that, where a single source of saturated steam was being used, steam jackets were hardly worth adopting because heat was merely transferred from one part of the cycle to another.[15] Superheating changed this because some of this extra heat carried by the steam could also be used through a jacket to reduce internal condensation without the steam itself condensing. Superheated steam might also contain sufficient extra heat to avoid initial condensation as it entered the cylinder and warmed up the cylinder walls again. Superheating at the turn of the century also changed the style of mill engine design. On most earlier horizontal engines, it was customary to run the main steam pipes under the floor of the engine house so that the steam was brought up around the cylinder under the lagging to the inlet valves. This cramped the layout of the pipes and must have restricted the steam flow through them although the jacket would have acted as a sort of small reservoir. On later engines, the steam pipes swept down into the tops of the cylinders in gentle curves and might even separate into two pipes, one for each end of the cylinder, to give the smoothest passage for the steam and to present the least restrictions. The planished lagging on these pipes could be particularly fine.

Triple expansion engines

The large four cylinder engines arranged as pairs of tandem compounds began to lose their dominance during the 1890s through the introduction of higher boiler pressures which led to the adoption of the triple expansion engine, and occasionally the quadruple expansion engine. As long as the initial steam pressure did not much exceed 100 p.s.i., it sufficed to expand the steam in ordinary compound engines. Elder had predicted that if ever the expansion of the steam could be made to exceed nine times the volume of the admitted steam before cut-off, it would be advisable to carry this out in three successive stages. With pressures of 150 p.s.i., there was economy in triple expansion and quadruple expansion had advantages with pressures of 200 p.s.i. or more.[16] The temperature difference would be spread out over a greater number of cylinders so that the fall in temperature, and therefore the amount of initial condensation when the steam entered, would be less at each stage. This had to be balanced against losses in the steam pipes connecting the cylinders and also in the extra costs of the additional cylinders and the friction of their moving parts.

Adamson took out a patent in 1861 for 'multiple expansion engines' with three or more cylinders connected to one beam or crankshaft and must have built a triple expansion engine before 1867 for the Victoria Mills. Elder took out a patent in 1862 but does not seem to have built any engines other than compounds. The first person to apply a triple expansion engine to a steam vessel was Charles Normand, a shipbuilder of Le Havre, France, who fitted one into a river boat on the Seine in 1871 and then some others in ocean-going steamships. The first British marine triple expansion engine has always been credited to A. C. Kirk in 1874. However this type did not become popular until the 1880s when boiler pressures had risen sufficiently to make their use economical.[17]

In 1850, in the Royal Navy, steam pressures in ships' boilers rarely exceeded 10 p.s.i. and in 1860, the limit was 20 p.s.i., the usual Admiralty pressure. In 1865 it became 30 p.s.i. and then in 1870 it was 60 p.s.i. By that time compound engines were beginning to be fitted into naval ships for the first one was installed in the *Constance* in 1863. For compounding to be really economical, let alone triple expansion, still higher pressures were needed. It was not until 1885 that triple expansion engines were being fitted in fighting ships when steam pressures had risen to 160 p.s.i. Corresponding pressures in the mercantile marine were in 1872, 52.4 p.s.i., in 1881, 77.4 p.s.i. and in 1891 158.5 p.s.i.[18] On land, Fredrick Colyer, writing in 1886, talked about high-pressure engines with steam at 40–60 p.s.i. The records of J. & E. Wood show the pressures of steam used in the engines they supplied and have been averaged for the years 1875 to 1900 (see Table 11). From these figures, it will be seen that, while a few brave souls ventured into pressures higher than 100 p.s.i. in the early 1880s, it can be said that these were not common until after 1890. William Fowler writing in about 1908 put the advance earlier for he said that, up to about 1880, steam pressure had risen to about 100 p.s.i. for general use.

> In less than five years after this it rose to about another 50 lbs, and in 10 years to about 200 lbs per square inch . . . For land purposes, however, 160 lbs to 180 lbs per square inch is about the maximum pressure in general use.[20]

All this suggests that it seems to have been only the last decade of the nineteenth century which really saw the introduction of pressures that made both compounding and triple expansion economic.

In 1872, Sir F. J. Bramwell gave a list of compounded marine engines with boiler pressures ranging from 45 to 60 p.s.i. which showed a coal consumption of between 2 and 2½ lb per hour per indicated horsepower. In a similar list published by F. C. Marshall in 1881, the boiler pressures averaged 77 lb and the consumption a trifle under 2 lb. Ten years later, with triple expansion engines, boiler pressures were 160 p.s.i. and the consumption averaged about 1½ lb of coal.[21] The advances from Savery's times were summed up in a list published in 1896 (see Table 12).

Table 11. *Boiler pressures for engines supplied by J. & E. Wood, 1875–1900*[19]

Year	Number of engines	Average pressure	Lowest pressure that year	Highest pressure that year
1875	1	75	75	75
76	4	92.5	70	150
77	10	72.5	60	80
78	7	71.4	60	80
79	11	62	35	80
1880	20	61.2	60	80
81	12	81.2	60	100
82	15	76	50	150
83	9	79.4	60	150
84	9	72.7	60	85
1885	2	80	80	80
86	13	75.3	60	100
87	11	78	50	100
88	18	88.8	70	100
89	14	87	70	100
1890	18	94.7	60	160
91	6	107.5	80	175
92	9	125.5	40	180
93	18	109	80	170
94	8	111.9	80	180
1895	18	130.5	80	180
96	13	123	30	180
97	10	123	90	160
98	13	125	100	180
99	13	132.7	90	180
1900	12	125.8	75	200

Table 12. *Coal consumption in pounds per horsepower per hour*[22]

Atmospheric engine, *Savery*, 1700	31
Low-pressure engine, *Watt*, 1768	8.8
High-pressure engine, *Evans*, 1804	6.7
Double cylinder engine, *Woolf*, 1804	4.5
Compound engine, *Elder*, 1850–1891	2.25
Triple expansion engine	1.76

The layouts of triple expansion engines on land were many and various. In a very few cases, three horizontal cylinders were placed in line, driving a common connecting rod onto a single crank. J. & W. McNaught built two of these in 1892 and another five years later.[23] The largest developed 650 h.p. Presumably, if a tandem compound engine gave a more even turning force than a single-cylinder engine, then a triple ought to have been smoother still. More usually in horizontal engines, there were two cranks, one either side of the flywheel. On one side there might be the high and intermediate cylinders and on the other the low and the air pump. To reduce the size of a single low-pressure cylinder and to distribute the power from three pistons more evenly, a popular layout was to divide the low-pressure cylinder into two separate ones so that there were four cylinders in all. In this case, there would be normally on one side the high-pressure and one low-pressure cylinder and on the other the intermediate and the other low. One such engine, which was built by J. & E. Wood in 1907 for the Trencherfield Mill in Wigan, has been preserved. It was rated at 2,500 h.p. with a boiler pressure of 200 p.s.i. There were 54 ropes round its 26 ft diam. flywheel.

Details in the available records show that Goodfellow built 3 three cylinder horizontal triple expansion engines in 1890. Between then and 1900, this firm had built a total of 12 three and four cylinder triples compared with 29 cross compounds and 19 pairs of tandem engines.[24] J. & E. Wood started to build horizontal triples in 1891 while Musgrave and Scott & Hodgson followed in the next year. In all cases, the most popular arrangement was the horizontal four cylinder type. J. & E. Wood supplied 16 of these in the next eight years and

Figure 66 The horizontal four cylinder triple expansion engine built by J. & E. Wood for Trencherfield Mill, Wigan in 1907. The high-pressure cylinder is on the left and the intermediate on the right and two low-pressure cylinders in front of them. (Courtaulds Ltd.)

Table 13. *Triple and quadruple expansion engines built by Musgrave, Bolton, 1889–1908*[25]

Year	Horizontal triple	Vertical triple	Quadruple expansion	Total engines that year
1889		1		33
1890		2		20
91	1	1	2	22
92			2	22
93	3	3	5	34
94	1	1	1	20
1895	4	6		30
96	13	6		40
97	3	5		27
98	4	4		23
99	6	2		32
1900	2	2		27
01	1	1		17
02				19
03	1			22
04	1			14
1905				19
06	1			36
07	4			24
08	3	1		23
	48	35	10	504

had a further flurry of activity around 1906 in which year six were built. The totals of triple expansion engines built by this firm up to 1907 were 27 horizontal four cylinder types, 7 in line, 7 two-crank types with three cylinders and two verticals. Table 13 shows Musgrave's output up to 1908 after which only three more triple expansion engines were built up to closure in 1926.

Although the horizontal triple layout proved to be the more popular, it was probably marine influence that introduced the inverted vertical inline triple. With the cylinders set above the crankshaft, although tall, this was a very compact design and reduced the ground space needed for large engines as well as the costs of the engine houses. Also, with three connecting rods, the cranks could be set at 120° which gave an exceptionally smooth running engine. Both Buckley & Taylor and Saxon were to build some of these engines after 1900 for the larger spinning mills. A 1,200 h.p. engine was supplied by Saxon to the Cedar Mill at Ashton under Lyne in 1905 which was a fine spinning mill and so required particular smoothness from its motive power. In 1966, the engine finally ceased working and the ropes were removed. It was decided to restart the engine so it could be filmed and the engineer was afraid that he might have

Figure 67 The inverted vertical triple expansion engine. (Nasmith, *Cotton Mill Construction*.)

trouble running it under no load. In fact, the engine ran very smoothly and there were no problems at all.

Quadruple expansion engines

A few quadruple expansion engines were built for textile mills. Following his 1861 patent for multiple expansion engines (see p. 241), Adamson built some before 1879 when W. H. Uhland published an account of one driving 48,096 spindles and all the preparation machinery. The boiler pressure was 100 p.s.i. and the steam was reheated as it passed between the final three cylinders. One advantage claimed for this engine was the lessening of steam leaking past the pistons as the fall in steam pressure in each cylinder was reduced. Uhland also claimed that

> The mechanical action of the quadruple engine, is admirably adapted for producing a uniform power, such as is required in a cotton, woollen, or flax mill, or for the grinding of corn, the force on the crankpins being nearly the same on every portion of a revolution.[26]

Six months of trials were carried out on a 12 year old triple expansion engine at Victoria Mill with 57,360 spindles and a quadruple expansion engine at Albert Mill with 48,240 spindles but the locations are not given. The coal consumption for the triple was 2½ lb per horsepower per hour and for the quadruple 1½ lb. At this same period, in 1862, Elder patented a quadruple expansion engine for marine use in which the steam could be reheated, or superheated as he phrased it, between the second and third cylinders.

Mention ought to be made of another instance of quadruple expansion being used in mill engines. The quadruple expansion engines listed above in the table of engines built by Musgrave had triangular connecting rods. The idea seems to have originated with W. Y. Fleming and P. Ferguson in 1887 and Musgrave patented further refinements in 1893.[27] The triangle pointed downwards and the lowest point was connected to the crank. The upper two corners were connected to piston rods from a pair of steam cylinders. This layout gave 'no dead centre' as the engine revolved and this was the name used for the type. Some were built with four cylinders, two on either side of the flywheel and so became quadruple expansion engines. Only ten were ever built like this but there were in addition some compounds with only two cylinders. In the early 1900s, there was a reversion away from both the four cylinder pair of tandem compound engines and all forms of multiple expansion engines to the straight cross compound layout. The additional gains from expanding the steam in more than two cylinders were not matched by the additional costs of multiple cylinders, especially when the steam was superheated.

Figure 68 Musgrave's 'no dead centre' engine with the triangular connecting rod. (Nasmith, *Cotton Mill Construction*.)

Superheating

Rankine defined superheated vapour as

> Vapour which has been brought to a temperature higher than the boiling point corresponding to its pressure, so as to be in the condition of a permanent gas.[28]

As long as steam was in contact with water in the boiler, it remained saturated and could not become superheated because the additional heat would turn more of that water into steam and not raise the temperature of the existing steam. Superheating was carried out after the steam had left the boiler and might occur in two ways. Either the steam might be passed through an additional source of heat, some of which it absorbed. Or when steam was expanded suddenly without doing work, for example when it passed through a reduction valve or was wire-drawn before it reached the cylinder of a steam engine, the pressure might fall faster than the temperature and so the steam became, in effect, superheated. In the discussion that follows, only the first type of superheating, that of passing the steam through an additional source of heat, will be considered. This gave the advantage that this additional heat could be imparted to the colder surfaces of the cylinder and warm them when the steam entered without causing condensation. Also superheating enabled a greater range of the fall in temperature to be covered in a steam engine but there were practical difficulties which had to be overcome first, such as finding a metal that would withstand the extra heat and also an oil that would not carbonise in the cylinder.

The origins of superheating are unknown. In 1826, a Mr Neville of Shadwell Thames designed a vertical tubular boiler for use with locomotives in which the steam was taken from the steam space down a large central tube to a chamber suspended in the firebox over the fire.[29] This box would soon have burnt out from the heat of the fire, particularly with the materials available at that date. In 1840, William Daubney patented a 'steam elevator' which was a heat exchanger filled with hot oil or other substances which boiled at high temperatures to impart heat to the steam passing through.[30] In 1845, J. A. Detmold patented a more conventional type in which the steam was taken through a coiled pipe and

> circulates through the whole length of the coil exposed, without the presence of water, to the great heat of the furnace, and being by this means greatly increased in bulk and temperature, it is then permitted to flow through the valve to the cylinder.[31]

This was, in essence, the form of later superheaters except copper piping was too soft for the higher temperatures and pressures.

Rankine gave no dates for the introduction of superheating. He wrote in 1861,

> It is difficult, if not impossible to specify any one as the first inventor of this process. Mr. Frost was at all events one of the first to recommend it and cause it to be put in practice. It was used many years ago in the engines of the American mail steamer "Arctic" with good effect, and has since been used by many makers in many engines, chiefly marine, with a great variety of forms of apparatus.[32]

Rankine mentioned that economies of over 15 per cent might be achieved, partly because the superheating apparatus took up heat which otherwise escaped wastefully through the chimney.[33] Hirn, in 1857, found that the consumption of steam was reduced from 19.4 to 16.2 lb per horsepower hour in a condensing engine by superheating the steam some 45 °C. Steam pressures at this time would have been less than 50 p.s.i. and 100° of superheat would have given an ultimate temperature of only 398 °F.[34]

However at this period, superheating was abandoned through the difficulties of finding materials for the superheating elements which would withstand the high temperatures and also the problems of lubricating the cylinders. This was well illustrated by Charles Normand, who outlined the disadvantages of earlier methods of superheating the steam.

> The practical realisation of these advantages has been prevented by various causes, the principle of which are, first, the rapid oxidation of the vessels in which the superheating is effected; second, the increased friction and rapid destruction of the cylinders, pistons, and valves, caused by the desiccation and decomposition of the oils or grease under the high temperature, which also burns the packing round the piston and rods.[35]

His method of avoiding this was to use what later became called 'reheating' the steam. In 1852, Adamson had patented 'superheating' steam in its passage between the high- and low-pressure cylinders of compound engines. In his high-pressure cylinders, Normand used steam straight from the boilers and so avoided the problem of carbonisation of the lubricating oil. The exhaust from the high-pressure cylinder was superheated by passing it through a heat exchanger in which high-pressure steam was employed as the heating medium. In this way he avoided burning out the superheating elements. Reheating was found occasionally in some later mill engines, for example the Galloway engine at Elm Street Mill in Burnley, but it never seems to have been common.

The effects of mineral oils for lubrication

The first lubricating oils used in steam engines were natural ones such as the oil from tallow, lard oil, or for general purposes, sperm oil. Sperm and lard oil flashed at about 500 °F (260 °C). Such animal and vegetable oils would decompose into acids in the steam cylinders and so would cause the metal to corrode rapidly.[36] A. Rigg confirms Normand's statements and tells how iron was converted 'into a sort of plumbago, so soft as to be readily cut with a knife'[37] and how the nuts and bolts on pistons inside cylinders were worn away. Worse happened in boilers where the water was supplied from the condensate with tallow in it. There was a rapid chemical action from the fatty acids and thick deposits from the tallow formed over the furnace crowns 'to their ultimate destruction'. Clark commented that

> Surcharged [superheated] steam is void of the quality of lubrication, and it carbonises any ordinary lubricant delivered into the cylinders, to the detriment of the working surfaces.[38]

Therefore the introduction of both higher pressures and superheating had to wait until mineral lubricants were available. Shale oil began to be made in France on a large scale in 1834 and 'paraffine-oils' at Glasgow by Young in 1850. From 1860, these oils began to be superseded by ones distilled from petroleum pumped out of the American wells. Clark did not mention mineral oils in 1862, but Rigg tentatively recommended them, which suggests that their use was being introduced during the later 1870s. In France, Hirn was a pioneer of mineral oils. A good quality mineral steam cylinder oil would have a flash point of 680 °F (360 °C), considerably higher than that of the natural oils.[39] Such oils contained no acids and did not decompose or vaporise with the increasing heat. Without them, the advent of the later high-pressure steam engine would have been impossible.[40]

Until the general introduction of mineral cylinder oils, the increase of boiler pressures above 60 p.s.i. led to the abandonment of superheating during the 1860s and 1870s. The British Board of Trade opposed it on the ground that there was a danger of the steam being broken down into its constituent elements and becoming dangerous, a quite erroneous idea. However, tests which were published in 1892 for the Alsatian Society of Steam Users showed a saving in coal of 20 per cent for superheaters installed integral with a boiler and 12 per cent where the superheaters were independently fired. The interest in superheated steam began to revive, partly through the introduction of the turbine where there were no problems with lubricants and where it was found that superheated steam produced a marked efficiency through the reduction in internal friction by preventing condensation.[41]

Some experiments with superheating were carried out on the turbines installed in the generating station of the Cambridge Electricity Supply Co. in 1891 but were soon abandoned.[42] A similar fate befell those undertaken in 1893 at the Charing Cross and Strand Electricity Supply Corporation in London.[43] One of the problems was the Lancashire boiler. The superheaters had to be placed beyond the boiler itself where the gases would normally have descended to pass through the return flues underneath. The disadvantage was that the temperature of the gases at this point was either too low to give a really effective superheat or, if the temperature were raised, there was an overall loss in efficiency.[44] In this position, the superheaters could be easily cut off from the flow of hot gases when not in use to save burning them. To overcome the disadvantages of superheaters in this position, around 1900, Galloway advertised superheaters in a separate bank heated by their own fire. But probably the best type was that fitted to the Babcock boilers. The elements were placed above the sloping watertubes under the main header drum. Here they were not exposed to the full force of the fire which was taken by the watertubes but were heated by the first pass of the gases through the tubes where they were still quite hot.[45] This company built its first superheaters in 1895 which were an immediate success.

The gains in economy with steam at 150 p.s.i. were stated to be

Degrees F of superheat	50	100	150	200	250	300
Gain in economy per cent	8	14½	21½	26½	31½	34½[46]

Put in a different way, in simple engines, the gain was about 1 per cent for each 10°F of superheat,[47] but 600 °F was about the safe limit for cast iron. In Britain, 100–150 °F of superheat was usual.[48] Superheating was quickly established in electricity generating stations. The first 10,000 h.p. engines planned in 1898 for the Manhattan transport system of New York were not superheated but similar engines ordered in 1902 for the New York Interborough Rapid Transit Co. were.[49]

Just how many textile mill engines were fitted with superheaters has been impossible to establish. When visiting the surviving mill engines running in the late 1960s, the impression gained was that superheating was rare. The inverted vertical cross compound engine at Monton Mill built in 1906, the little Musgrave tandem compound of 1907 at Bamford Mill, the 1909 Carel engine at Moston Mill, Moston, Urmson & Thompson's engine at Ace which started in 1920, Galloway's Elm Street Mill engine, 1926, were all superheated. On the other hand, the Buckley & Taylor cross compound installed at Wye No. 2 mill in 1926 was not superheated at a date when the advantages one would have thought would have been well recognised.

In an engine using superheated steam in a series of cylinders, the effect of the superheating was confined basically to the first cylinder. Stumpf claimed that, in an ordinary counterflow cylinder engine, the superheat was excessive for the first cylinder and too little in the subsequent ones.[50] The steam usually became saturated in the high-pressure cylinder before release and so entered the receiver in a wet state. While initial condensation was eliminated in the high-pressure cylinder, it occurred in subsequent ones. In this way the use of superheated steam partially obviated the desirability of compounding and multiple expansion.

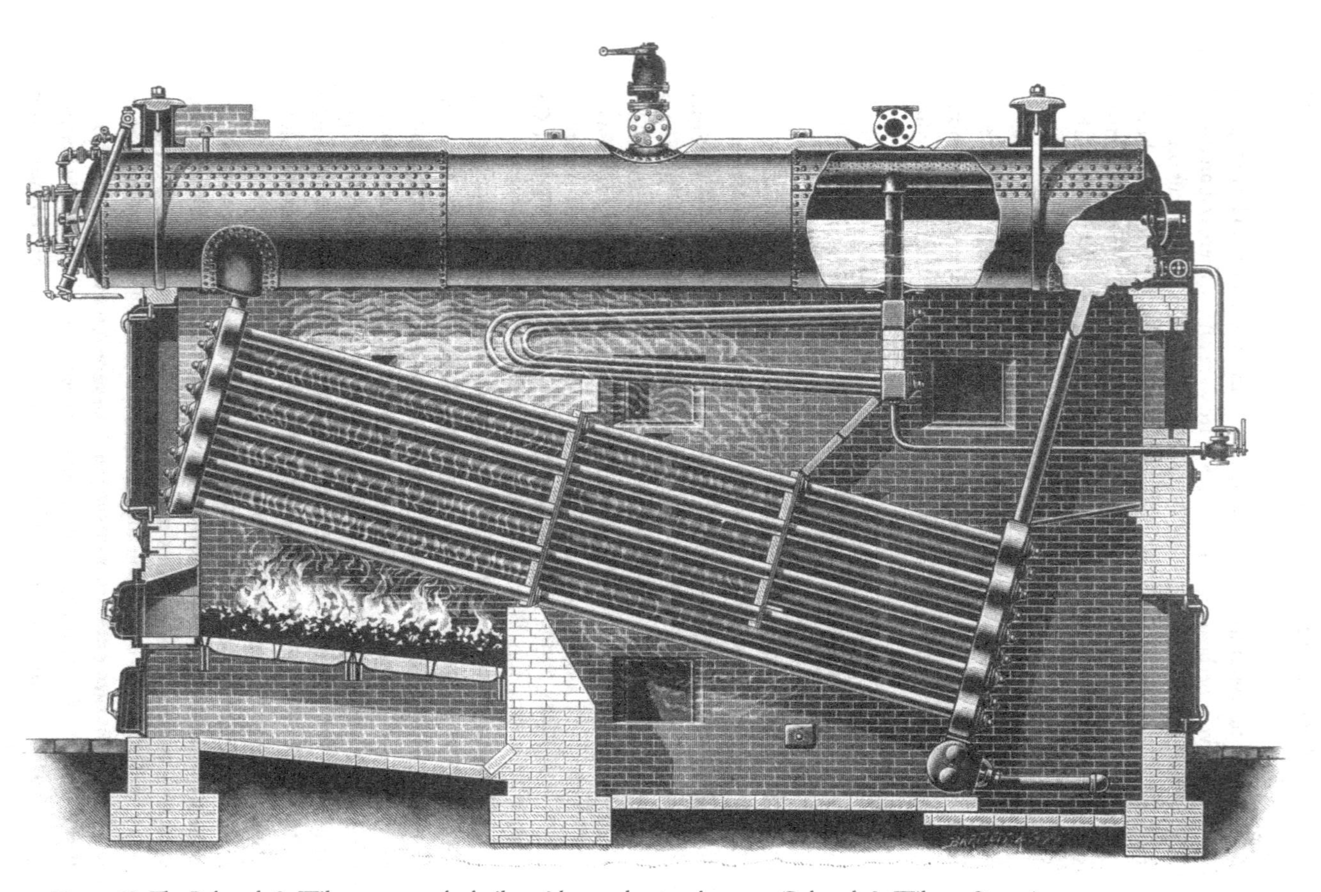

Figure 69 The Babcock & Wilcox water tube boiler with superheater elements. (Babcock & Wilcox, *Steam*.)

> When superheated steam is used, its excess heat prevents any immediate condensation and may keep the steam dry until cut-off. Moreover, the lesser heat conductivity of superheated steam results in a lesser transfer of heat to and from the cylinder walls. Hence it follows that the economy of superheating is not so great in compound and triple-expansion engines as in simple engines.[51]

This was another of the reasons for the return to the cross compound layout of mill engines in the decade before the First World War.

The introduction of steel

Without the introduction of steel, the advance in mill engine design around 1900 would have been impossible. Yet, having made this claim, it has proved to be very difficult to establish when the engine builders and boiler manufacturers actually changed to this material. As far back as 1861, Rankine wrote,

> *Steel Boilers* Recent improvements in the manufacture of steel have so far diminished its cost as to render it commercially available as a material for boilers. Its tenacity is on an average about 1.6 times that of iron; and hence, by its use, boilers of a given strength may be made much lighter than heretofore. In the steel steamer "Windsor Castle", lately built by Messrs. Caird & Co., the shell of the boiler is made of steel plates, with steel rivets. It has to withstand a working pressure of about 40 lbs. on the square inch; while its thickness is only 5⁄16 inch, or little more than 5⁄8 of the thickness of an iron boiler of the same strength.[52]

Where there was no necessity to save weight, as in the boilerhouse of a textile mill, there was no incentive to risk this new material which to begin with proved dangerous because proper annealing techniques had not been developed to prevent embrittlement during bending.

Adamson was one of the first to introduce steel for Lancashire boilers. By the beginning of 1867, he was using Bessemer steel plates from Bolton Forge in preference to iron. Hick Hargreaves had followed by the following year. In boilers for railway locomotives, steel was probably accepted later and became common only during the 1880s.[53] By 1890, Clark could write,

> Of recent years steel of a mild quality has been extensively and successfully used; and is now almost exclusively used in the best class of work . . . The fitting method of treatment is now thoroughly understood, and steel of uniform quality in boilers can be ensured. There are manufacturers of boilers who use steel exclusively in their construction.[54]

It would seem that wrought iron was retained for the Babcock boilers supplied to Ferranti's Deptford Power Station in 1888 which were built to withstand a pressure of 200 p.s.i.[55] This could be done safely in this type of boiler owing to the small diameter of the pipes and drums, giving it another advantage over the Lancashire boiler. The introduction of strengthened fire tubes, again by Adamson, helped to increase the pressure of steam which could be generated in these boilers but it was only after J. Thompson, Wolverhampton, brought out one with dished ends that Lancashire boilers became suitable for pressures up to 250 p.s.i.

Steel gradually replaced wrought iron in parts of the mill engines themselves. Colyer wrote in 1886,

> Since steel has been so much improved and cheapened, many of the moving parts of these engines are made in this material, such as the piston-rods, slide-rods, connecting rods, cross-heads, and guide blocks.[56]

Scott & Hodgson were using steel on their engines by 1889. Crankshafts were some of the larger parts to be made of steel first and were common in this material by 1896.[57] America probably took the lead for in 1900, the Harrisburg Foundry could advertise that 'The connecting rod is made of specially selected steel of tough and homogeneous character, having the greatest tensile strength possible'.[58] They also made the crossheads and crankpins of steel. Wrought iron was more amenable to forging and the horizontal tandem compound engine supplied by J. & W. McNaught in 1907 to Firgrove Mill, Rochdale, was made almost entirely from wrought iron with cast iron for the castings. Wrought iron was still used in 1926 for many of the smaller forgings on the later Elm Street Mill engine, but the larger ones such as the connecting rods were steel and showed how much smaller and lighter they could be than the earlier wrought iron.

Steel castings for pistons were being recommended by 1888,[59] and were adopted on the small high-speed generating sets such as the Ashworth & Parker. So far as is known, no mill engine builder ever installed a steel foundry as did the private locomotive manufacturer, Beyer, Peacock, in 1899.[60] The castings used in mill engines even for the cylinders on the last engines built in the 1920s were poured from iron and not steel. In 1902, Parsons tried some steel castings for the high pressure cylinders for the turbines of H.M.S. *Amethyst*, one of which cracked under steam tests and so he reverted to cast iron. However he found that cast iron began to creep with the higher-pressure steam being used in turbines after 1910 and had to introduce steel castings for the casings of the high-pressure cylinders.[61]

Other improvements

Although not directly affecting the design of the engines themselves, one important addition was the introduction of the pass-out or extraction engine. Here steam was taken out of the pipe joining the high- and low-pressure cylinders for use in process work or heating in other parts of the mill. Sulzer Bros were pioneers in this field and had built an engine with steam extraction on a cross compound with automatic pressure regulation in 1887. They supplied another one to a Munich brewery in 1904.[62] The McNaught engine at Firgrove Mill in 1907 and the Galloway engine at Elm Street Mill in 1926 were fitted with pass-out systems for supplying steam to drying cylinders on the sizing plant in the weaving sections. This arrangement gave a very economical source of low-pressure steam which could be generated more economically at high pressure and reduced through the high-pressure cylinder of the steam engine where it would produce useful work. Weaving sheds and papermills were among the main users of pass-out engines and it was in such places that steam power survived longest through this dual use of the steam. Any steam that was not taken out for process work went into the low-pressure cylinder where it helped drive the mill.

One small but vital improvement for the introduction of higher pressures and superheated steam was the replacement of yarn and asbestos materials in the glands for piston rods by metallic packings in which metal segments were held against the moving rods by springs. These had appeared before 1900. As well as using improved mineral oils for lubrication, gradually the whole system of lubricating the moving parts was changed from drip feeds with the total loss of the oil. For main bearings a circulating system with gravity feed into the bearings and return of the oil by a small pump was introduced which finally evolved into a full pressure system. For most of the larger mill engines running at 60 r.p.m., the system where the oil flowed out of an 'aquarium' or tank above the bearing and down into the bearing itself sufficed. However the higher-speed vertical engines needed full pressure lubrication.[63] The Ferranti engines described earlier were among the pioneers and F. E. Musgrave patented a system for forced lubrication on vertical engines in 1899.[64] It was really the introduction of the Uniflow with its higher speeds that led to the adoption of forced lubrication on horizontal engines after 1910. With forced lubrication, the moving parts had to be totally enclosed to prevent the oil splashing everywhere.

The same developments occurred with balancing. The large slow-speed engines do not appear ever to have been balanced, although the necessity for this had been recognised on railway locomotives for many years.[65] While Elder had calculated in 1866 that from 10 to 15 per cent of the indicated power of a marine engine could be wasted in friction by neglecting to balance the forces on the driving shaft,[66] the practice was never followed in mills except at the very end of the stationary steam engine era. It was only the higher speeds of rotation that compelled its adoption on mill engines, particularly the vertical and Uniflow types.

A look at the developments in condensers and airpumps will complete the improvements to mill engines up to the end of their history. Porter had used a horizontal condenser and airpump situated behind the single cylinder of his engine displayed in 1867. A few other makers followed this practice including Pollit & Wigzell who patented their own type in 1877.[67] One advantage of this arrangement was that the engine foundations were not weakened by the hole for the condensing apparatus. However, most manufacturers preferred to place the condensers below the level of the main cylinders both so that the condensate formed in the low-pressure cylinder could flow by gravity into the condenser and so that there would be less danger of the injecting water backing up through the condenser and into the engine cylinders.

In the airpumps, rubber disc valves replaced the earlier flap valves. There was a considerable shock when the piston descended into the condensate because the area of the valve openings was considerably less than the surface area of the column of water in the bottom of the airpump. This meant that the water had to be forced through the restricted apertures. In 1894 F. Edwards patented an airpump which had valves only at the outlet.[68] Near the bottom of the pump barrel were two rings of ports. The condensate flowed through the lower ones into the bottom of the airpump which had a conical shape. The airpump piston was a solid cone which almost fitted the cone in the cylinder. In the lowest position of the piston, the condensate was forced back through the lowest ports by the piston and it was flung through the higher ports onto the top of the piston. The piston, rising again, trapped this water on top of itself as it moved above the ports and closed them. This water was lifted and ejected through valves in the top of the airpump in the normal way.[69] These airpumps were

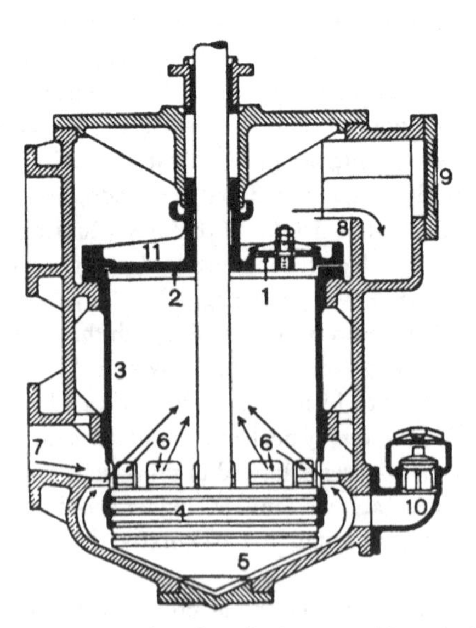

Figure 70 The Edwards airpump with conical bucket. (Low, *Heat Engines*.)

claimed to draw a better vacuum than earlier ones and, with their few valves, were more suited for high speed engines such as the Uniflow types. Only in the 1920s did a few makers introduce rotary extraction pumps or sometimes steam ejectors on the condensers.

14
An economical source of motive power

Right at the end of the reciprocating steam engine era appeared one based on different principles from earlier designs which enabled it to compete for a while not only with the steam turbine but also the diesel. In fact development work on it continued in the United States of America right up to the 1940s. It was called the 'Uniflow', or on the Continent and United States 'Unaflow', through the way in which the steam was used in the cylinder. Normally this type was double-acting with only inlet valves fitted at either end of the cylinder. The steam was exhausted through a central ring of ports in the middle of the cylinder which were closed by the movement of the piston. The piston had to be made almost as long as the length of the stroke, about 10 per cent less was customary. The steam entered at one end of the cylinder and pushed the piston along. Most of it escaped through the exhaust ports and what was left was compressed as the piston returned, raising its temperature. The attractive feature of the Uniflow engine was the good thermodynamic layout because the inlet end always remained hot and the centre with the exhaust stayed cold. The residual steam was reheated by the compression back to the temperature of the incoming steam so there was no heat loss through condensation.

Leonard Jennett Todd

The Uniflow type of steam engine was first used by Jacob Perkin in England in 1827 and patented in America in 1857 by Eaton when engines with piston valves were placed on two small steamers operating on Lake Chautauque but there appears to have been no further interest.[1] Likewise there was no follow-up from the work of the Englishman Leonard Jennett Todd who took out a patent for a 'Terminal-exhaust' engine in 1885. The title of this patent reads 'Improvements in or relating to producing and maintaining gradations of temperature in the cylinders of double-acting Steam Engines'.[2] Todd's aim was

> to produce a double acting steam engine which shall work more efficiently, which shall produce and maintain within itself an improved gradation of temperature extending from each of its two Hot Inlets to its common central Cold Outlet, which shall cause less

> condensation of the entering steam, and which shall work with greater economy than has hitherto been the case.

On the working stroke, the steam expanded normally so that the temperature fell according to an adiabatic curve. During the return stroke, Todd was fully aware of the importance and advantage of compressing the residual steam. Compression occurred for about 90 per cent of the stroke after the 'terminal exhaust' ports had been closed by the piston. It therefore followed that a prolonged and continual rise in temperature was produced from the outlet temperature back up to, or nearly up to, the temperature of the incoming steam. Todd realised that the temperature of the interior metal skin of the cylinder was also raised back to approximately the same temperature as that of the fresh steam, that there was little condensation of the entering and expanding steam and also that there was little transfer of heat either from the steam to the metal or the reverse. In other words, all the heat lost during expansion would be turned into work. In this patent, Todd covered virtually all the theoretical points of the Uniflow engine yet for some reason his engines were unsuccessful.

The drawings in this and two further patents taken out in 1886 and 1887 show some of the problems Todd tried to overcome. If he did not build an engine on his principles then, he certainly carried out some practical experiments in the middle of the 1890s on a locomotive with 'dual' exhaust cylinders, presumably with the ring of ports and exhaust valves.[3] The drawings in Todd's patent of 1885 show that his valve design followed the contemporary British practice. His first type consisted of separate slide valves for each end of the cylinder, linked together, with cut-off valves on their backs. These cut-off expansion valves had additionally on their backs, piston cut-off valves controlled by the governor. There is no indication of the speed at which his engines were run, but, while positively driven valves could have been set to give quick opening and closing, the layout is exceedingly complex with tortuous steam passages. Todd's next idea simplified the layout at the expense of omitting the variable cut-off feature, for he used a short piston valve with long steam passages. His reason for using a piston valve was to avoid it being lifted off its face by excessive compression, a problem which will be examined later.

The drawings in Todd's 1886 patent[4] show that he realised the necessity for the inlet of the steam to be controlled by the governor through variable cut-off valves. This would allow the steam to enter the cylinder with the least amount of wire-drawing and give the best economy. He reverted to slide valves, again the double type with cut-off valves on their backs, but this time he placed them closer to the ends of the cylinders with shorter steam passages. He developed a reversing gear which he claimed,

> possesses many advantages over the ordinary 'link motion' or similar mechanism for altering the direction of rotation. By its means the steam can be cut off at the very earliest point in the stroke of the piston

> whether the engine be running forwards or backwards. Thus steam can be expanded many times in a single cylinder, irrespective of direction of rotation, and without altering either the Lead, the Exhaust or the Compression.[4]

The wording of this patent with reference to reversing gears may infer that Todd was considering placing his engine on a railway locomotive at that time. When the Uniflow engine was tried on locomotives on the Continent after 1910, it was found that the weight of the long cylinders with the heavy pistons was excessive.

Todd was moving in the right direction and he went even further in his next patent[5] when he proposed using valves of 'the vibrating, oscillating, or rotary type, transversely arranged, and by preference controlled by the action of the governor'. By this date, the Corliss valve gear was becoming popular in Britain, for, as Todd remarked,

> Governor-controlled valves of the type described [Corliss] are the most efficient known, and cylinders of the type described [his mid-cylinder exhaust type] are the most efficient known, and the combination of these valves and these cylinders produces a new and improved and useful effect not previously attained.

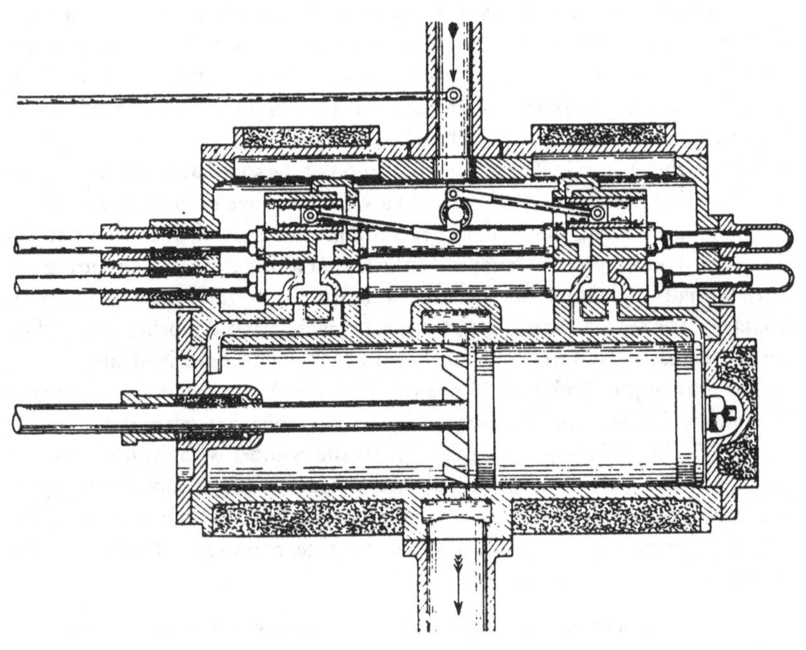

Figure 71 The valve gear on Todd's Uniflow engine from his 1885 patent. (Patent drawing.)

One of Todd's ideas, that of having a single oscillating valve in the middle with long passages to each end of the cylinder, would not have worked well, but the other, with Corliss valves below the bore, one close to each end of the cylinder, should have worked satisfactorily at the speeds his engines were likely to attain. By 1886, the Corliss trip gear and dashpots had been developed far enough to permit speeds in excess of 80 r.p.m. and it is doubtful if Todd would have tried to run his engines much faster.

Todd realised that his single inlet valves at each end of the cylinder had a great thermodynamic advantage over the 'Wheelock' arrangement where the hot incoming steam passed over the back of the cold exhaust valve. He claimed that his inlet valves

> can be placed in a new and improved and more useful position than has hitherto been possible.
>
> And thus a steam cylinder is produced possessing great thermodynamic advantages over the ordinary "Corliss" or "Wheelock" engine, and yet at the same time a cylinder which requires only one, at the most two steam valves, in place of the usual four steam valves of these engines.[6]

If, as would seem to have been the case, Todd had produced a Uniflow engine with an inlet valve system that would have worked well enough, where or why did he fail?

His failure probably lay in his inability to solve the problem of reducing the excessive compression, for, as he said,

> When starting or stopping a "terminal exhaust" engine, the maximum compression may under certain circumstances become temporally excessive; as for instance when the vacuum is imperfect, or when a reversing engine gives a half turn forward and then back.[7]

For normal running, Todd had devised a method of proportioning the clearance space at the ends of the cylinders and in the steam passages to the difference between the initial and the final steam pressures. By determining this ratio correctly, the maximum compression could be contained to equal the initial steam pressure. In practice, Todd pointed out that, when the correct ratio was used, the maximum compression would not be reached because, if the cylinder were not steam-jacketed, some of the heat from compression would be transferred to the skin of the enclosing metal

> and thus the compression of the waste steam in my improved cylinders is always kept, in normal working, a very convenient and economical distance within the initial steam pressure.[8]

This system was satisfactory when the initial and terminal pressures remained constant, but Todd does not say how he overcame the abnormal pressures generated when starting a condensing engine before the vacuum had been fully created or when the vacuum failed. His only suggestion seems to have been to use piston valves which would not lift from their seats.[9] In other words, he failed to solve a critical feature of Uniflow engine design, because, to gain the greatest economy with any steam engine, it was necessary to run it condensing.

Professor Johann Stumpf

It was Professor Johann Stumpf of Charlottenburgh, Germany, who really solved the technical problems of the Uniflow engine and its history may be said to start with his first one constructed in 1908. Stumpf made the point that it was only after his invention that his critics pointed out to him the earlier work of others in this field, and he claimed that 'Probably if I had advised myself fully of the work done by these gentlemen, I might have been led astray. My investigations, however, have been entirely untrammeled'.[10] While he was in America in the late 1890s, he had been working on various types of steam turbine design and this prompted him to wonder whether a similar unidirectional flow of steam could be achieved in a reciprocating engine. He realised that, in a turbine, the steam went in hot at one end, had its energy extracted axially, and emerged at the other end in the cold exhaust.[11] Just as Watt, many years before, had wanted to keep his steam cylinder hot to avoid initial condensation of steam and to keep his condensing chamber cold, so Stumpf became more and more convinced of the necessity of keeping the hot end of a reciprocating steam engine cylinder hot and the exhaust cold. But this was impossible in the usual double-acting engine of his day, for, once the steam had pushed the piston along to the end of the cylinder, had expanded, done work and lost heat, the piston pushed it back along the cylinder to an exhaust port at the same end that it had entered. Therefore the steam cooled the rest of the cylinder as it was exhausted.

With the Uniflow design, Stumpf realised that he could make a double-acting engine by, in effect, placing two single-acting cylinders end to end with a common ring of exhaust ports in the middle and having inlet valves at the outer ends. This layout enabled Stumpf to expand the steam in one cylinder, instead of in two or three stages, so the number of working parts was reduced and the cost of building, lubricating and maintaining Uniflow engines was lessened. But there were snags. A large ratio of expansion was necessary to produce the most work from a given quantity of steam. In some cases, this ratio could be as high as 35 or 45 in a condensing Uniflow with a good vacuum,[12] so the single cylinder had to be quite large. One way of reducing the size was to make the engine run more quickly but then the engine had to be built more strongly with forced lubrication. Because the steam was being expanded in a single stage, the inlet period was very brief. Often the cut-off at full load was only 10 per cent of the stroke, compared with 25–50 per cent in an ordinary engine.[13] So control of the cut-off valve became much more critical and one has to speculate whether

Todd encountered such difficulties and failed to overcome them. Then the reciprocating parts had to be made heavy and more accurate to withstand the force of the steam from full boiler pressure down to final exhaust pressure (which is not the case in a compound engine)[14] and there was the large mass of the piston to be controlled.

Stumpf realised that the steam left in the cylinder after the exhaust ports had been closed by the piston was a thermodynamic advantage because, during its compression, it beame hot. His compatriot, Rudolf Diesel, had taken advantage of this in 1896 to create an engine in which air was compressed to such an extent that fuel injected into it became ignited. Stumpf compressed the residual steam to bring the cylinder walls back to the temperature of the incoming steam which thereby was not wasted in heating up the cylinder and piston but could do its work immediately when starting.[15] However, there was a hazard when the pressure in the exhaust varied, which happened with starting or when the vacuum failed during running. Then the Uniflow could be turned into a potential diesel engine, when it would compress a cylinder full of steam at atmospheric pressure instead of the vacuum pressure.

No doubt Stumpf soon found that he had been preceded and so could not patent the basic concept of the Uniflow engine, but his six patents sealed in 1908/9 help to show the course of his experiments and investigations. In his first patent,[16] he applied steam-jacketing to the cylinder head or cover and over that part of the cylinder uncovered by the piston during the admission of the steam. This was based on his principle of keeping the hot end as hot as possible. On ordinary engines, the cylinder cover, the largest area exposed to the incoming steam, was not jacketed and so remained cold until heated by that steam. Stumpf claimed that, by passing the fresh steam through the cylinder covers before reaching the inlet valves, the covers would remain hotter than the steam inside the cylinder which diminished the initial condensation. In addition, extra heat from this source would be given to the steam as it expanded during the working stroke which again reduced condensation. After the cold steam had passed out through the exhaust ports, the remaining steam would be reheated both by compression and by heat from the cylinder cover. In this way, none of the heat from the jacket would be lost to exhaust as it was in counterflow engines. Stumpf claimed that the extra thermodynamic gains justified the extra complexity of the cylinder head design and the greater difficulty in stripping the engine down for maintenance or in cases of breakdown. In addition, Stumpf patented a special design of piston with a lengthened head and reduced diameter so that it would fit into the jacketed area. In these ways, this first patent had very little to do with the principle of the Uniflow but with the advantages of steam-jacketing a Uniflow.

At the end of 1908, Stumpf patented another idea of steam-jacketing.[17] This time, the steam passed first through the cylinder head and then round part of the barrel of the cylinder itself, either in a spiral-shaped passage or a series of rings, so that, as it made its way towards the centre of the cylinder, it would become cooler, following what was happening inside the cylinder itself. This steam would not subsequently drive the engine but would be taken away for use elsewhere. There was a gap between these steam jackets and the exhaust ring round the exhaust ports so the latter remained cool. Any source of heat, such as the flue gases from the boiler, might be used instead of steam. This patent was not much used, if at all, in later engines but the idea of jacketing the cylinder cover was used on practically all ordinary Uniflow engines with the exception of the low-pressure cylinder of the Galloway heat extraction cross compound engine at Elm Street Mill, Burnley. Two of Stumpf's next patents were for the application of Uniflow cylinders to railway locomotives and for extracting steam out of the cylinder at various stages to be used for other processes.[18] His last two patents were crucial to the success of the Uniflow. They were for an improved form of valve gear[19] and for what may be termed 'expansion chambers'.[20]

The problem with the valve gear lay in the combination of higher rotating speeds and shorter valve opening periods which necessitated quick and precise closing if the engine were to be governed by the cut-off of the valve gear. Although by 1908 the speeds of engines with Corliss valve gear had risen above 100 r.p.m., the return springs for shutting such valves had to be very powerful to overcome the sliding resistance for closing them quickly. This, together with the inertia of the heavy Corliss valve, meant that this valve gear was not suitable for giving periods of cut-off less than 1/6 of the stroke and much less was required with Uniflow engines.[21] The increase in speed of the Uniflow engine to keep the size of its single cylinder down to reasonable proportions meant that the cut-off was reduced too to give the greatest possible expansion so that sometimes the steam inlet valve would be open for less than 1/20 of a second.[22]

Stumpf's patent described how he would use drop valves which were operated through a shifting eccentric and a common cam rod controlled by the governor. He may have been following Continental practice where Sulzers had been fitting such valves on their mill engines since the middle 1860s and many other manufacturers did so too. Not only were these valves much lighter than

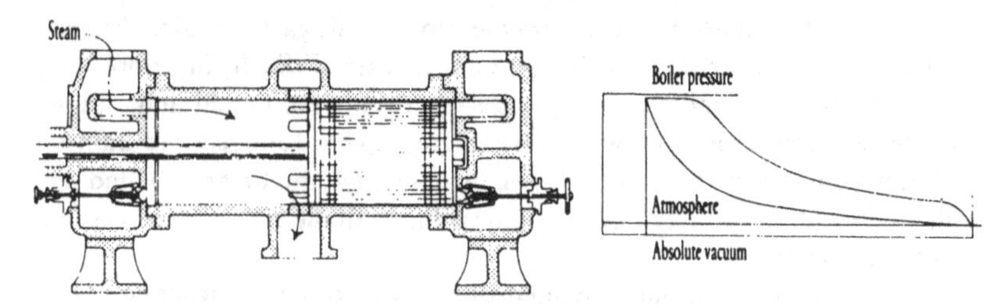

Figure 72 Stumpf's Uniflow cylinder with expansion chambers at either end. (Smithsonian Institution.)

the Corliss type but they could cope both with high-pressure and high-temperature superheated steam. When well designed, the seats remained steam tight, a vital point when operating against the range of steam pressure in a Uniflow cylinder. It was essential that the valves did not leak during the return stroke of the piston, otherwise the increased volume of steam would cause the pressure inside the cylinder to rise too much. The double-beat drop valve with its small stem needed little force to raise it off its seat except that to compress the return spring, and virtually none to close it. Nearly all Uniflow engines were fitted with various forms of these valves.[23] On some later designs, one seating surface was sprung so that both would remain steam-tight under varying ranges of temperature.[24]

Stumpf's other patent provided a solution to the problem of excessive compression particularly when starting a condensing engine. He proposed to make another chamber in the cylinder head which would be connected to the cylinder by a valve so it could act as an additional clearance space or expansion chamber. If the steam jacket were situated between the cylinder and the expansion chamber, not only would the chamber be kept warm by the steam but the space would help to act as a heat insulator for the cylinder end. There could be two or three chambers, all inter-connected by their own valves, one beyond the other, if necessary. Self-acting spring loaded valves might be fitted, or, on many of the early engines in Britain, it was the job of the engine driver on starting up to turn handwheels and open these valves. Of course the man had to remember to close them when proper running conditions were reached. If the vacuum were lost for some reason during normal running, the consequences might be disastrous because this type of valve was not automatic. At least Stumpf had found a solution to make a workable Uniflow engine.

The first Uniflow engines

Stumpf's first engine was built in 1908 by the Erste Brünner Maschinenfabrik-gesellschaft in Brünn, then in Austria but now Czechoslovakia. It was in fact a conversion of an existing 80 h.p. single-cylinder condensing engine fitted with a Uniflow cylinder having Lentz drop valves.[25] By the end of July 1911, Stumpf claimed that there were engines with a total output of over half a million horsepower working or in actual construction.[26] The largest one built up to 1914 was a mill engine with a normal load of 4,000 h.p. and a maximum continuous load of 6,000 h.p. This had a single-cylinder 1,700 mm bore by 1,700 mm stroke running at 120 r.p.m. but its location is not given.[27]

In Britain, J. Musgrave & Sons, Bolton, were the first to construct one in 1909.[28] Possibly they investigated this market when they saw that large engines were no longer required for generating electricity. They took out a licence from Stumpf and, on 5 May 1910, supplied a single cylinder 500 h.p. engine to J. Nuttall & Sons. In the next ten years, they built 54 Uniflow engines ranging from 150 to 1,000 h.p. (see Table 14) which was nearly half their entire output.

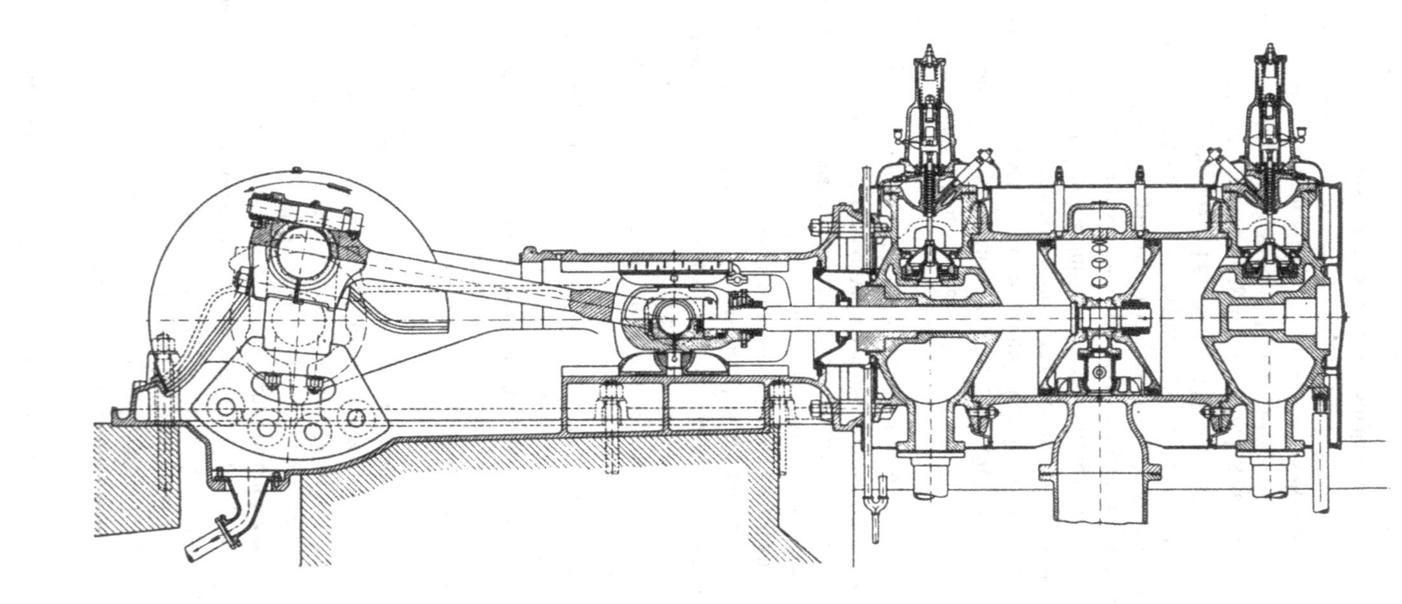

Figure 73 Layout of a Sulzer Uniflow engine which was typical of most. (Ewing, *Steam Engine*.)

Table 14. *Output of Musgrave engines, 1901–26*[29]

Year	No. Uniflow	Total no. engines
1901	—	17
02	—	19
03	—	22
04	—	14
1905	—	19
06	—	36
07	—	24
08	—	23
09	—	13
1910	8	21
11	13	28
12	4	9
13	6	15
14	7	15
1915	4	9
16	1	4
17	5	6
18	1	2
19	3	5
1920	2	9
21	—	2
22	—	3
23	—	3
24	—	1
1925	—	2
26	—	2

All ran at over 100 r.p.m. with the fastest 200 r.p.m. It is impossible to say whether this was creating new markets at a time when the production of mill engines was declining or whether people were ordering Uniflows in preference to other types. In 1926, Musgrave closed and the drawings and patterns were taken over by Galloway. Table 14 does not include in the Uniflow column at least two tandem compound engines which were built as pass out engines with ordinary high-pressure cylinders and the Uniflow type for the low-pressure ones. There may have been others with different valve gear so Musgrave may have built a total of 66 Uniflow engines.

At least 16 other manufacturers in Britain built Uniflows. It is not known when Robey started but Cole, Marchent & Morley and Hick Hargreaves both began in 1911. Hick Hargreaves supplied Robert Walker with a 200 h.p. one running at 75 r.p.m. Hick Hargreaves had built nine by 1914 ranging from 180 to 1,000 h.p. with speeds from 47 to 130 r.p.m.[30] They displayed a 325 h.p. one at the British Empire Exhibition at Wembley in 1924 and built their last in 1937.[31] Galloway had entered the market by 1914 when one was supplied to a firm in London.[32] Later they were to become leaders in this field, also exhibiting one at Wembley. Scott & Hodgson produced designs in 1926 but probably never built any.

Developments in design

These different manufacturers all contributed their own developments to improve the design of the Uniflow engine. Amongst the most important were improvements to the compression relief valves because these proved to be one of the features which conditioned the acceptance of the Uniflow. For some years, Cole, Marchent & Morley engines were fitted with cam-operated, spring-loaded relief valves which had three settings. While warming up, a lever was placed in one position which kept the valves fully open so they acted as drain valves. When starting, the lever was moved to a second position which allowed the cams to hold the valves open during the return stroke but to close them during the power stroke. In the third position, the cams were inoperative for running with the vacuum established. The valves then acted as spring-loaded relief valves which operated if the vacuum failed.[33] Not only did this mechanism rely upon the memory of the engineman, but the problem with spring-loaded valves was that they had to be set to release the steam at a pressure higher than that of the incoming steam. For maximum economy, the compression pressure in the cylinder should approach that of the incoming steam only at the final part of the stroke or power would be wasted.

What was needed was a device that would release excess steam or air in starting before compression commenced. In this country, Robey and Galloway produced mechanisms which worked by the vacuum. The Robey gear consisted of small auxiliary exhaust valves in each cylinder head, operated by cams fixed on a common shaft. There was a clutch on the end of this shaft which was engaged through a small cylinder and piston which had the underside connected to the exhaust passage and the upper side to atmospheric pressure. The necessary balance of pressure was achieved by means of a coil spring underneath the piston. When the vacuum was low, e.g. when starting or running non-condensing, the action of the spring engaged the clutch and the auxiliary exhaust valves came into operation. Upon resumption of normal vacuum, the piston descended and the relief valves stopped working as the clutch was disengaged.[34]

Galloway had a similar but neater arrangement. In their case, the cylinder, with its spring-loaded piston linked up to the exhaust pipe, was placed horizontally at the end of the camshaft on which were fixed two eccentrics for the inlet valves and two cams to work auxiliary relief valves. These cams were free to slide along the camshaft and were pushed or pulled along it depending upon the degree of vacuum in the cylinder at the end. The cams were tapered with maximum lift at one end for use when there was no vacuum, to no lift at all when the vacuum had been created. This mechanism was developed around 1920 and on the Elm Street Mill engine, built in 1927, it discharged into the exhaust passage round the centre of the cylinder.[35] Galloway had developed a compression release gear which not only gave full opening during starting but also a graduated opening related to the state of the vacuum should the vacuum begin to fail at any time during running.

Such systems worked satisfactorily in Europe where practically every steam plant operated condensing but, in America, fully 90 per cent of the steam engines installed operated non-condensing.[36] The reason for this lay in the different methods of heating the mills. In Britain, it was the custom to use steam either directly at boiler pressure or to run extraction or pass-out engines and take off some steam at about 30 p.s.i. as it passed from the high to the low pressure cylinders of compound engines. In America, it seems to have been usual to take steam at even lower pressures from the exhaust of a non-condensing engine and pass that round the mill at only 4 or 6 p.s.i. Some engines were run non-condensing during the winter and, when during the summer there was no use for the exhaust steam, were run condensing. The Uniflow engine was most economical and efficient when run condensing but it could not be worked with two different exhaust pressures unless it were either to have large clearance spaces at the ends of the cylinders, which would be uneconomic when running condensing, or to have special auxiliary exhaust valves fitted.

It was the Skinner Engine Co. of Erie, Pennsylvania, which found the best solution to this problem in the 'Universal' Unaflow engine. A. D. Skinner, Vice-President of the company, had read about the Uniflow engine and decided to tour Germany in about 1910 to see it being manufactured. On returning to America, he used the drawings, descriptions and impressions he had gained to build a Stumpf Uniflow engine without a licence.[37] The Stumpf representatives

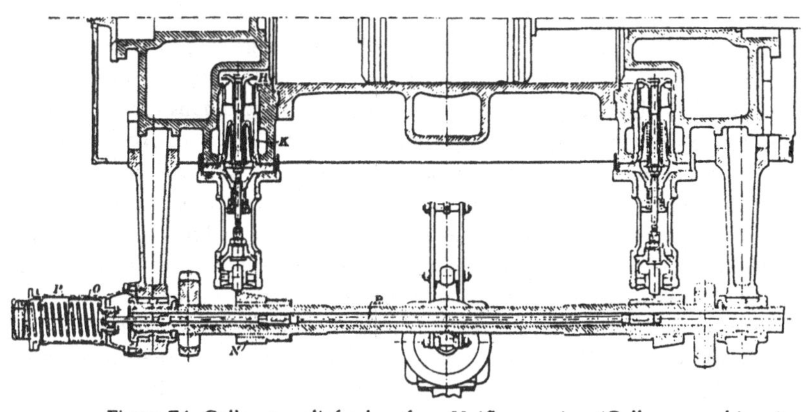

Figure 74 Galloway relief valves for a Uniflow engine. (Galloway archives.)

brought a law suit which they won, but presumably the Skinner Engine Co. must have obtained a licence for, in 1911, its engineering department under L. G. Skinner undertook to adapt the Uniflow engine to non-condensing service, producing the first one in 1913. Stumpf had tried auxiliary valves inside the piston itself. These were worked by a lever from the connecting rod and exhausted through the middle of the piston into the central ring of exhaust ports. Being mechanically operated, the opening of these valves could not be varied according to differing pressures.[38]

Skinner realised that, if auxiliary relief valves were placed at the ends of the cylinders, the engine would become partially a counterflow type with the cold exhaust taking away heat from the cylinder walls and the jacketed head. When the valves were being used normally only for starting, this loss was not very significant, but it became serious when working non-condensing for long periods. The central ring of exhaust ports gave a much larger area through which steam could pass than with ordinary valves so the working vacuum inside the cylinder of a Uniflow engine was greater than on ordinary engines. However, a diagram was produced showing that, at light loads, 95.2 per cent of the exhaust steam might pass through the auxiliary exhaust valve and only 4.8 per cent through the ring of exhaust ports. At full load, the figures were 45.6 per cent through the valve and 54.4 per cent through the ports.[39] This showed that such engines had a large 'counterflow' element in the passage of steam through them and lost a great deal of the economy which should have been gained from a Uniflow.

Therefore the Skinner designers located their relief exhaust valves away from the hot ends of the cylinders. It was seen that, if the valves were located a sufficient distance from the ends of the cylinder so that compression would take place at about the same point as was usual in Corliss engine practice, no cold exhaust steam would be washed past the hot cylinder walls that had been heated by the boiler steam before cut-off had taken place or past the steam-heated ends. Also the clearance spaces at the ends of the cylinder could be reduced so that the compression would bring the pressure and temperature of the steam back up to that of the incoming steam, thus retaining this vital Uniflow principle.

Skinner's auxiliary exhaust valves were single-seated poppet valves set in pockets beneath the cylinder. They discharged into the exhaust pipe and were operated by rocking levers from the engine valve gear. The timing of their opening did not need to take place until the end of the stroke when most of the steam had been exhausted so they opened against no pressure. This is why they could be the poppet type and not double-seated drop valves, and had the advantage of being more steam tight. The timing of their closure was not critical because the piston itself covered the holes in the pockets as it moved back and so effectively shut them. The rockers for opening the valves were enclosed in oil boxes and worked through a double system of levers. One of these levers was connected directly to a piston in a cylinder and the cylinder in its turn was connected by a pipe to the condenser. When there was a vacuum in the condenser, the second lever was drawn out of engagement with the valve stem so that the valve was not lifted and the engine became a true Uniflow. If the vacuum failed, or was not being used during the winter, the lever engaged the valve again and the engine worked non-condensing.[40] All that was needed to make the change was a valve to divert the exhaust steam either to the condenser or to the steam-heating pipes and the engine could be set to change itself automatically. Therefore these 'Universal' Uniflow engines could be operated condensing during the non-heating season and non-condensing during the heating months, the engines giving the maximum economy under both conditions. Skinner claimed that it was the only engine ever built which could accomplish this economical feat. This system also would work at starting and in the event of the vacuum failing. Skinner continued to build engines of this type for driving mills or generating electricity until the early 1950s.

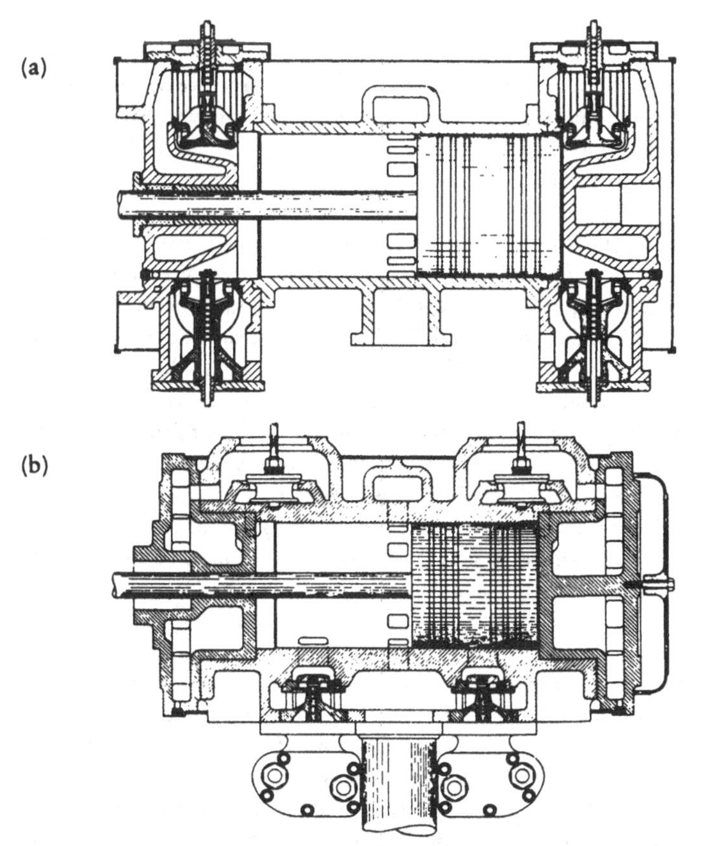

Figure 75 Skinner positioned his supplementary exhaust valves in the central portion of each part of the cylinder. (*a*) The wrong way – the usual construction; (*b*) the right way – 'Unaflow Universal' design.

Engineering problems

The higher speeds, the higher pressure range within the cylinder and the stronger, heavier reciprocating parts meant that Uniflow engines had to be much better engineered than their slower speed counterparts. In fact, it was commented that they had to be built upon oil engine lines rather than the old fashioned steam engine style.[41] For the higher speeds, adequate lubrication was essential. On the Elm Street Mill engine, Galloway fitted a forced lubrication system with oil supplied at 30 p.s.i. to all main bearings, crossheads, crankpins, etc. and all the other moving parts had a pressure drip feed arrangement. Improved cylinder oiling became vital because, as superheated steam became more common, there were no water droplets to assist with lubrication. Then it was found that the way the steam passed through the cylinder in a Uniflow did not allow any oil carried in the steam itself to touch the cylinder walls because it passed straight through. Therefore lubrication points had to be fitted in the cylinder walls.

Considerable difficulties were experienced in early designs of Uniflow engines owing to the seizing of the piston followed by fracture of the cylinder. This may have been due partly to the oiling deficiency but it was also caused by the unequal expansion of the cylinder through the ends remaining hotter than the centre. Fracture usually occurred across the exhaust ports at the centre of the cylinder, where it was coolest and weakest. An improvement was effected by boring the cylinder with a double taper, the largest diameter being in the middle.[42] Similar precautions had to be taken with the large piston too because its ends became hotter than its centre.[43] Allowance had to be made for the expansion of the cylinder longitudinally through the heat. The front end was bolted firmly to the target plate on the crosshead guides and so the rear part had to be allowed to expand. On the Elm Street Mill engine, Galloway supported the rear end of the cylinder on steel stays which had rounded ends so the cylinder could rock on them. The tail-rod guides could slide on their supports so the whole of the back of the cylinders was flexible. On some large engines, the weight of the pistons was carried by the crossheads and the tail-rod slides so that wear inside the cylinder was reduced. These rods might be cambered to allow for deflection.

Superheating

While Stumpf realised that his engine was suitable for running with superheated steam, he was against its use because he claimed that 'the una-flow steam engine must give almost the same economic efficiency with superheated as with saturated steam'.[44] The chief advantage with superheated steam in ordinary engines was the lessening of condensation when the steam first entered the cylinder, which of course did not apply in Uniflow engines. Also, in the Uniflow engine, the temperature of the steam inside the cylinder could not be raised by compression back to that of superheated steam without increasing the pressure

Table 15. *Consumption of steam at different degrees of superheat*[46]

Condition of supply		Steam in lb per hour per i.h.p.
Saturated		12.9
Superheated	13 °C	12.4
	42	11.6
	98	10.4
	155	9.6

beyond that of the boiler pressure. So Stumpf stated,

> Stages and superheating are therefore "make shifts" used to correct the injurious results of the principal error in the ordinary form of steam engine construction which . . . is . . . the counter-flow type. *If now these fundamental mistakes are avoided, that is to say, if the counter-flow is replaced by a uni-directional flow, the cooling of the surfaces is avoided so that the extraction of the energy of the steam in stages and superheating become superfluous.*[45]

Superheating was still in course of being accepted in 1908 but oddly Stumpf seems to have forgotten that, for maximum efficiency, the temperature of the working fluid must be raised as close as possible to that of the heat source. The efficiency of an engine cannot be considered on its own, that of the boiler must be included too. So in fact it was found advantageous to have a moderate degree of superheat. Although losses due to initial condensation were greatly reduced in the Uniflow engine, they still occurred, as tests on one showed. The steam supplied was at 170 p.s.i. and, with a cut-off at 5 per cent, the consumption was found to be as shown in Table 15. In another Uniflow engine, working without a condenser, the consumption was reduced from 21.6 to 13.4 lb per i.h.p. per hour by superheating the steam 110 °C.[46] Musgrave's catalogue of the Uniflow engine published in 1910 stated,

> In taking up the manufacture of this Engine our chief aim is to supply our clients with an Engine *in which highly superheated steam of high pressure can be most favourably and economically utilized.*[47]

Later developments

During the 1920s, the Uniflow engine was applied to a variety of uses. While generally considered a high-speed type, rotating at over 100 r.p.m., it was used also for driving blast furnace blowers, rolling mills and water pumping.[48] Galloway designed one which was coupled directly to sheet rolling mills which ran at only 28 r.p.m. They also built a set of seven 1,500 h.p. engines for blowing air for blast furnaces which had a steam consumption of 10½ lb of steam per i.h.p. per hour compared with 16 lb for an ordinary condensing engine.[49] But the main line of development was to take the Uniflow engine away from Stumpf's original concept of an engine with all the expansion carried out in a single cylinder.

As boiler pressures rose and superheated steam became more common, it became more difficult to accommodate the expansion within a single cylinder. Not only did the size of the cylinder increase in proportion but the cut-off had to be shortened. The pressure in the cylinder, and therefore the turning force on the crank, fluctuated widely from full pressure at the beginning of the stroke to very little at the end. Once again the heaviness of the parts needed to withstand the great pressures was acclaimed as a virtue as their inertia would help to even out the turning force on the crank. What was involved may be judged by the weight of one Uniflow piston which was 5 ft 6 in long and, with its rod, weighed about 5 tons in a cylinder 60 in diam.[50] F. B. Perry claimed,

> The early cut-off in a cylinder large enough to give the required ratio of expansion implies a high initial load upon the piston and consequently heavy working parts. The weight of these parts, however, is not without advantage, in as much as their inertia reduces the stresses on them and tends to equalize the turning effort throughout the revolution.[51]

It is surprising that the old argument should have surfaced once more at this late date. It was pointed out that a single-cylinder Uniflow would still have two dead points and that no advantage of inertia of the parts could get over that.[52]

Valve gears were developed to cope with these extreme conditions. Among them was a cut-off mechanism operated by hydraulic pressure designed by Henry Pilling of Galloway's which he patented in 1919.[53] On the layshaft at the side of the cylinder, an eccentric for each valve operated a plunger pump so oil pressure lifted the valve. The timing was set to give a constant point of opening while variable closure was achieved by the governor controlling a release valve. A spring in an oil-filled dashpot closed the valve in the usual way. This hydraulic valve gear gave remarkably effective control of the cut-off from zero to about 60 per cent of the stroke and needed very little power to operate it.[54] It was fitted to the Elm Street Mill engine where it gave over 40 years continuous service with no trouble and is still operating satisfactorily in the museum in Manchester.

Heat-extraction engines

Right at the end of the reciprocating mill engine era appeared a type with the most thermodynamically efficient layout that had been devised. Superheated steam entered a high-pressure counterflow cylinder from where it could be either passed out to process work in the mill or be used in a Uniflow low-pressure cylinder. In many mills, low-pressure steam was needed for process work which could not be obtained from a single cylinder Uniflow engine except with the pass-out valves on the cylinder itself mentioned in Stumpf's patent. It is interesting to note that the East Lancashire Paper Co. was supplied with a 1,000 h.p. Uniflow engine by Musgrave in 1910, one of the earliest they built. However in 1926, this same mill received a 2,500 h.p. tandem compound drop valve engine, also from Musgrave, which may have been the last built by them. This engine drove a d.c. dynamo on the crankshaft and also the beaters through shafting and, in addition, steam was taken off between the high- and low-pressure cylinders for the steam drying cylinders. Musgrave used the Uniflow layout for the low-pressure cylinder which Galloway were doing as well by this time. Stumpf mentions an instance of an extraction or pass-out engine with a low-pressure Uniflow cylinder[55] but it is interesting to see that even this application was anticipated by Todd. In his second patent of 1866,[56] he recommended that, in compound, triple or quadruple expansion engines, the last cylinder should be constructed according to his 'mid-cylinder-exhaust' plan.

> The lowest pressure cylinder of a compound engine is always the most important in an economic sense, as heat which escapes any higher pressure cylinder may afterwards be partly intercepted by the lowest pressure cylinder, but heat which escapes the lowest pressure cylinder is entirely lost to the engine. And as my improved mid-cylinder-exhaust cylinder is less subject to initial condensation of the entering steam than ordinary cylinders, and as the low temperature of the final exhaust has less cooling effect upon it than usual, it intercepts and uses it more beneficially than ordinary lowest pressure cylinders do. And therefore the application of such a cylinder, as the final or lowest pressure cylinder of a compound engine, becomes economically advantageous.[57]

It was advantageous also to use a high-pressure cylinder as a sort of reducing valve to bring the steam pressure down to that needed for the processing. Power generated in this way cost very little more than raising the process steam directly.[58] When process steam was needed at 30 p.s.i., as was the case at Elm Street Mill, a Uniflow engine exhausting at this pressure would have needed large clearance spaces to cope with such a terminal pressure and would not have been as successful as a counterflow cylinder with superheated steam. Steam at 150 p.s.i., superheated between 150 and 200 °F, as was the case at Elm Street Mill, eliminated condensation problems in the high-pressure cylinder of this cross compound engine. By using a Uniflow cylinder on the low-pressure side, which exhausted into a condenser, this second cylinder could be designed to take maximum advantage of Uniflow principles with small clearance spaces, good compression and easy exhaust to the condenser.

Galloway evolved their compound high-pressure counterflow and low-pressure Uniflow engines in two stages. First they produced a tandem compound layout. The problem with the pass-out or heat extraction engine has always been that the need for power might not balance the need for process steam. Some steam might be wasted if the engine exhausted too much or more had to be added if the engine did not pass enough. Also the steam being used for processing needed to be kept at a constant pressure however much the consumption might vary. In the tandem compound layout, the high-pressure cylinder could exhaust at a constant pressure and what was not needed for processing would go into the low-pressure cylinder and through that into the condenser. The combined power from the two cylinders would turn the single crank.

In a cross compound engine, the turning force on the two cranks needed to be approximately equal. In the ordinary type of cross compound in a textile mill, the load for power would be fairly constant and so the inlet valves on the low-pressure side were set with a fixed cut-off suitable for driving half of that load. Changes in load were met by the governor altering the cut-off on the high-pressure cylinder which meant that the steam pressure in the transfer pipe between the cylinders would vary. This, of course, was unacceptable when steam was being taken away for other uses in a mill. Therefore a dual governing system had to be devised, one for speed against load and the other to control the speed and intermediate pressure against varying extraction of steam.

For their cross compound steam extraction engine, Galloway developed a special type of pressure regulator.[59] When the pressure in the transfer pipe

Figure 76 The Elm Street Mill engine. The high-pressure counterflow cylinder is on the left and the Uniflow low-pressure cylinder on the right. (Author.)

between the cylinders fell through too much steam being taken off, the regulator operated a servo-motor which reduced the cut-off of the inlet valves on the low-pressure cylinder. This reduced the steam consumption of the engine and diverted it to the mill. At the same time, it also reduced the engine speed. At that point, the governor of the high-pressure cylinder operated to let more steam into the engine and so the balance was restored once more. Such engines might be employed in paper-making, bleaching, dyeing, brewing, as well as in the case of the Elm Street Mill engine, calico weaving. It was claimed that it was possible in the majority of cases with these combined steam extraction engines to obtain overall thermal efficiencies exceeding 50 per cent, whereas, in the most favourable cases, the efficiency of ordinary condensing Uniflow engines rarely exceeded 20 per cent.[60] Therefore steam extraction engines with counterflow superheated high-pressure cylinders and Uniflow low-pressure cylinders may be considered the ultimate development of the reciprocating mill engine designs but it is ironical that they appeared only as the last examples of reciprocating engines installed in mills.

A comparison of the steam consumption of Uniflow condensing engines with that of other types showed that it was as low as triple expansion engines and lower than two cylinder compounds. This economy was attained not only in very large engines but also in small ones. The steam consumption per indicated horsepower was more even over the whole range of loads between a quarter and full load than with ordinary engines in which fuel consumption rapidly became excessive if the load varied either upwards or downwards from a very narrowly designed area.[61]

Multi-cylinder Uniflow engines

For the final development of the Uniflow engine, we must cross the Atlantic to America. While in 1912 Stumpf mentioned a Uniflow marine engine with two cylinders and a rolling mill engine with three,[62] it was not until the middle 1920s that multi-cylinder double-acting vertical Uniflow engines really began to make their appearance. Three firms might be mentioned who exploited this type, the Ames Iron Works, Oswego, New York, the Mattoon Engine Works, Illinois, and the Skinner Engine Co. So far as is known, firms like Belliss, Allen, Browett & Lindley, Ashworth & Parker never tried to follow this lead in Britain. The multi-cylinder type retained all the thermodynamic advantages of the Uniflow. In addition, it operated exceptionally smoothly because all the reciprocating parts in each cylinder, the piston, the crosshead, connecting rod, etc., were all the same size which resulted in the reciprocating forces on each crank being the same. This contrasted with a compound, triple or quadruple expansion engine where the increasing sizes, and therefore the weights, had to be balanced separately and differed on each crank.[63] The only way of avoiding this was to make the pistons all the same weight so that those in the high-pressure cylinders became needlessly heavy, or the low-pressure pistons were cast in lightweight steel.

These multi-cylinder Uniflow engines were used extensively for generating electricity and, because of their smooth running, became very popular in ships. The Skinner Engine Co. built its first marine Uniflow engine in 1929 with three cylinders, 10 in diam. by 20 in stroke, developing 800–1,000 h.p. at 170–185 r.p.m., using steam at 250 p.s.i. with 200 °F superheat and a 26 in vacuum. Such engines could be operated at higher speeds than was practicable with other types. With multiple high-pressure cylinders, there was no dead centre and it was recorded that a change was made from *full speed ahead* to *full speed astern* in *five seconds*, or as fast as the operator could handle the control levers.[64] A great many of these engines were built for the American Navy during the Second World War. In ten years, the Skinner Engine Co. built engines totalling over 1,250,000 h.p.

The final evolution of the Uniflow engine during the late 1930s and 1940s took it even further away from Stumpf's concept as boiler pressures increased and superheating temperatures rose still further. To meet these extreme conditions, the Skinner Engine Co. developed the 'Compound Unaflow Marine Steam Engine'[65] which had a single-acting high-pressure cylinder exhausting into a single-acting low-pressure cylinder placed underneath it. A common cylinder head served both cylinders. This steeple compound could be built with any number of cranks and, because each crank had equal reciprocating weights, there was an optimum balance of the inertia forces. The cycle of the flow of steam was this. Near bottom dead centre, full steam pressure was admitted under the high-pressure piston to begin the power stroke upwards.

> Before this stroke is half completed, the steam valve closes and cuts off the pressure, allowing expansion to continue to the top of the stroke. In that position, the piston uncovers ports in the HP cylinder liner, and the expanded steam is by-passed around the HP piston to its top side and into the annular space that surrounds the liner.
>
> It is at this point in the cycle that the transfer valve opens, and the steam from the HP cylinder passes through the cylinder head to the top of the LP cylinder . . . After closure of the transfer valve, the steam continues expansion in both cylinders [this is referring to the top of the HP cylinder and the LP cylinder] on the down stroke, and the steam remaining under the HP piston is compressed to the initial pressure. Near the end of the down stroke, the LP piston uncovers the large unaflow ports, and the steam is exhausted to the condenser through the exhaust manifold.[66]

The low-pressure piston compressed the remaining steam on its upward dead stroke after an auxiliary exhaust valve had closed.

Not only did this engine expand the steam in two stages contrary to Stumpf's

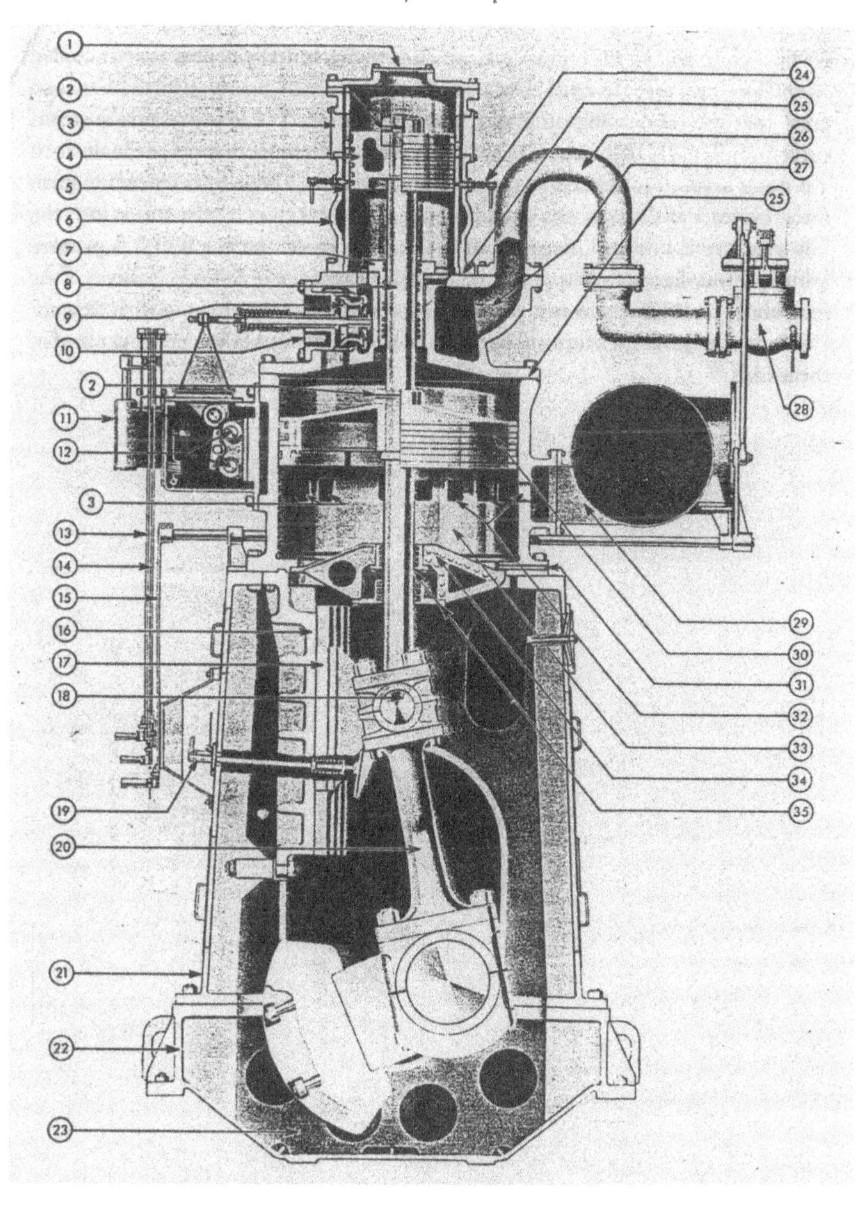

Figure 77 Skinner's steeple 'Compound Unaflow Marine Steam Engine', which was used with high-pressure superheated steam. (Smithsonian Institution.)

original concept, but the high-pressure cylinder was actually deliberately cooled by the low-pressure steam. The temperature of the high-pressure steam was so great that the lubricating oil might have carbonised. The low-pressure steam surrounding the high-pressure cylinder reduced the temperature of the walls to a point where lubrication was readily accomplished. The heat extracted from the cylinder walls was not wasted because it was taken by the steam into the low-pressure cylinder. Engines with four cranks developing 3,000 h.p. were built to this design for the train ferries of the Chesapeake & Ohio Railway.[67] As recently as 1985, there was a proposal to build coal-fired ships with Skinner Unaflow engines for working between Poland and Sweden but nothing came of the idea.[68]

15
The most economical mode of obtaining power

Immediately after the First World War, the cotton textile industry entered a period of boom as manufacturers tried to catch up on orders that had been delayed by that war. However this was short-lived because some countries to which Britain had been accustomed to export cloth had developed their own capacity for spinning and weaving. In Lancashire, some mills in the course of construction when the war started, like the Ace Mill at Chadderton, were completed when it finished. But then others which it was intended should be doubled like the Hare Mill (renamed Mons after the war) at Todmorden, and the Pear Mill, Stockport, never had the second section added. The last three traditional style Lancashire cotton spinning mills were finished in 1926 and 1927 and each was driven in a different way.

For the Wye Mill Co.'s No. 2 Mill, Shaw, Buckley & Taylor built their largest and last engine in the traditional style with 50 ropes round a 24 ft diam. flywheel weighing 90 tons driving the shafting. It was a cross compound designed to develop 2,500 h.p. with cylinders 32 and 70 in bore by 5 ft stroke. The speed was a stately 66 r.p.m. The parts were massive to take such power at such a slow speed with the enormous single slipper crossheads. Even these huge castings proved to be inadequate when the foundations at the back of the low-pressure cylinder settled and the crosshead guides cracked. However, this was repaired and the engine continued to drive the mill until 1965 when it was stopped because current from the electricity grid was cheaper.

At their new Elk Mill, Shiloh Spinners installed in 1926 a Parsons turbine to provide power through a rope drive for this mill and through an alternator for the nearby Shiloh Mill. This turbine developed 2,600 h.p. at 5,000 r.p.m. in high- and low-pressure cylinders placed side by side which were coupled together through a gearbox. Out of the gearbox, one drive turned a 5 ft diam. rope pulley at 333 r.p.m. from which 50 ropes distributed the power to the various floors of Elk Mill. Another drive drove a 728 kW alternator to power Shiloh Mill. The turbine and gearbox were only 16 ft long and so would have fitted between the cylinders of the Wye No. 2 engine. Superheated steam was supplied from four Daniel Adamson boilers at a pressure of 260 p.s.i., which must have been the highest in any cotton mill.[1] It was a highly economical plant and continued to drive the mules until they stopped in 1973.

The last traditional type of multi-storey cotton spinning mill was Sir John

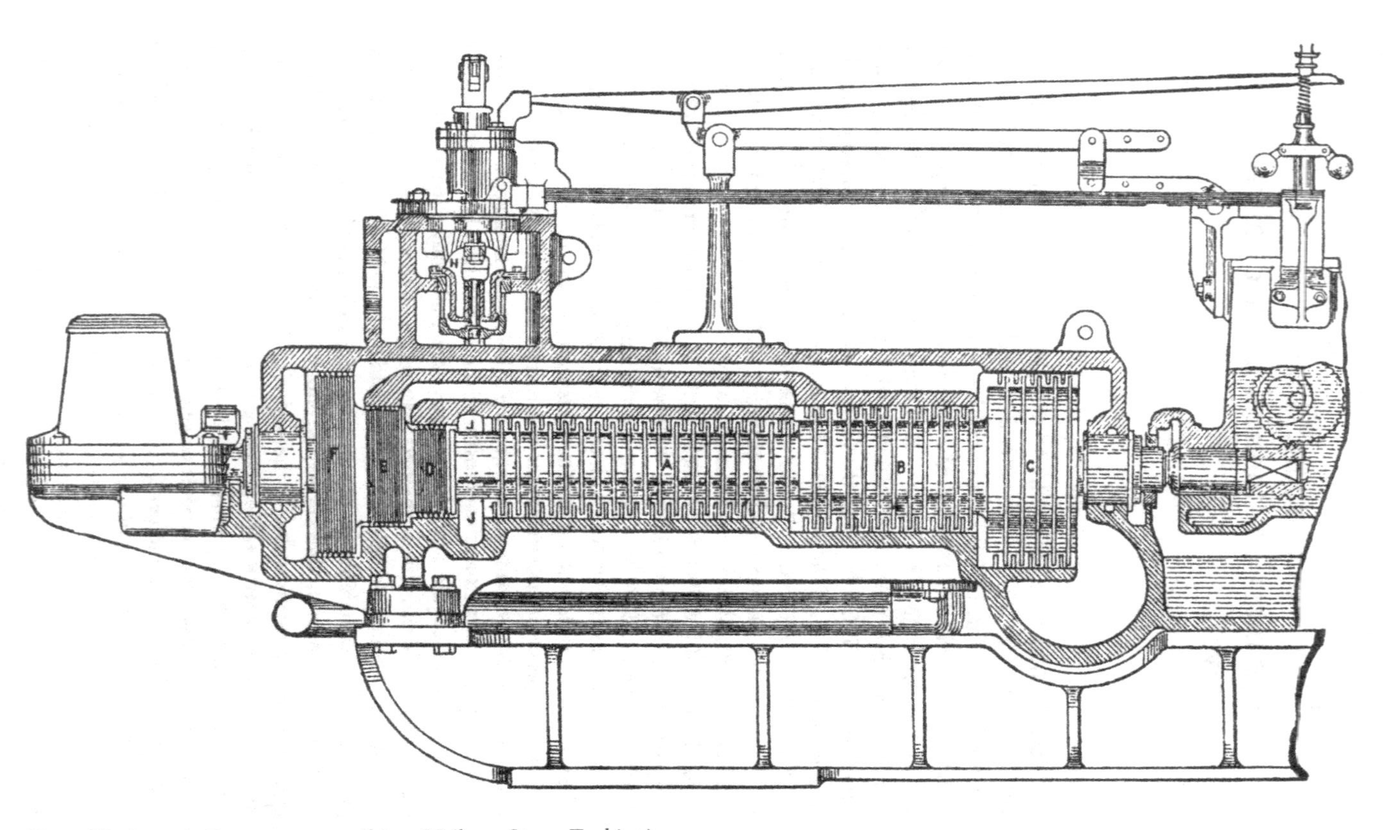

Figure 78 An early Parsons steam turbine. (Neilson, *Steam Turbine*.)

Holden's Mill in Bolton finished in 1927 but it had no tall chimney and only a small boiler to provide heating in the spinning rooms. Its power came from the nearby Bolton Back-o-th-Bank electricity generating station for this was an all electric mill. The line-shafts on each floor were driven by their own individual motors. At one end of the building, the floors of the final bay were lower than the rest so the motors could be placed on them and drive the shafting directly just under the ceilings of the floors of the mill. Individual electric motors fitted to each machine were still really in the future. The electric current at Bolton power station was generated by steam turbines and so this installation marked the beginning of the end of individual mills generating power themselves by whatever means.

In a remarkably short space of time, the steam turbine had risen from nothing to a place of complete dominance. Although its history may be traced back to Hero of Alexandria in AD 100 or Branca of Loretto, Italy, in 1629, it really started with the patents of Parsons in 1884.[2] The essential feature which distinguishes steam turbines from steam engines of other types is that, in the turbine, the action of the steam is kinetic, depending upon its kinetic energy. In other engines, it acts by pressing against a moving piston to develop the motive force. In the turbine, it acts as a mass that is set in motion in consequence of its own power to expand. The force which is utilised is produced by the impulse of the jets on moving vanes or by their reaction on moving holes from which the jets issue. The heat energy in the steam is employed to set the steam in motion and the moving steam drives the mechanism through its momentum by impulse or reaction or both.

Parsons was seeking to develop a high-speed motor more suitable for driving electric generators than the reciprocating engines available at that time. His first turbine on the reaction principle was directly coupled to a generator which it turned at 18,000 r.p.m., an unheard of speed, and produced 10 h.p. One was shown at the Inventions Exhibition in London in the following year which the *Engineer* disregarded but *Engineering* commented,

> If the motor invented by the Hon. Charles Algernon Parsons accomplishes one half that is ascribed to it by rumour . . . viz, that it will run steadily at many thousands of revolutions per minute with an expenditure of steam that is not extravagant, it will constitute by far the most noticeable feature in the electric lighting department of the exhibition and will even rank among the foremost novelties of the entire collection on the ground.[3]

It was to take many years of development before the turbine began to equal the reciprocating steam engine in either size or efficiency. In fact, when Parsons was applying for an extension for his patents when they were about to expire in 1898, he claimed that, without charging for his time but assuming 7 per cent interest on the capital employed, he made a loss of £1,107.13.10.[4]

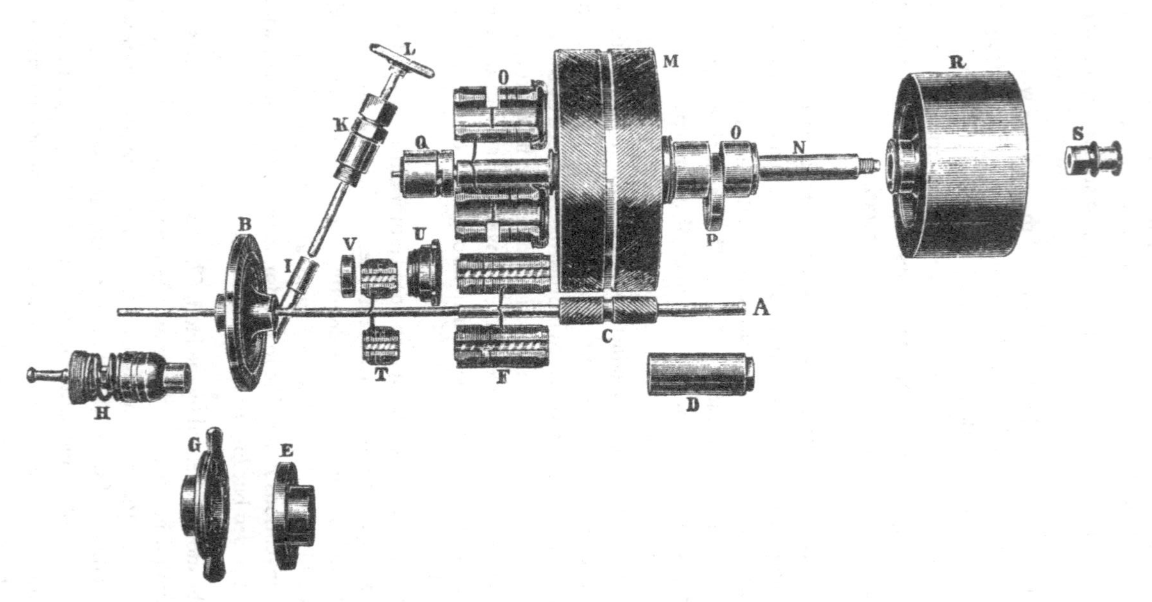

Figure 79 The de Laval steam turbine. (Neilson, *Steam Turbine*.)

The early turbines gained a reputation for guzzling steam. By 1888, the steam consumption for a 32 kW generator, which originally had been 200 lb per kW hour, was reduced to 34½ lb, when working non-condensing with steam at 90 p.s.i. Such high steam consumption was tolerable in ships where space and weight were the main considerations and in the first four years, 360 of these turbo-dynamos were supplied for ships' lighting, ranging from 1 to 75 kW.[5] So far as is known, none found their way into textile mills but during these years, Gustaf de Laval, of Stockholm, had been working on a type of impulse turbine.

A year or two before 1884, de Laval had designed a small turbine on the Hero principle[6] but his name is really associated with the type he had brought to a practical form by 1889. He passed the steam through a jet with a divergent nozzle so that it impinged on blades set all round the circumference of a wheel designed to be capable of very rapid rotation. The steam was expanded in a single stage from the initial to the final pressure.[7] To secure economy with this extremely high-velocity steam issuing from the nozzles, the blades had to rotate at high peripheral speeds. In a small de Laval turbine developing about 5 h.p., the wheel made some 30,000 r.p.m. and in one developing 600 h.p. with a mean diameter of 37 in, the revolutions were 9,500. Later trials of de Laval turbines fitted with condensers produced an average consumption of about 20 lb per hour per brake horsepower in a 60 h.p. turbine and 15 or 16 lb in one of 300 h.p.[8] At these speeds, it was necessary to introduce gearing between the turbine and its generator and de Laval was the first to do this. The turbine went into production in Sweden in 1893 and licences for Britain were obtained by Greenwood & Batley of Leeds. Bamford Mill in Derbyshire was lit by one of their generating sets and presumably there were others elsewhere but how many is not known.

To give a detailed account of all the inventions which went into improving steam turbines would take up too much space here but some are necessary to understand how it was able to compete with the reciprocating engine. Parsons originally had concentrated on the axial-flow pattern in which the steam passed through a series of small turbines one after the other to form the rotor. Between each set of turbine blades was a ring of fixed blades which guided the steam onto the next moving blades. Each ring of moving blades took up the energy of the jets directed against it by the preceding ring of fixed blades and to that extent this turbine acted by impulse. But in passing through the channels of the moving blades, the steam became accelerated. Its velocity relative to the blades increased and also altered direction. Consequently it exerted pressure on the blades by reaction, giving a greater driving force than it would have done if it had passed over them without any increase of relative speed. There was also a drop in pressure and of course a temperature drop at each set of blades. Because the steam was always flowing in the same direction, a steady rate of fall in the temperature and pressure could be designed to give the maximum thermodynamic efficiency at each stage.

In this layout, the whole of the steam pressure drop was distributed throughout all the wheels in a series of small stages which enabled the wheels to be kept at a moderate radius. The peripheral velocity could be kept moderate and so good efficiency could be obtained with the moderate dimensions needed for strength. Because the blades on the rotor of a turbine did not actually touch the outer casing, the friction of the internal parts was very small indeed when compared with a reciprocating engine. The only rubbing parts were the main bearings which could be supplied with oil under pressure and other moving parts such as the governor and condenser pumps. Another advantage was that, because none of the internal parts through which the steam passed needed lubricating, the steam, and so the condensate, was not contaminated with oil and this would lead eventually to the use of steam at much higher pressures and superheat. However, in 1889, Parsons lost control of his original patents and had to design a radial type of turbine in which the steam passed outwards from the shaft through alternate rings of moving and fixed blades. In 1894 he regained possession of his patents and was able to return to developing the axial-flow type.

In the earliest of Parsons's turbines, the metal of the rotor was machined to form the blades which were cut straight. Experiments on the shapes of blades to determine the best form for maximum efficiency led to the adoption of a curved blade in 1887. This was further improved and strengthened in 1896 and remained the standard pattern for many years. In 1887, shrouding was fitted around the ends of the blades which helped to reduce leakage and increase efficiency.[9] Controlling the speed of the de Laval turbine was achieved by varying the number of nozzles. In this way, each nozzle could be operated at full boiler pressure and at maximum efficiency. This was not possible with the Parsons turbine and so in 1890 the 'gust' governor was patented. A series of puffs of steam was admitted which could be increased to obtain more power. Each puff was at full pressure in order to give the maximum efficiency. There was some condensation when each puff was admitted but this was found to be preferable to throttle governing because all the moving parts of the governor could be kept free and hunting avoided.[10]

The early Parsons turbines were non-condensing but in 1891 the Cambridge Electric Supply Co. ordered two condensing turbines which were running in 1892. Parsons wrote about these machines,

> Their Committee carefully tested the merits of the question before deciding and they were of opinion that it was the most economical altogether, and much cheaper in first cost. It seems that when condensing, and with intermittent admission of steam, it is much more economical at *average* loads than the best compound engines and dynamos – partly because there is less friction, and partly because

> there is no condensation when steam-jacketed, and the steam gets fairly expanded at all loads.[11]

Professor Ewing tested these turbines in August, 1892, and wrote,

> The general result of the trials is to demonstrate that the condensing steam turbine is an exceptionally economical heat engine. The efficiency under comparatively small fractions of the full load is probably greater than in any steam engine, and is a feature of special interest in relation to the use of the turbine in electric lighting from central stations.[12]

Work continued to improve the condensers because they proved to be so vital to the efficiency of turbines. The freer passage from the last set of turbine blades to the condenser compared with the restrictions caused by the exhaust valves and ports enabled the vacuum within the turbine to be less than in the cylinder of a reciprocating engine. An advance from 26 in to 28 in vacuum reduced the steam consumption by only 2 per cent in the case of a reciprocating engine but by 10 per cent in a turbine. In 1903, Parsons first applied his vacuum augmenter to condensers which was considered to be one of the most important improvements to the turbine at that period.[13] Through tests on this, it was shown that, with the great range of expansion possible with compound turbines, the effect of an increase in vacuum of 1 in at 26 in was to decrease steam consumption by 4 per cent, at 27 in by 4½ per cent, at 28 in by 5½ per cent and between 28 in and 29 in by 6–7 per cent.[14]

Superheating was first applied to a Parsons turbine at the Cambridge power station. The equipment was placed in the main flue, on the up-stream side of the Green's economiser, but the operating conditions must have been too severe for it because it was abandoned after a few months. It seems that no more Parsons turbines in power stations were fitted with superheaters until after 1900.[15] Yet it was the turbine which seems to have helped to revive superheating. Superheating caused a marked improvement in efficiency particularly in the turbine through its effect in reducing internal friction by preventing condensation and so the formation of water droplets which scoured the blading. This lessening of the internal fluid friction was one reason and the fact that there was a gradual reduction in temperature along the barrel of the turbine was another. These reasons accounted for the results obtained in tests on Parsons turbines which showed a greater percentage increase in efficiency with superheating than was due to thermodynamic reasons alone.[16] It was found that each 10 °F of superheat reduced the steam consumption by 1 per cent. Table 15 shows the steadily increasing size and efficiency of turbines used for generating electricity.

In 1900, two Parsons turbines were sent to Elberfeld in Germany, the first of any type in that country. At 1,000 kW they were the largest constructed so far

Table 16. *Performance of Parsons turbo-generators*[17]

Date	Power kW	Steam per kW hour, lb	Vacuum in	Superheat °F	Steam pressure p.s.i.
1885	4	200	0[a]	0[a]	60
1888	75	55	0[a]	0[a]	100
1892	100	27	27	50	100
1900	1,250	18.22	28.4	125	130
1902	3,000	14.74	27	235	138
1907–10	5,000	13.2	28.8	120	200

[a]These were non-condensing turbines using saturated steam.

and their consumption of steam at 19 lb per kW hour (also given as 18.22 kWH) was then unexcelled by any engine. They confirmed Parsons's prediction that, with increase in size, turbines would possess additional advantages.[18] In the same year, 1,500 kW Parsons turbines were installed in the Neptune Bank Station of the Newcastle-upon-Tyne Electric Supply Co. which was already equipped with the Wallsend Slipway marine four cylinder triple expansion reciprocating engines of the most efficient type at that period. Numerous tests were carried out to compare the efficiency of both types of prime-movers during 1901 and 1902. At full load, the steam consumption of the triple expansion engine was 20.7 lb per kW hour and of the turbine, 18.4 lb in April 1902 and 18.1 lb in June 1902. At less than full load, the reciprocating engine compared more favourably. Additional tests were made later in 1904 by the Commission appointed to consider whether reciprocating engines or turbines should be adopted in the Cunard express liners, *Mauretania* and *Lusitania*, when the turbine appeared even more favourably. The figures at full load were for the reciprocating engine 24.22 lb per kW hour and the turbine 18.67 lb and, at three-quarter load, 23.48 lb and 19.65 lb respectively.[19] As a result of these tests and the performance of turbines elsewhere, 1905 may be said to have been the turning point in the education of engineers as to the efficiency of the turbine.

The attention of the world had been drawn to the outstanding powers of the steam turbine in 1897 when Parsons's ship the *Turbinia* sailed through the famous Naval Review held at Spithead in honour of the Diamond Jubilee of Queen Victoria. Proceeding at over 30 knots, there was no other ship which could catch her.[20] Plans to build a ship powered by a turbine had started in 1894 but it took many years of trials to evolve propellers which would drive a boat both at such high speeds through the water and also to match the high rotational speed necessary with direct turbine drive. *Turbinia* at Spithead was fitted with three propeller shafts, and the turbine was divided into three sections, high-, intermediate- and low-pressure cylinders, so each one drove its own shaft.

In spite of the evident success of the *Turbinia*, it was not until 1901 that the first mercantile ship, the Clyde passenger ferry *King Edward*, was built with turbines driving three shafts. A sister ship, the *Queen Alexandra*, was completed in the following year. In the 1904 season, not only did these two vessels cover a greater mileage than their compound engined paddle steamer rivals but they had a speed 2–3 miles per hour faster and the coal consumption over the years proved to be nearly 20 per cent less.[21] The successful trials of the turbine driven cruiser H.M.S. *Amethyst* in that same year demonstrated the superiority of steam turbines so conclusively that they were adopted subsequently in the design of all new ships for the Royal Navy.[22] The trials of the turbines in the Neptune Bank power station at Newcastle-upon-Tyne resulted in orders for the 40,000 ton *Mauretania* and *Lusitania* to be fitted with turbines producing 70,000 h.p. in each ship. The *Mauretania* was launched in 1907 and could sail at 26 knots. Final mention might be made of the battlecruiser H.M.S. *Hood*, which was launched in 1918 and completed in 1920. She was designed with turbines of 144,000 h.p. to propel her at 31 knots and on trials recorded 151,000 shaft h.p. and 32.07 knots.[23] Once again this showed how quickly the turbine had increased in size, far outstripping reciprocating engines.

But for the history of the textile mill engine, another development for turbines used to propel ships had greater significance. The early turbines driving propellers directly were suitable only for vessels where high speed was the main criterion. Parsons saw that there would be a large market for ordinary ships if the propeller speed could be reduced. The use of gearing in connection with steam turbines originated with de Laval and his type of turbines around 1889. In 1896, Parsons constructed a generating unit of 150 kW capacity for the Forth Banks Power Station consisting of a turbine running at 9,600 r.p.m. driving an alternator at half this speed by means of double-helical gearing. Their efficiency, deduced from heat losses, was about 98 per cent.

Such efficiency was the result of years of development in many areas since the times of Boulton & Watt. The crude gearing of those days was improved through better theories and more accurate knowledge about the behaviour of gear wheels so that stronger teeth were designed which created less friction. Automatic gear cutting machine tools began to appear in the 1860s, but their accuracy depended upon inventions in other fields. First there was the milling machine which could generate tooth profiles with specially shaped milling cutters. This particular application was pioneered by Joseph Brown in America around 1864. Accurate milling cutters needed to be ground and high precision was achieved after 1877 when F. B. Norton fabricated a grinding wheel from clay and emery or later feldspar. Finally, after 1880 and especially after 1900, there was the introduction of special alloy steels for the cutters and tools for machine tools. F. W. Taylor from the Midvale Steel Co. in Pennsylvania carried out experiments from 1880 onwards which he showed in some spectacular demonstrations at the Paris Exhibition of 1900. Chrome tungsten alloy, followed by the introduction of the more exotic materials such as titanium, molybdenum and especially vanadium, accelerated even more the revolution that he had started and helped to create an era of much greater precision in engineering which assisted the wider spread of the steam turbine.[24]

Parsons had carried out tests in 1897 on a small launch with a turbine at 33,000 r.p.m. which through helical gearing drove two propeller shafts at 1,500 r.p.m. Although the machinery worked well, no further attempts were made until, in 1909, the tramp steamer, *Vespasian*, 4,350 tons displacement, was purchased and trials carried out on her existing triple expansion reciprocating engines on a voyage to the Mediterranean and back. Then she was reengined with turbines which drove the existing propeller shaft through double-helical reduction gearing with the high- and low-pressure turbines coupled to a common gearwheel. The steam consumption of the turbines ranged from 13½ to 19 per cent less than the previous reciprocating engines.[25] A shaft speed of around 63 r.p.m. gave the vessel a speed of 8–10 knots. So successful were these trials that many geared turbines were quickly ordered for other ships and by 1918 direct coupling had become obsolete at sea. The use of gearing allowed the turbine to rotate at its optimum speed and the machinery, whether on land or sea, could be matched to it.

It was these geared turbines which proved to be suitable for driving textile mills. The India Mill of Messrs Kershaw, Leese & Co., Stockport, was driven through ordinary gearing and line-shafting by a four cylinder McNaughted beam engine of 2,000 h.p. In 1913, it was decided to modernise this and the old engine was replaced by a Parsons turbine with separate high- and low-pressure cylinders arranged side by side, rated at 2,300 b.h.p. but capable of developing 2,900 b.h.p. with a steam pressure of 110 p.s.i., superheated to 494 °F. The turbines drove at 3,500 r.p.m., through common reduction gearing at 14 to 1 to a rope pulley turning 44 ropes at 250 r.p.m.[26] The rearrangement of the rope drive was probably installed by Scott & Hodgson.[27] It was this type of turbine which was supplied to the Elk Mill in 1926.

The development of successful turbines by Parsons encouraged many other people such as Charles Curtis, Birger Ljungström and Auguste Rateau to experiment and produce rival types but their history barely impinges on the story of reciprocating power in textile mills. While Scott & Hodgson seems to have worked with Parsons, various other mill engine builders, such as Daniel Adamson, Hick Hargreaves and Musgrave, took up the construction of turbines themselves. By 1914, there was a multitude of different designs from which to choose. Turbines could be easily adapted as pass-out engines with steam being taken out at a particular section of the blading where the pressure was appropriate. Many such turbines were installed in papermills where steam was needed to heat the cylinders for drying the paper but few found their way into textile mills. Where they did, it was often to drive a generator directly and then the electric current would be distributed to motors throughout the mill.

One of the earliest conversions to electric transmission in Lancashire was Ashworth Hadwen's at Droylsden where, in 1907, the 1,000 h.p. beam engine with gear drives comprising 90 pairs of bevel wheels, was replaced by a turbo-alternator and numerous motors. In the spinning mill, one motor was arranged to drive two floors by ropes reaching upwards and downwards.[28]

The turbine continued to outstrip the reciprocating engine in efficiency. To give but one example, in 1913, Parsons supplied a condensing turbo-alternator to the Fisk Street Power Station of the Commonwealth Edison Co. at Chicago which became known as 'Old Reliability'. Its performance caused the same company in 1923 to order a 50,000 kW turbo-alternator for their Crawford Avenue power station. The steam consumption of 'Old Reliability', at 200 p.s.i. and 29 in vacuum, was 11.25 lb per kW hour. The later machine had a pressure of 550 p.s.i. and 29¼ in vacuum and, with reheating the steam between cylinders, it showed a steam consumption of only 7.35 lb per kW hour.[29] The range of heat which was being utilised in these later power stations was far greater than anything achieved in a textile mill. The higher boiler pressures could be achieved because, with no parts of the turbine rubbing where there was steam, there was no need for oil in the steam and so there was no fear of contamination in the boiler feed water. Against such performances, the large reciprocating steam engine could not compete although it still could in small sizes while the Diesel engine had not been developed above 1,000 h.p.

Yet the mill engine lingered on until the late 1960s. In some of the older mills, such as the Hawthorne Mill, Chadderton, the Saxon engine of 1878, modified to a triple expansion layout in 1909, continued to drive the whole mill until its closure in 1970. Sometimes, where ring frames replaced mules, these new machines were driven by ropes or belting from the existing line-shafting. One such mill was Magnet where once again the engine drove all the machines in that mill from 1902 until it was closed in 1966. In a few cases, the original rope drives to all the floors were scrapped and new ropes were fitted to drive an alternator which supplied current to electric motors on the different machines throughout the mill. This was the case at Brooklands Mill which was powered by the same 1,200 h.p. engine from 1893 to 1967. It is surprising that such a system should have been competitive with current supplied from the grid. More often, when new ring spinning machines were installed, they were fitted with their own electric motors and took current from the grid so gradually the power supplied by the engine through its shafting was reduced. Of course the engines became less and less efficient as the load became smaller and finally the mill closed or the engine was scrapped. Shift working too was another reason for the demise of the reciprocating steam engine because often the night shift had different power requirements from the day and the engine was less economic at light loads. In a spinning mill, the last machines to be driven by steam were usually the preparatory and opening machines because the least modernisation took place here.

Figure 80 The pair of tandem compound engines at Magnet Mill, Chadderton, built by Saxon in 1902 and which ran until 1966. (Courtaulds Ltd.)

It was in the weaving sheds that steam engines lingered longest. Owing to the peculiar nature of the power cycle in weaving, should all the looms pick and send the shuttle across at the same time, the power required would peak suddenly. This might send an electricity maximum demand meter into a higher range of charges while, with a steam engine, the peak would be absorbed in a slight slowing of the shafting and flywheel. As the numbers of mill engines declined, so the whole infrastructure of skills and knowledge needed to keep them operating gradually disappeared. In particular, firms who could make and repair the Lancashire boilers closed or turned to other types. However efficient electric motors may be, they never have the same attraction as the reciprocating steam engine which usually was the pride and joy of the men who attended it so it was kept in immaculate condition, as befitted a machine on which the whole of the mill depended.

Notes

Chapter 1

1 M. Reynolds, *Stationary Engine Driving. A Practical Manual for Engineers in Charge of Stationary Engines* (London 1885) p. 1.
2 C. Hadfield & G. Biddle, *The Canals of North West England* (Newton Abbot 1970) pp. 26 & 163.
3 For a full account of cotton spinning, see W. E. Morton, *An Introduction to the Study of Spinning* (London 1937).
4 W. J. M. Rankine, *A Manual of the Steam Engine and Other Prime Movers* (London 1861) p. 299.
5 O. Reynolds, 'The Transmission of Energy', Society for the Encouragement of Arts, Manufactures and Commerce, *Cantor Lectures*, 1883, p. 5.
6 W. Ripper, *Heat Engines* (London 1909) p. 9.
7 Reynolds, 'Transmission', p. 4.
8 C. H. Peabody, *Thermodynamics of the Steam-Engine and Other Heat-Engines* (London 1889) p. 23, Rankine, *Manual*, p. 306 and R. H. Thurston, *A Manual of Steam-Boilers: Their Design, Construction and Operation* (New York 1888) p. 240.
9 M. A. Alderson, *An Essay on the Nature and Application of Steam, with an Historical Note of the Rise and Progressive Improvement of the Steam Engine* (London 1834) pp. 43–4. Part of this is quoted from Belidor, see A. Rees, *The Cyclopaedia; or Universal Dictionary of Arts, Sciences and Literature* (London 1819, reprint Newton Abbot 1972) Vol. V, p. 106.

Chapter 2

1 L. T. C. Rolt & J. S. Allen, *The Steam Engine of Thomas Newcomen* (Hartington 1977) p. 17.
2 H. W. Dickinson, *Sir Samuel Morland, Diplomat and Inventor, 1625–1695* (Cambridge 1970) p. 76 and J. Farey, *A Treatise on the Steam Engine, Historical, Practical and Descriptive* (London 1827, reprint Newton Abbot 1971) Vol. I, p. 91.
3 R. Jenkins, *The Collected Papers of . . .*, see 'A Contribution to the History of the Steam Engine, (i) The Note Books of Roger North, (ii) The Work of Sir Samuel Morland', Lecture to the Newcomen Society, 22 January 1936, p. 42.
4 Rolt & Allen, *Steam Engine*, p. 18.
5 T. K. Derry & T. I. Williams, *A Short History of Technology from Earliest Times to A.D. 1900* (Oxford 1960) p. 313 and Farey, *Steam Engine*, Vol. I, p. 88.
6 Farey, *Steam Engine*, Vol. I, p. 93.
7 Patent 356, 25 July 1698. For further details about Savery's life and his engine, see R. Jenkins, 'Savery, Newcomen and the Early History of the Steam Engine', *Transactions of the Newcomen Society*, Vol. 3, 1922–3, pp. 96–118 and J. S. P. Buckland, 'Thomas Savery, His Steam Engine Workshop of 1707', *T.N.S.*, Vol. 56, 1984–5, pp. 1–20.

8 T. Savery, *The Miners Friend or an Engine to Raise Water by Fire* (London 1702) Intro. p. A2.
9 *Ibid.* and *Philosophical Transactions*, No. 253, Vol. 21 and H. W. Dickinson, *A Short History of The Steam Engine* (Cambridge 1938) p. 23.
10 See R. L. Hills, 'A One Third Working Model of the Newcomen Engine of 1712', *T.N.S.*, Vol. 44, 1971–2, p. 66.
11 Dickinson, *Short History*, Plate I and Buckland, 'Savery', pp. 2–5.
12 Savery, *Miners Friend*, p. 10.
13 *Ibid.*, p. 19.
14 *Ibid.*, p. 17.
15 Buckland, 'Savery', pp. 7–9.
16 Dickinson, *Short History*, p. 24 and R. Stuart, *A Descriptive History of the Steam Engine* (London 1824) pp. 40 & 42.
17 Savery, *Miners Friend*, p. 5.
18 J. T. Desaguliers, *A Course of Experimental Philosophy* (3rd edn, London 1743) Vol. II, p. 467.
19 *Ibid.*, p. 472 and Buckland, 'Savery', pp. 17 & 19.
20 Desaguliers, *Experimental Philosophy*, p. 485.
21 Dickinson, *Short History*, p. 27.
22 See L. T. C. Rolt, *Thomas Newcomen, The Prehistory of the Steam Engine* (Dawlish 1963) and Rolt & Allen, *Steam Engine*.
23 Hills, 'Working Model', pp. 63–77 and K. Cookson, 'An Experimental Determination of the Power and Efficiency of an Early Newcomen Engine' (U.M.I.S.T. 1970–1).
24 See V. Zonca, *Novo Teatro di Machine et Edificii* (Padoua 1607).
25 For a biographical study of Triewald, see S. Lindqvist, *Technology on Trial, The Introduction of Steam Power Technology into Sweden, 1715–1736* (Sweden 1984).
26 M. Triewald, *Short Description of the Atmospheric Engine* (Stockholm 1734, trans. Cambridge 1928) p. 2.
27 Rolt, *Thomas Newcomen*, pp. 58–9.
28 A. Rees, *The Cyclopaedia; or Universal Dictionary of Arts, Sciences and Literature* (London 1819, reprint Newton Abbot 1972) Vol. V, p. 75.
29 Desaguliers, *Experimental Philosophy*, p. 473.
30 *Ibid.*, p. 473.
31 Triewald, *Short Description*, pp. 2–3.
32 Desaguliers, *Experimental Philosophy*, p. 471.
33 *Ibid.*, p. 477 and Rees, *Cyclopaedia*, p. 82.
34 C. O. Becker & A. Titley, 'The Valve Gear of Newcomen's Engine', *T.N.S.*, Vol. 10, 1929–30, p. 3.
35 G. Birkbeck & H. & J. Adcock, *The Steam-Engine Theoretically and Practically Displayed* (London 1827) p. 85, Desaguliers, *Experimental Philosophy*, p. 476 and Rees, *Cyclopaedia*, pp. 76 & 78.
36 A. W. Skempton, ed., *John Smeaton, F.R.S.* (London 1981), pp. 179ff.
37 Cookson, 'Experimental Determination', p. 56.
38 Rees, *Cyclopaedia*, p. 79.
39 *Ibid.*, p. 92.
40 See Rolt, *Thomas Newcomen*, pp. 73ff. and Rolt & Allen, *Steam Engine*, pp. 58ff.
41 Rolt & Allen, *Steam Engine*, p. 88.

Chapter 3

1 L. T. C. Rolt & J. S. Allen, *The Steam Engine of Thomas Newcomen* (Hartington 1977), p. 68.
2 'An Account of Mr. Tho. Savery's Engine for Raising Water by the Help of Fire', *Philosophical Transactions*, Vol. 21, 1699, p. 228.

3 T. Savery, *The Miners Friend or an Engine to Raise Water by Fire* (London 1702) p. 40.
4 M. Triewald, *Short Description of the Atmospheric Engine* (Stockholm 1734, trans. Cambridge 1928) p. 47.
5 S. Lindqvist, *Technology on Trial, The Introduction of Steam Power Technology into Sweden, 1715–1736* (Stockholm 1984) p. 249.
6 *Ibid.*, pp. 247–9.
7 Patent 571, 1740.
8 H. P. Spratt, *The Birth of the Steam Boat* (London 1958) p. 49.
9 A. Rees, *The Cyclopaedia, or Universal Dictionary of Arts, Sciences and Literature* (London 1819, reprint Newton Abbot 1972) Vol. V, p. 122.
10 J. Farey, *A Treatise on the Steam Engine, Historical, Practical and Descriptive* (London 1827, reprint Newton Abbot 1971) Vol. I, p. 109.
11 J. T. Desaguliers, *A Course of Experimental Philosophy* (3rd edn, London 1743) Vol. II, p. 467 and Rees, *Cyclopaedia*, p. 69.
12 J. Bourne, *A Treatise on the Steam Engine in its Application to Mines, Mills, Steam Navigation and Railways* (London 1846) p. 7.
13 R. Stuart, *Historical and Descriptive Anecdotes of Steam Engines and their Inventors and Improvers* (London 1829) Vol. I, p. 185.
14 Desaguliers, *Experimental Philosophy*, p. 467.
15 *Ibid.*, p. 488.
16 *Ibid.*, pp. 489–90.
17 R. Stuart, *A Descriptive History of the Steam Engine* (London 1824) p. 85.
18 *Phil. Trans.*, Vol. 47, p. 436 and T. Tredgold, *The Steam Engine, Its Invention and an Investigation of its Principles for Navigation, Manufactures and Railways* (London 1838) p. 18.
19 Farey, *Steam Engine*, Vol. I, p. 120.
20 A. W. Skempton, ed., *John Smeaton F.R.S.* (London 1981) p. 179.
21 E. Straker, *Wealden Iron* (1931, reprint Newton Abbot 1969) pp. 72–3.
22 Rolt & Allen, *Steam Engine*, p. 122.
23 Skempton, *Smeaton*, pp. 186–7.
24 Rolt & Allen, *Steam Engine*, p. 123.
25 Patent 848, 10 June 1766.
26 W. Blakey, *A Short Historical Account of the Invention, Theory and Practice of Fire-Machinery* (London 1793) p. 5 and H. W. Dickinson, *A Short History of the Steam Engine* (Cambridge 1938) p. 69.
27 J. Bootsgezel, 'William Blakey – A Rival to Newcomen', *Transactions of the Newcomen Society*, Vol. 14, 1935–6, pp. 97–106 for a fuller discussion on the work of Blakey, and Rees, *Cyclopaedia*, p. 71.
28 Rees, *Cyclopaedia*, p. 72.
29 Blakey, *Short Account*, pp. vii, viii & 17 and Stuart, *Descriptive History*, p. 88.
30 H. W. Dickinson, *Matthew Boulton* (Cambridge 1936) p. 75.
31 *Ibid.*, p. 76.
32 *Ibid.*, p. 76.
33 For accounts of the early cotton industry, see R. L. Hills, *Power in the Industrial Revolution* (Manchester 1970) and A. P. Wadsworth & J. de Lacey Mann, *The Cotton Trade and Industrial Lancashire* (Manchester 1931).
34 For an account of the early Industrial Revolution, see T. S. Ashton, *The Industrial Revolution, 1760–1830* (Oxford 1948) and for technical developments in textile machines, see W. English, *The Textile Industry* (London 1969).
35 C. S. Aspin & S. D. Chapman, *James Hargreaves and the Spinning Jenny* (Helmshore 1964).
36 R. S. Fitton & A. P. Wadsworth, *The Strutts and the Arkwrights, 1758–1830* (Manchester 1958) and R. L. Hills, *Richard Arkwright and Cotton Spinning* (London 1973).
37 B. Cooper, *Transformation of a Valley, The Derbyshire Derwent* (London 1983).

38 Dobson & Barlow Ltd, *Samuel Crompton, The Inventor of the Spinning Mule* (Bolton 1927), G. J. French, *The Life and Times of Samuel Crompton* (London 1859) and H. Catling, *The Spinning Mule* (Newton Abbot 1970).
39 E. Baines, *History of the Cotton Manufacture in Great Britain* (London 1835) p. 215.
40 J. Priestley, *The Historical Account of the Navigable Rivers, Canals and Railways, throughout Great Britain* (1831, reprint Newton Abbot 1969) p. 447.
41 H. Malet, *The Canal Duke* (Dawlish 1961).
42 W. S. Jevons, *The Coal Question: an Inquiry concerning the Progress of the Nation, and the Probable Exhaustion of our Coal-mines* (London 1865).
43 Baines, *History*, p. 226 and A. E. Musson & E. Robinson, *Science and Technology in the Industrial Revolution* (Manchester 1969) p. 395.
44 J. Aikin, *A Description of the Country from Thirty to Forty Miles Round Manchester* (London 1795) p. 174.
45 Farey, *Steam Engine*, Vol. I, p. 125.
46 Musson & Robinson, *Science & Technology*, p. 403.
47 Farey, *Steam Engine*, Vol. I, p. 122.
48 Bolton & Watt Collection. J. Watt to J. Rennie, 20 August 1785.
49 Hills, *Power*, pp. 142 & 168.
50 E. Kilburn Scott, ed., *Matthew Murray, Pioneer Engineer, Records from 1765 to 1826* (Leeds 1928) p. 8. For a fuller discussion, see Musson & Robinson, *Science and Technology*, pp. 399ff.
51 B. & W. Coll., from P. Ewart, 7 December 1791.
52 *Ibid.*, from Beverley Cross & Co., 23 June 1792.
53 For a fuller discussion, see Hills, *Power*, pp. 140ff.
54 G. Birkbeck & H. & J. Adcock, *The Steam Engine Theoretically and Practically Displayed* (London 1827) pp. 59–72.
55 *Ibid.*, p. 68.
56 *Ibid.*, p. 71.
57 E. Galloway, *History and Progress of the Steam Engine* (London 1830), p. 137 and Stuart, *Descriptive History*, p. 161.
58 Dobson & Barlow, *Samuel Crompton*, p. 95.
59 See P. N. Wilson, 'Waterwheels of J. Smeaton', *T.N.S.*, Vol. 30, 1955–7, p. 48.
60 B. & W. Coll., J. Southern to P. Ainsworth, 28 June 1790.
61 *Ibid.*, P. Ewart to J. Southern, 20 October 1791.
62 *Ibid.*, P. Ewart to J. Southern, 30 April 1792.

Chapter 4

1 Some of this information is based on a communication from Professor D. S. L. Cardwell, see also H. W. Dickinson, *James Watt, Craftsmen and Engineer* (Cambridge 1935) p. 32 and J. Farey, *A Treatise on the Steam Engine, Historical Descriptive and Practical* (London 1827, reprint Newton Abbot 1972) Vol. I, pp. 310 & 313.
2 J. T. Desaguliers, *A Course of Experimental Philosophy* (3rd edn, London 1743) Vol. II, p. 536.
3 R. J. Law, *James Watt and the Separate Condenser* (London 1969) p. 17.
4 For a much fuller discussion, see D. S. L. Cardwell, *From Watt to Clausius, the Rise of Thermodynamics in the Early Industrial Age* (London 1971) pp. 40ff.
5 T. Tredgold, *The Steam Engine, Its Invention, and An Investigation of the Principles, for Navigation, Manufactures, and Railways* (London 1838) Vol. I, p. 49.
6 Cardwell, *From Watt*, pp. 46–7 and Farey, *Steam Engine*, Vol. I, p. 311.
7 Dickinson, *James Watt*, p. 36 and J. P. Muirhead, *The Origin and Progress of the Mechanical Inventions of James Watt* (London 1854) Vol. I, p. lxxix.
8 Law, *James Watt*, p. 27.
9 Boulton & Watt Collection, Portfolio 1381.

10 Patent 913, 5 January 1769.
11 Muirhead, *The Origin*, Vol. I, p. 34.
12 *Ibid.*, Vol. I, p. 34.
13 H. W. Dickinson & R. Jenkins, *James Watt and the Steam Engine* (1927, reprint Ashbourne 1981) pp. 98f.
14 Muirhead, *The Origin*, Vol. I, p. 47, see also letter to Roebuck, 16 March 1769, p. 48.
15 Dickinson & Jenkins, *James Watt*, pp. 113ff.
16 N. P. Burgh, *Practical Treatise on the Condensation of Steam* (London 1871) p. 8.
17 A. Rees, *The Cyclopaedia; or Universal Dictionary of Arts, Sciences and Literature* (London 1819, reprint Newton Abbot 1972) Vol. V, p. 105.
18 H. W. Dickinson, *Matthew Boulton* (Cambridge 1936) p. 81 and Dickinson, *James Watt*, pp. 71–2.
19 Farey, *Steam Engine*, Vol. I, pp. 291 & 326 and Rees, *Cyclopaedia*, Vol. V, p. 96.
20 Dickinson & Jenkins, *James Watt*, p. 89.
21 Farey, *Steam Engine*, Vol. I, pp. 333 & 339 and Rees, *Cyclopaedia*, Vol. V, p. 99.
22 M. Arago, *Historical Eloge of James Watt* (London 1838) p. 88.
23 Rees, *Cyclopaedia*, pp. 91 & 110.
24 Patent 571, 1740.
25 Muirhead, *The Origin*, Vol. III, p. 38, note.
26 Patent 1213, 1779. See also Farey, *Steam Engine*, Vol. I, p. 409.
27 Rees, *Cyclopaedia*, p. 93.
28 Patent 1,263, 1780.
29 Farey, *Steam Engine*, Vol. I, p. 410.
30 J. Aikin, *A Description of the Country from Thirty to Forty Miles Round Manchester* (London 1795) p. 177.
31 B. & W. Coll., from P. Drinkwater, 3 April 1789.
32 J. Farey, *Steam Engine* (London 1827) Vol. I, p. 422.
33 R. L. Hills, *Power in the Industrial Revolution* (Manchester 1970) pp. 150ff.
34 Aikin, *Description*, p. 177.
35 Dickinson, *Boulton*, p. 113.
36 Patent 1,306, 1781, enrolled 23 February 1782.
37 Muirhead, *The Origin*, Vol. III, p. 42.
38 *Ibid.*, p. 45.
39 *Ibid.*, p. 47.
40 *Ibid.*, p. 49.
41 *Ibid.*, p. 51.
42 Patent 1,321, 4 July 1782.
43 Farey, *Steam Engine*, Vol. I, p. 361.
44 Muirhead, *The Origin*, Vol. III, p. 42.
45 *Ibid.*, p. 63.
46 Dickinson & Jenkins, *James Watt*, p. 139.
47 Muirhead, *The Origin*, Vol. III, p. 75.
48 *Ibid.*, p. 81.
49 Dickinson & Jenkins, *James Watt*, p. 159.
50 Farey, *The Steam Engine*, Vol. I, p. 426.
51 *Ibid.*, Vol. I, p. 664.
52 A. E. Musson & E. Robinson, *Science and Technology in the Industrial Revolution* (Manchester 1969) p. 427.
53 Farey, *Steam Engine*, Vol. I, p. 658, Hills, *Power*, pp. 147ff. and F. Nixon, 'The Early Steam Engines in Derbyshire', *Transactions of the Newcomen Society*, Vol. 31, 1958–9, pp. 6ff.
54 B. & W. Coll., from Davison & Hawksley, 3 November 1797.
55 Dickinson, *James Watt*, p. 136.
56 Farey, *Steam Engine*, Vol. I, p. 430.

57 Patent 1,432, 25 August 1784.
58 Dickinson, *James Watt*, p. 139.

Chapter 5

1 M. Arago, *Historical Eloge of James Watt* (London 1839) p. 145.
2 J. Farey, *A Treatise on the Steam Engine, Historical, Practical and Descriptive* (London 1827, reprint Newton Abbot 1971) Vol. I, p. 677.
3 D. B. Barton, *The Cornish Beam Engine* (Truro, New edn 1969) p. 27 and H. W. Dickinson, *A Short History of the Steam Engine* (Cambridge 1938) p. 88.
4 J. Moses, 'The Albion Mills, 1784–1791', *Transactions of the Newcomen Society*, Vol. 40, 1967–8, pp. 47ff. and Farey, *Steam Engine*, Vol. I, pp. 438 & 442.
5 See Farey, *Steam Engine*, Vol. I, p. 434, and L. T. C. Rolt, *James Watt* (London 1962) p. 95.
6 Boulton & Watt Collection, from G. & J. Robinson, 13 August 1785.
7 *Ibid.*, to Messrs Robinson, 17 August 1785.
8 *Ibid.*, from G. Robinson, 9 October 1785.
9 *Ibid.*, to G. Robinson, 11 October 1785.
10 *Ibid.*, from John Robinson, 14 December 1785.
11 *Ibid.*, from John Robinson, 27 February 1786.
12 *Ibid.*, from John Robinson, 26 March 1786.
13 *Ibid.*, from G. & J. Robinson, 18 December 1786.
14 S. D. Chapman, 'The Midlands Cotton and Worsted Industry, 1769–1800', Ph.D. Thesis, London 1966, p. 206; see *Royal Exchange Registers*, Vol. XVIII, 119287.
15 B. & W. Coll., from John Robinson, 7 September 1790.
16 *Ibid.*, from James Robinson, 29 July 1797.
17 *Ibid.*, from J. Southern to John Robinson, 17 April 1791.
18 *Ibid.*, from H. Pearson, 12 April 1786.
19 *Ibid.*, from J. Kendrew, 13 August 1790.
20 *Ibid.*, from P. Drinkwater, 3 April 1789.
21 R. L. Hills, *Power in the Industrial Revolution* (Manchester 1970) pp. 165ff.
22 B. & W. Coll., to C. B. & A. Burden, 11 August 1785.
23 *Ibid.*, to R. Dennison, 23 June 1791.
24 *Ibid.*, from J. Southern to P. Drinkwater, 22 April 1789.
25 *Ibid.*, to T. Harris, 27 November 1785.
26 *Ibid.*, from T. Harris, 5 November 1786.
27 H. W. Dickinson, *James Watt, Craftsman and Engineer* (Cambridge 1935) p. 186 and see A. Rees, *Cyclopaedia; or Universal Dictionary of Arts, Sciences and Literature* (London 1819, reprint Newton Abbot 1972) Vol. V, p. 140 for an account of Fenton, Murray & Wood.
28 B. & W. Coll., to J. & S. Simpson, 23 November 1792.
29 H. W. Dickinson and R. Jenkins, *James Watt and the Steam Engine* (London 1927, reprint Ashbourne 1981) pp. 375 & 399.
30 B. & W. Coll., to J. Rennie, 20 August 1785.
31 T. Balston, *William Balston, Paper Maker, 1759–1849* (London 1954, reprint New York 1979) p. 41.
32 B. & W. Coll., from R. Gorton, 7 June 1788.
33 *Ibid.*, from T. Gorton & Son, 13 June 1788.
34 *Ibid.*, from Gorton & Thompson, 19 September 1788.
35 *Ibid.*, from J. Southern to T. Gorton & Son, 21 September 1788.
36 *Ibid.*, from Gorton & Thompson, 18 April 1791.
37 *Ibid.*, from Gorton & Thompson, 20 April 1791.
38 *Ibid.*, from Gorton & Thompson, 24 August 1791.
39 *Ibid.*, from Gorton & Thompson, 10 September 1791.

40 *Ibid.*, from Gorton & Thompson, 2 September 1791.
41 *Ibid.*, from J. Southern to Gilpin, 21 October 1791 and from Gorton & Thompson, 19 February 1793.
42 *Ibid.*, from P. Ewart, 18 August 1792.
43 Rees, *Cyclopaedia*, Vol. V, p. 132 and E. K. Scott, ed., *Matthew Murray, Pioneer Engineer, Records from 1765 to 1826* (Leeds 1928) p. 9.
44 Rees, *Cyclopaedia*, Vol. V, p. 134.
45 Farey, *Steam Engine*, Vol. I, pp. 169 & 473.
46 B. & W. Coll., Southern to Pearson & Grimshaw, 24 March 1791.
47 *Ibid.*, from P. Drinkwater, 21 November 1789.
48 *Ibid.*, to P. Drinkwater, 25 November 1789 and Farey, *Steam Engine*, Vol. I, pp. 436, 458 & 465ff.
49 For a full history of early fire-proof construction, see H. R. Johnson & A. W. Skempton, 'William Strutt's Cotton Mills', *T.N.S.*, Vol. 30, 1956. See also Farey, *Steam Engine*, Vol. I, pp. 274 & 434.
50 J. Banks, *On the Power of Machines* (Kendal 1803) pp. 90–7.
51 *Ibid.*, p. 97.
52 A. J. Pacey, *The Maze of Ingenuity: Ideas and Idealism in the Development of Technology* (London 1974) p. 228.
53 Farey, *Steam Engine*, vol. I, pp. 473 & 612 and Rees, *Cyclopaedia*, p. 129.
54 B. & W. Coll., M. R. Boulton to J. Watt Junior, 17 January 1799.
55 Scott, *Matthew Murray*, pp. 39–40 and T. Turner, 'History of Fenton Murray & Wood', M.Sc. Thesis, U.M.I.S.T., 1966, Ch. 2, p. 11.
56 Dickinson, *Short History*, p. 75.
57 B. & W. Coll., Indenture of A. Hunt, 1 March 1792.
58 *Ibid.*, from S. J. & B. Churchill, 6 June 1793.
59 Hills, *Power*, pp. 168ff.
60 *Ibid.*, pp. 217–19.
61 For a full discussion on determining the horsepower in early steam engines, see R. L. Hills & A. J. Pacey, 'The Measurement of Power in Early Steam-driven Textile Mills', *Technology and Culture*, Vol. 13, No. 1, January 1972, pp. 25–43.
62 T. Savery, *The Miner's Friend or An Engine to Raise Water by Fire* (London 1702) Ch. 2.
63 M. Triewald, *Short Description of the Atmospheric Engine* (Stockholm 1734, trans. Cambridge 1928) pp. 29–31.
64 *Reports of the Late John Smeaton* (2nd edn, London 1837) Vol. II, p. 77.
65 R. L. Hills, *Machines, Mills and Uncountable Costly Necessities. A Short History of the Drainage of the Fens* (Norwich 1967) p. 170 and Rees, *Cyclopaedia*, p. 128.
66 B. & W. Coll., Portfolio 56.
67 *Ibid.*, from J. James, 27 February 1786.
68 *Ibid.*, from J. Southern to J. Cartwright, 17 November 1788.
69 *Ibid.*, from J. Watt to W. Carr, 19 March 1790.
70 R. Buchanan, *Practical Essays on Millwork and Other Machinery* (2nd edn, London 1823) Vol. I, pp. 156–7.
71 B. & W. Coll., from J. Watt to J. Cartwright, 6 September 1788.
72 Dickinson, *Short History*, p. 85 and Farey, *Steam Engine*, Vol. I, pp. 481ff.
73 B. & W. Coll., from G. Lee to J. Lawson, 12 April 1796.
74 *Ibid.*, from G. Lee to J. Lawson, 12 April 1796.
75 Anon., 'Account of a Steam-Engine Indicator', *Quarterly Journal of Science*, Vol. 13, 1822, p. 95.
76 B. & W. Coll., from G. Lee to J. Lawson, 6 April 1798.
77 Strutt Letters, Derby Public Library, G. Lee to W. Strutt, 29 December 1803.
78 Farey, *Steam Engine*, Vol. I, p. 7 gives a note with an appropriate number of engines working in the main manufacturing towns; these include London with 290; Manchester, 240; Leeds, 130 and Glasgow, 80 or 90.

Chapter 6

1 M. Arago, *Historical Eloge of James Watt* (London 1839) pp. 149–50.
2 J. Farey, *A Treatise on the Steam Engine, Historical, Practical and Descriptive* (Newton Abbot 1971) Vol. II, p. 72 and T. Tredgold, *The Steam Engine, Its Invention and an Investigation of Its Principles for Navigation, Manufactures and Railways* (London 1838) p. 41.
3 Patent 913, 5 January 1769.
4 Patent 1,432, 25 August 1784.
5 A. Elton and B. Drummond, 'The Life and Times of William Murdock, Frederic Albert Winsor and Philippe Lebon' (typescript) p. 36.
6 G. Bathe, *An Engineering Miscellany* (*c.* 1940) p. 56 and H. W. Dickinson, *A Short History of the Steam Engine* (Cambridge 1938) p. 94.
7 *Relfs Philadelphia Gazette and Daily Advertiser*, Saturday 13 July 1805.
8 *Emporium of the Arts and Sciences* (Boston, U.S.A.) Vol. 4, 1813, pp. 201 & 211.
9 For an early illustration, see V. Zonca, *Novo Teatro di Machine et Edificii* (Padua 1607) p. 103. This was the type of pump used by the Romans.
10 Dickinson, *Short History*, p. 38 and Farey, *Steam Engine*, Vol. II, pp. 140 & 146.
11 F. Trevithick, *Life of Richard Trevithick with an Account of his Inventions* (London 1872) Vol. I, p. 69.
12 H. W. Dickinson and A. Titley, *Richard Trevithick, the Engineer and the Man* (Cambridge 1934) p. 39.
13 Trevithick, *Trevithick*, Vol. I, pp. 78, 80 & 87.
14 *Ibid.*, Vol. I, p. 92.
15 Dickinson & Titley, *Trevithick*, p. 60.
16 Farey, *Steam Engine*, Vol. II, p. 9.
17 Trevithick, *Trevithick*, Vol. II, p. 132.
18 *Ibid.*, Vol. II, p. 143.
19 A. Rees, *The Cyclopaedia; or Universal Dictionary of Arts, Sciences and Literature* (London 1819, reprint 1972) Vol. V, p. 139.
20 Farey, *Steam Engine*, Vol. II, pp. 240–1, Trevithick, *Trevithick*, Vol. II, p. 157 and D. S. L. Cardwell, *Technology and Change, 1750–1914* (Milton Keynes 1984) p. 76.
21 T. Lean, *On the Steam Engines in Cornwall* (Truro 1839, reprint 1969) p. 10.
22 *Ibid.*, pp. 19–20 and Rees, *Cyclopaedia*, p. 111.
23 Farey, *Steam Engine*, p. 336, Trevithick, *Trevithick*, Vol. II, pp. 70–2 and D. B. Barton, *The Cornish Beam Engine* (Truro, new edn 1969) p. 33.
24 Barton, *Beam Engine*, p. 37, note 1.
25 Trevithick, *Trevithick*, Vol. II, p. 75.
26 *Ibid.*, Vol. II, p. 94.
27 *Ibid.*, Vol. II, pp. 96 & 100.
28 Patent 2,726, 21 January 1804.
29 Patent 2,772, 1804.
30 *Ibid.*
31 Farey, *Steam Engine*, Vol. II, p. 57.
32 *Ibid.*, Vol. II, p. 65.
33 T. R. Harris, *Arthur Woolf, The Cornish Engineer, 1776–1837* (Truro 1966) pp. 49 & 52.
34 Farey, *Steam Engine*, Vol. II, p. 70.
35 E. Vale, *The Harveys of Hayle, Engine Builders, Shipwrights and Merchants of Cornwall* (Truro 1966) pp. 106–7.
36 Barton, *Beam Engine*, p. 33.
37 Harris, *Woolf*, p. 53.
38 *Ibid.*, pp. 55–6.
39 Barton, *Beam Engine*, p. 37.

40 Lean, *Cornwall*, p. 31.
41 Harris, *Woolf*, p. 57.
42 Barton, *Beam Engine*, pp. 42–4 and Farey, *Steam Engine*, Vol. II, pp. 206, 216 & 297.
43 Z. Colburn, *Locomotive Engineering and the Mechanism of Railways* (London 1871) pp. 49–50 and Farey, *Steam Engine*, Vol. I, p. 468.
44 R. L. Hills and A. J. Pacey, 'The Measurement of Power in Early Steam-driven Textile Mills', *Technology and Culture*, Vol. 13, No. 1, January 1972, pp. 36ff.
45 Farey, *Steam Engine*, Vol. II, pp. 317–18.
46 Harris, *Woolf*, pp. 85–7.
47 Farey, *Steam Engine*, Vol. II, p. 146.
48 *Ibid.*, Vol. II, pp. 147 & 198.
49 *Ibid.*, Vol. II, p. 198.
50 Barton, *Beam Engine*, p. 57 and Rees, *Cyclopaedia*, Vol. V, p. 117.
51 Farey, *Steam Engine*, Vol. I, p. 589.
52 Barton, *Beam Engine*, pp. 49–50.
53 T. M. Goodeve, *Text-Book on the Steam Engine* (London, 6th edn 1884) p. 93.
54 Lean, *Cornwall*, pp. 146–7.
55 N. P. Burgh, *A Practical Treatise on the Condensation of Steam* (London 1871) p. 12.
56 R. Buchanan, *Practical Essays on Mill Work and Other Machinery* (London, 3rd edn 1841) p. xix.
57 W. Fairbairn, *Treatise on Mills and Millwork* (London 1861) Vol. II, p. 72.
58 Buchanan, *Essays*, p. xix.
59 R. Fitzgerald, 'Albion Mill Manchester', *Industrial Archaeology Review*, Vol. 10, No. 2, Spring 1988, pp. 204ff.
60 Fairbairn, *Millwork*, Vol. I, p. 180.
61 B. Hick, 'First Order Book', quoted by kind permission of Hick Hargreaves Ltd.
62 *Ibid.*, 23 February 1835.
63 See Patent 1,432, 1784, Patent 2,599, 1802 and Patent 1,637, 1802.
64 Farey, *Steam Engine*, Vol. I, pp. 502 & 508.
65 J. Montgomery, *The Theory and Practice of Cotton Spinning or the Carding and Spinning Master's Assistant* (Glasgow, 3rd edn 1836) p. 20.
66 Hick, Order Book, 4 March 1836.
67 W. English, *The Textile Industry* (London 1969) p. 197.
68 E. Baines, *History of the Cotton Manufacture in Great Britain* (London 1835) pp. 235 & 237 and W. Fairbairn, *Useful Information for Engineers* (London, 4th edn 1864) p. 17.
69 Baines, *History*, p. 240.
70 H. Catling, *The Spinning Mule* (Newton Abbot 1970) p. 63 and English, *Textile Industry*, p. 178.
71 R. L. Hills, *Power in the Industrial Revolution* (Manchester 1970) pp. 127ff. and Montgomery, *Cotton Spinning*, p. 193.
72 Fairbairn, *Millwork*, Vol. I, pp. 236–7.
73 C. F. Beyer, '1841 Note Book', p. 114. Mr L. C. Davenport kindly allowed me to read this.
74 Farey, *Steam Engine*, Vol. II, p. 306, note. 'It has been stated, at p. 192 that Mr. Woolf's engines have never been tried, and are scarcely known in the great manufacturing districts in the North of England, and in Scotland. It should be mentioned, that Mr. Hick of Bolton, in Lancashire, has of late taken up the making of Mr. Woolf's compound rotative engines, and has made two engines of a larger size than any previous of the kind. They are excellent specimens, and on improved proportions of the parts, with every perfection of execution which has hitherto been attained in the construction of steam engines; and

although both have been sent abroad, one to France, and the other to Spain, they will probably lead to the introduction of such engines in the manufactories of Lancashire'.

75 Beyer, Note Book, p. 38.

Chapter 7

1 W. Fairbairn, *Treatise on Mills and Millwork* (London 1861) Vol. I, p. 253.
2 R. H. Thurston, *A Manual of Steam-Boilers, Their Design, Construction and Operation* (New York 1888) pp. 11–12.
3 H. W. Dickinson, *A Short History of The Steam Engine* (Cambridge 1938) pp. 12 & 25, and J. Farey, *A Treatise on the Steam Engine, Historical, Practical and Descriptive* (London 1827, reprint Newton Abbot 1971) Vol. I, p. 84.
4 Dickinson, *Short History*, Plate I and A. Rees, *The Cyclopaedia; or Universal Dictionary of Arts, Sciences and Literature* (London 1819, reprint 1972) Vol. V, p. 70.
5 J. T. Desaguliers, *A Course of Experimental Philosophy* (London, 3rd edn 1743) p. 472 and R. Stuart, *A Descriptive History of the Steam Engine* (London 1824) p. 76.
6 N. P. Burgh, *A Practical Treatise on Boilers and Boiler-Making* (London 1881) p. 2.
7 Farey, *Steam Engine*, Vol. I, p. 117 and C. F. Partington, *An Historical and Descriptive Account of the Steam Engine* (London 1822) p. 136 and Rees, *Cyclopaedia*, Vol. V, p. 83, a hogshead equalled 63 gallons.
8 H. R. Kempe, *The Engineer's Year Book, 1929* (London 1929) p. 2850.
9 Desaguliers, *Philosophy*, p. 467.
10 Rees, *Cyclopaedia*, Vol. V, p. 69.
11 T. Tredgold, *The Steam Engine, Its Invention and An Investigation of Its Principles for Navigation, Manufactures and Railways* (London 1838) p. 124.
12 L. T. C. Rolt, *Thomas Newcomen. The Prehistory of the Steam Engine* (Dawlish 1963) p. 96.
13 J. S. Allen, 'The 1715 and Other Newcomen Engines at Whitehaven, Cumberland', *Transactions of the Newcomen Society*, Vol. 45, 1972–3, p. 251 and L. T. C. Rolt & J. S. Allen, *The Steam Engine of Thomas Newcomen* (Hartington 1977) pp. 56 & 90.
14 Farey, *Steam Engine*, Vol. I, p. 230 and Rolt, *Newcomen*, p. 83 for Mr Wauchope's engine in 1727 and Dickinson, *Short History*, p. 118 for the one at Jesmond Colliery, Newcastle-upon-Tyne, erected in 1773.
15 R. Jenkins, *The Collected Papers of* ... (Cambridge 1936) see 'Boiler Making', *The Engineer*, 19 July 1918, p. 127.
16 M. Triewald, *Short Description of the Atmospheric Engine* (Stockholm 1734, English translation, Cambridge 1928) pp. 7–8.
17 Farey, *Steam Engine*, Vol. I, pp. 145–6, Partington, *Historical Account*, p. 136 and Rees, *Cyclopaedia*, Vol. V, p. 81.
18 J. Bourne, *A Treatise on the Steam Engine in Its Application to Mines, Mills, Steam Navigation and Railways* (London 1846) p. 9.
19 Farey, *Steam Engine*, Vol. I, pp. 168 & 203.
20 J. Leupold, *Theatrum Machinarum Hydraulicam* (Leipzig, 1725) para. 203.
21 J. Curr, *The Coal Viewer, and Engine Builder's Practical Companion* (Sheffield 1797) p. 43.
22 Tredgold, *Steam Engine*, p. 251.
23 Rees, *Cyclopaedia*, Vol. V, p. 81 and Rolt & Allen, *Steam Engine*, p. 107.
24 This information was communicated to me by the late T. Althin, Royal Institute of Technology Library, Stockholm.
25 Farey, *Steam Engine*, Vol. I, p. 410.
26 T. H. Hair, *Sketches of the Coal Mines in Northumberland and Durham* (London 1844) p. 40.

27 R. Armstrong, *A Rudimentary Treatise on Steam Boilers; their Construction and Practical Management* (London, 3rd edn 1857) p. 27.
28 H. W. Dickinson & R. Jenkins, *James Watt and the Steam Engine* (London 1927, 2nd edn Ashbourne 1981) p. 234.
29 Boulton & Watt Collection, Sketch of Mr. Harris Engine, 8 November 1785.
30 Farey, *Steam Engine*, Vol. I, p. 489.
31 Rolt, *Newcomen*, pp. 96–7.
32 Farey, *Steam Engine*, Vol. I, pp. 580–1.
33 See *ibid.*, p. 453, Rees, *Cyclopaedia*, Vol. V, p. 126 and Boulton & Watt Collection, Drawing for Samuel Oldknow's Engine, 15 April 1791.
34 Rees, *Cyclopaedia*, Vol. V, p. 107 and Stuart, *Descriptive History*, p. 159.
35 Farey, *Steam Engine*, Vol. I, p. 581, note.
36 B. Hick, 'First Order Book', see for example order from Messrs Hill & Smith, Liverpool, 30 May 1834.
37 T. Wicksteed, *Cornish and Boulton and Watt Engines Erected at the East London Waterworks* (London 1842) Plate IV.
38 R. Armstrong, *A Practical Essay on Steam Engine Boilers as now used in the Manufacturing District around Manchester* (Manchester 1838) p. 12.
39 Armstrong, *Rudimentary Treatise*, p. 27.
40 Patent 1,351, 1783 and R. A. Mott, ed. P. Singer, *Henry Cort, The Great Finer* (London 1983) pp. 27ff.
41 Jenkins, *Collected Papers*, p. 127.
42 Curr, *Coal Viewer*, p. 88.
43 Partington, *Historical Account*, pp. 97–8 & 116.
44 Farey, *Steam Engine*, Vol. II, p. 14.
45 *Ibid.*, p. 17 and Partington, *Historical Account*, p. 162.
46 F. Trevithick, *Life of Richard Trevithick, with an Account of His Inventions* (London 1872) Vol. II, pp. 138ff. and H. W. Dickinson & A. Titley, *Richard Trevithick, The Engineer and the Man* (Cambridge 1934) p. 73.
47 Trevithick, *Trevithick*, p. 155, Dickinson, *Short History*, p. 120 and Dickinson & Titley, *Trevithick*, p. 127.
48 G. Bathe, 'The First High Pressure Steam Engine in America', *An Engineering Miscellany* (no date) p. 56, Dickinson, *Short History*, p. 94 and E. Galloway, *History and Progress of the Steam Engine* (London 1830) p. 147.
49 Patent 2,937, 1806.
50 Trevithick, *Trevithick*, Vol. II, p. 156.
51 *Ibid.*, Vol. II, p. 16.
52 Farey, *Steam Engine*, Vol. I, p. 370, Vol. II, p. 31, Armstrong, *Practical Essay*, p. 38 and Trevithick, *Trevithick*, Vol. II, p. 143.
53 Patent 2,726, 21 January 1804.
54 Galloway, *History*, p. 156, Rees, *Cyclopaedia*, Vol. V, p. 119 and Tredgold, *Steam Engine*, p. 129.
55 Patent 2,726, 21 January 1804.
56 Partington, *Historical Account*, p. 112.
57 *Ibid.*, p. 112.
58 T. R. Harris, *Arthur Woolf, The Cornish Engineer, 1766–1837* (Truro 1966) pp. 59–60.
59 *Ibid.*, pp. 75–6.
60 T. Lean, *Historical Statement of the Improvements Made in the Duty Performed by Steam Engines in Cornwall* (London 1839) p. 10.
61 W. S. Jevons, *The Coal Question: an inquiry concerning the progress of the nation, and the probable exhaustion of our coal-mines* (London 1865) p. 108.
62 Armstrong, *Practical Essay*, p. 45.
63 D. B. Barton, *The Cornish Beam Engine* (Truro 1965, new edn 1969) pp. 37 & 47.
64 Armstrong, *Rudimentary Treatise*, p. 5.
65 Farey, *Steam Engine*, Vol. II, p. 305.

66 E. Alban, *The High-Pressure Steam Engine. An Exposition of Its Comparative Merits and an Essay Towards an Improved System of Construction* (London 1848) p. 6.
67 Trevithick, *Trevithick*, Vol. I, p. 214.
68 E. L. Cornwall, ed., *The Pictorial Story of Railways* (London 1974) p. 49.
69 Burgh, *Practical Treatise*, p. 106.
70 R. D. Hoblyn, *A Manual of the Steam Engine* (London 1842) p. 85.
71 Armstrong, *Practical Essays*, p. 101 and Burgh, *Practical Treatise*, p. 105.
72 J. G. H. Warren, *A Century of Locomotive Building by Robert Stephenson & Co., 1823–1923* (1923, reprint Newton Abbot 1970) p. 140.
73 Armstrong, *Practical Essay*, p. 100.
74 Bourne, *Treatise*, p. 55.
75 Patent 10,166, 1844.
76 C. W. Williams, *The Combustion of Coal and the Prevention of Smoke Chemically and Practically Considered* (London, 2nd edn 1841) p. 154 and J. F. Petrie, 'Charles Wye Williams, A Pioneer in Steam Navigation and Fuel Efficiency', *Transactions of the Newcomen Society*, Vol. 39, 1966–7, pp. 35ff.
77 *The Mechanics Magazine, Museum, Register, Journal and Gazette*, Vol. 42, January–June 1845, London, p. 122.
78 Fairbairn, *Mills*, Vol. I, p. 258.
79 Patent 13,552, 11 March 1851 and D. K. Clark, *The Steam Engine: A Treatise on Steam Engines and Boilers* (London 1890) pp. 705ff.
80 Armstrong, *Rudimentary Treatise*, pp. 72–3.
81 Patent 10,986, 1845.
82 Fairbairn, *Mills*, Vol. I, p. 271 and M. McLean, *Modern Electric Practice* (London no date) p. 203.
83 Dickinson & Titley, *Trevithick*, p. 61 and Hoblyn, *Manual*, p. 81.
84 W. H. Chaloner, *National Boiler, 1865–1964. A Century of Progress in Industrial Safety* (Manchester 1964) pp. 1–2.
85 J. P. Muirhead, *The Origin and Progress of the Mechanical Inventions of James Watt* (London 1854) Vol. III, p. 115.
86 Armstrong, *Rudimentary Treatise*, p. 144 and G. Bauer & L. S. Robertson, *Marine Engines and Boilers, Their Design and Construction* (London 1905) pp. 456ff.
87 Trevithick, *Trevithick*, Vol. II, p. 143.
88 Armstrong, *Rudimentary Treatise*, p. 144, Bourne, *Treatise*, pp. 49ff. and Williams, *Combustion*, p. 127.
89 W. Fairbairn, *Report on the Consumption of Fuel and the Prevention of Smoke* (London 1845) p. 117.
90 See Williams, *Combustion*, and a series of lectures he delivered at the Victoria Gallery Conversazione, Manchester, 1841, 'On the Combustion of Smoke' and Clark, *Steam Engine*, pp. 157ff.
91 Thurston, *Manual*, pp. 189–91.
92 Armstrong, *Rudimentary Treatise*, p. 144.
93 R. L. Hills & D. Patrick, *Beyer, Peacock, Locomotive Builders to the World* (Glossop 1982) p. 32.
94 D. K. Clark, *Railway Locomotives and their Progress, Mechanical Construction and Performance* (Glasgow 1860) Vol. I, p. 126.
95 Armstrong, *Rudimentary Treatise*, p. 42 and Tredgold, *Steam Engine*, Vol. I, p. 125. "It was justly remarked by Mr Watt, that the sole object of the arrangement of his boilers 'was to economise the fuel as much as possible. It is not the shallowness or depth of the boiler that produced this effect; but the making of the boilers of such a shape that the air which passes through the fire shall be robbed of almost all its heat before it can make its escape' ".
96 Hoblyn, *Manual*, p. 90.
97 Fairbairn, *Mills*, Vol. I, p. 255.
98 Armstrong, *Rudimentary Treatise*, p. 42.

99 Thurston, *Manual*, p. 219.
100 Cornwall, *Railways*, p. 34.
101 Jenkins, *Collected Papers*, p. 129.
102 Armstrong, *Rudimentary Treatise*, p. 97.
103 Bourne, *Treatise*, p. 229.
104 Armstrong, *Rudimentary Treatise*, p. 97 and Jenkins, *Collected Papers*, p. 130.
105 Armstrong, *Rudimentary Treatise*, p. 97 and E. J. A. Kenny & R. L. Hills, 'The Steam Pumping Engine at Stretham, Cambridgeshire', *T.N.S.*, Vol. 36, 1963–4, p. 19.
106 R. J. Law, *The Steam Engine* (London 1965) p. 31.
107 Jevons, *Coal Question*, p. 295.

Chapter 8

1 E. Alban, *The High-Pressure Steam Engine. An Exposition of Its Comparative Merits and an Essay towards an Improved System of Construction* (London 1848) p. xii.
2 A. W. Skempton, 'Engineering in the Port of London, 1789–1808', *Transactions of the Newcomen Society*, Vol. 50, 1978–9, p. 93 and W. J. M. Rankine, *A Manual on the Steam Engine and Other Prime Movers* (London 1861) p. xxvi.
3 M. R. Bailey, 'Robert Stephenson & Company, 1823–1836', M.Sc. Thesis, Newcastle-upon-Tyne, 1984, p. 22.
4 *Ibid.*, p. 64.
5 J. Walker, *Report to the Directors on the Comparative Merits of Loco-motive and Fixed Engines, as a Moving Power* (London 1829).
6 Bailey, 'Stephenson', p. 117.
7 *Ibid.*, Appendix V.
8 *Ibid.*, p. 311.
9 I. Brunel, *I. K. Brunel* (London 1870) p. 245.
10 E. Corlett, *The Iron Ship: The History and Significance of Brunel's 'Great Britain'* (Bradford-on-Avon 1974) pp. 69ff.
11 Brunel, *Brunel*, pp. 308–9.
12 L. T. C. Rolt, *Isambard Kingdom Brunel. A Biography* (London 1957) pp. 240 & 248.
13 J. Guthrie, *A History of Marine Engineering* (London 1971) p. 119.
14 Sir J. A. Ewing, *The Steam-Engine and Other Heat-Engines* (Cambridge 1926) p. 566 and D. S. L. Cardwell, 'Some Further Thoughts on the History of the Heat Engine' (typescript).
15 H. Evers, *Steam and the Steam Engine, Land, Marine and Locomotive* (London, 4th edn 1880) p. 147.
16 J. G. Winton, *Modern Steam Practice and Engineering* (London 1885) p. 491 and D. S. L. Cardwell & R. L. Hills, 'Thermodynamics and Practical Engineering in the Nineteenth Century', *History of Technology*, Vol. 1, 1976, p. 10.
17 R. H. Thurston, *A History of the Growth of the Steam Engine* (London 1878, 1939 edn) p. 394.
18 D. K. Clark, *The Steam Engine. A Treatise on Steam Engines and Boilers* (London 1890) p. 647.
19 D. K. Clark, *The Exhibited Machinery of 1862* (London 1864) p. 358.
20 Winton, *Steam Practice*, p. 494.
21 Clark, *Exhibited*, p. 358.
22 D. S. L. Cardwell, *From Watt to Clausius. The Rise of Thermodynamics in the Early Industrial Age* (London 1971) p. 78 and J. Farey, *A Treatise on the Steam Engine, Historical, Practical and Descriptive* (London 1827, reprint Newton Abbot 1971) Vol. I, p. 384.
23 Patent 1,298, 1781 and H. W. Dickinson, *A Short History of the Steam Engine* (Cambridge 1938) p. 78.

24 Farey, *Steam Engine*, Vol. I, p. 384 and J. Tann, 'Mr. Hornblower and His Crew, Steam Engine Pirates in the Late 18th. Century', *T.N.S.*, Vol. 51, 1979–80, pp. 96 & 105.
25 Farey, *Steam Engine*, Vol. I, p. 390.
26 A. Rees, *The Cyclopaedia; or Universal Dictionary of Arts, Sciences and Literature* (London 1819, reprint Newton Abbot 1972) Vol. V, p. 114.
27 Patent 2,772, 1804. Another patentee for a compound engine, with cylinders superimposed, was Willis Earle, Patent 2,836, 1805, in which he claimed double the power of a Boulton & Watt engine with steam eight times as strong as common steam.
28 Patent 2,772, 1804.
29 T. R. Harris, *Arthur Woolf, The Cornish Engineer, 1766–1837* (Truro 1966) pp. 26ff.
30 *Ibid.*, pp. 40ff.
31 Patent 2,772, 1804, Farey, *Steam Engine*, Vol. II, p. 55 and Rees, *Cyclopaedia*, Vol. V, pp. 116–17.
32 M. A. Alderson, *An Essay on the Nature and Application of Steam, with an Historical Note on the Rise and Progressive Improvement of the Steam Engine* (London 1834) p. 56.
33 Ewing, *Steam Engine*, p. 215.
34 Farey, *Steam Engine*, Vol. II, pp. 90 & 304.
35 *Ibid.*, pp. 94 & 123.
36 Patent 1,620, 5 May 1873.
37 See Galloway Drawings, Order 4880.
38 W. Fairbairn, *Treatise on Mills and Millwork* (London 1861) p. 69.
39 Patent 11,001, 1845.
40 *Ibid.*
41 R. S. Burn, *The Steam Engine, Its History and Mechanism* (London 1854) p. 69.
42 A. Rigg, *A Practical Treatise on the Steam Engine* (London 1888) p. 292.
43 Patent 12,988, 1850.
44 Patent 862, 13 March 1868.
45 See also E. Galloway, Patent 13,545, 10 March 1851 and J. Bourne, Patent 2,872, 9 December 1853.
46 T. W. Croft, *Steam-Engine Principles and Practice* (New York 1922) p. 260.
47 G. C. V. Holmes, *The Steam Engine* (London 1888), p. 480.
48 Clark, *Exhibited*, p. 313.
49 Ewing, *Steam-Engine*, p. 216 and Holmes, *Steam Engine*, pp. 461 & 464.
50 J. Rose, *Modern Steam Engines* (Philadelphia 1893) p. 239.
51 Ewing, *Steam-Engine*, p. 360.
52 W. H. Chaloner, *National Boiler, 1864–1964. A Century of Progress in Industrial Safety* (Manchester 1964) pp. 1 & 11.
53 W. Minchinton, 'The Energy Basis of the British Industrial Revolution' (typescript 1985) from J. Kanefsky, 'The Diffusion of Power Technology in British Industry 1760–1870', Ph.D. Thesis, University of Exeter, 1979, p. 126.

Chapter 9

1 R. S. Burn, *The Steam Engine, Its History and Mechanism* (London 1854) p. v.
2 A. Rees, *The Cyclopaedia; or Universal Dictionary of Arts, Sciences and Literature* (London 1819, reprint Newton Abbot 1972) Vol. V, p. 114.
3 J. Farey, *A Treatise on the Steam Engine, Historical, Practical and Descriptive* (London 1827, reprint Newton Abbot 1971) Vol. I, pp. 71–4.
4 W. Fairbairn, *Treatise on Mills and Millwork* (London 1861) p. 237.
5 Farey, *Steam Engine*, Vol. I, p. 399.
6 G. de Pambour, *A New Theory of the Steam Engine* (London 1838) p. 12.

7 Fairbairn, *Mills*, p. 237.
8 E. Alban, *The High Pressure Steam Engine. An Exposition of Its Comparative Merits and an Essay towards an Improved System of Construction* (London 1848) p. 34.
9 *Ibid.*, p. 34.
10 M. A. Alderson, *An Essay on the Nature and Application of Steam, with an Historical Note of the Rise and Progressive Improvement of the Steam Engine* (London 1834) pp. 10–14.
11 T. M. Goodeve, *Text-Book on the Steam Engine* (London 1884) p. 6.
12 D. S. L. Cardwell, 'Power Technologies and the Advance of Science, 1700–1825', *Technology & Culture*, Vol. 6, No. 2, Spring 1965, p. 198 and P. Lervig, 'Sadi Carnot and the Steam Engine: Nicolas Clément's Lectures on Industrial Chemistry, 1823–28', *British Journal for the History of Science*, Vol. 18, Part 2, No. 59, July 1985, p. 150.
13 S. Carnot, *Reflections on the Motive Power of Fire* (1824, E. Mendoza, ed., New York 1960) pp. 48–9.
14 D. S. L. Cardwell, *From Watt to Clausius, The Rise of Thermodynamics in the Early Industrial Age* (London 1971), p. 193 and Lervig, 'Sadi Carnot', p. 148.
15 Carnot, *Reflections*, p. 50.
16 E. Claperyon, *Memoir on the Motive Power of Heat* (1834) (see Carnot, *Reflections*, E, Mendoza, ed., New York 1960) p. 103.
17 Goodeve, *Text-Book*, p. 52 and A. Jamieson, *Elementary Manual on Steam and the Steam Engine* (London 1898) pp. 51–2.
18 Goodeve, *Text-Book*, p. 52 and Jamieson, *Manual*, p. 52.
19 Cardwell, *Watt*, p. 233 and G. W. Sutcliffe, *Steam Power and Mill Work, Principles and Modern Practice* (London 1895) p. 16.
20 Goodeve, *Text-Book*, p. 61.
21 W. J. M. Rankine, *A Manual of the Steam Engine and Other Prime Movers* (London 1861) p. 299.
22 Cardwell, *Watt*, p. 235.
23 A. J. Pacey, 'Some Early Heat Engine Concepts and the Conservation of Heat', *British Journal for the History of Science*, Vol. 7, Part 2, No. 26, July 1974, pp. 138 & 40.
24 E. E. Daub, 'The Regenerator Principle in the Sterling and Ericsson Hot Air Engines', *British Journal for the History of Science*, Vol. 7, Part 3, No. 27, November 1974, p. 272.
25 D. S. L. Cardwell & R. L. Hills, 'Thermodynamics and Practical Engineering in the Nineteenth Century', *History of Technology*, Vol. 1, 1976, pp. 5–6.
26 D. S. L. Cardwell, 'Towards the Comprehension of the Steam Engine' (typescript), p. 16.
27 Sir J. A. Ewing, *The Steam-Engine and Other Heat-Engines* (Cambridge 1926) p. 35.
28 R. Fox, 'Science, Industry and the Social Order in Mulhouse, 1798–1871', *British Journal for the History of Science*, Vol. 17, Part 2, No. 56, July 1984, p. 157.
29 Cardwell, 'Comprehension', p. 19.
30 Cardwell & Hills, 'Thermodynamics', p. 12.
31 Cardwell, 'Comprehension', p. 17.
32 Goodeve, *Text-Book*, p. 1.

Chapter 10

1 J. Farey, *A Treatise on the Steam Engine, Historical, Practical and Descriptive* (London 1827, reprint 1971) Vol. I, p. 486.
2 Patent 9,571, 22 December 1842.
3 W. H. Uhland, *Corliss-Engines and Allied Steam Motors . . .* (London 1879) Vol. II, p. 99.

4 E. Leigh, *The Science of Modern Cotton Spinning* (Manchester, 3rd edn 1875) Vol. II, p. 196.
5 R. H. Thurston, *A History of the Growth of the Steam-Engine* (New York 1878, Centennial edn 1939) p. 311.
6 J. H. White, *A History of the American Locomotive* (New York 1968, reprint 1979) p. 203.
7 G. B. Williamson, 'Steam Engine Building in Rochdale', *Transactions of the Rochdale Literary & Scientific Society*, Vol. 22, 1944–46, pp. 11 & 13.
8 Patent 10,193, 22 May 1844 and Patent 12,988, 1850.
9 Patent 12,244, 17 August 1848.
10 Patent 12,105, 22 March 1848.
11 Patent 12,960, 1850.
12 Patent 1,194, 14 May 1853.
13 Patent 3,841, 7 November 1874.
14 R. S. Burn, *The Steam-Engine, Its History and Mechanism* (London 1854) p. 110.
15 A Rees, *The Cyclopaedia; or Universal Dictionary of Arts, Sciences and Literature* (London 1819, reprint 1972) Vol. V, p. 76.
16 Farey, *Steam Engine*, Vol. II, p. 277.
17 *Ibid.*, Vol. I, p. 469.
18 Hick Hargreaves Records, 'Typical Engines Built by Hick Hargreaves & Co. Ltd'.
19 Farey, *Steam Engine*, Vol. I, p. 530.
20 W. Fairbairn, 'On the Expansive Action of Steam and a New Construction of Expansion Valves for Condensing Steam Engines', *Proceedings of the Institution of Mechanical Engineers*, 1 July 1849, p. 22.
21 *Ibid.*, p. 28.
22 W. Fairbairn, *Treatise on Mills and Millwork* (London 1861) Vol. I, pp. 233 & 235.
23 Leigh, *Science*, Vol. I, p. 26.
24 Uhland, *Corliss Engines*, Vol. I, p. 43.
25 E. Alban, *The High-Pressure Steam Engine. An Exposition of its Comparative Merits and an Essay towards an Improved System of Construction* (London 1848) p. 236.
26 J. Rose, *Modern Steam Engines* (London, 3rd edn 1893) p. 144.
27 D. K. Clark, *The Steam Engine: A Treatise on Steam Engines and Boilers* (London 1890) Vol. II, pp. 39ff., G. G. Phillips, 'The Early History of the Corliss Engine in the United States', paper read to the *Engine Builders' Association of the United States*, 1 December 1902, p. 3 and J. L. Wood, 'The Introduction of the Corliss Engine to Britain', *Transactions of the Newcomen Society*, Vol. 52, 1980–1, p. 1.
28 U.S. Patent 6,192, 10 March 1849.
29 *Ibid.*, p. 1.
30 *Ibid.*, p. 3 and Clark, *Steam Engine*, Vol. II, p. 40.
31 Smithsonian Steam Engine Archives, photograph.
32 D. K. Clark, *The Exhibited Machinery of 1862* (London 1864) p. 318.
33 U.S. Patent 8,253, 29 July 1851.
34 *Ibid.*, p. 2.
35 Clark, *Exhibited*, p. 318.
36 U.S. Patent 24,618, 5 July 1859.
37 Clark, *Exhibited*, p. 318.
38 Clark, *Steam Engine*, Vol. II, p. 42.
39 *Ibid.*, p. 42 and Uhland, *Corliss Engines*, Vol. I, p. 2.
40 Wood, 'Introduction', p. 4.
41 Uhland, *Corliss Engines*, Vol. I, pp. 26 & 27.
42 Clark, *Exhibited*, p. 318.
43 J. Bourne, *Recent Improvements in the Steam-Engine* (London 1869) Fig. 38.

44 *Ibid.*, p. 110.
45 Wood, 'Introduction', pp. 3–5.
46 Patent 652, 9 March 1863.
47 Patent 82, 11 January 1865.
48 *Ibid.*
49 *Ibid.*
50 Wood, 'Introduction', p. 7.
51 H. W. Dickinson, *A Short History of the Steam Engine* (Cambridge 1938) p. 138 and A. Rigg, *A Practical Treatise on the Steam Engine* (London 1888) pp. 289–90.
52 J. T. Henthorn, *The Corliss Engine* (London 1897) p. 4.
53 Patent 2,713, 26 September 1867, see also Patent 481, 17 February 1869. The earliest record details a double-crank compound engine, No. 103, supplied in 1875 to C. Shorrock & Co., with Corliss valves on the HP cylinder and slide valves on the low, speed 56 r.p.m.
54 Patent 1,377, 5 May 1869 and Patent 1,972, 1876, Patent 2,085, 17 May 1876 and Patent 4,293, 6 September 1883.
55 B. Goodfellow Order 800, 2 May 1877, for a horizontal compound engine for Robert McClure & Sons. See Clark, *Steam Engine*, Vol. II, p. 255 for drawings of an engine by Goodfellows.
56 J. & W. McNaught Order 873, 1892, Ellen Road Spinning Co.
57 Galloway Order 4,075, 1896.
58 Patent 2,965, 25 July 1878 and J. Nasmith, *Recent Cotton Mill Construction and Engineering* (Manchester, 2nd edn *c.* 1900) p. 180.
59 Patent 11,295, 22 September 1885 and Patent 17,135, 29 October 1889.
60 H. W. Fowler, *Stationary Steam Engines* (Manchester *c.* 1908) p. 40.
61 H. Haeder, *A Handbook on the Steam Engine with Especial Reference to Small and Medium-Sized Engines* (London, 2nd edn 1896) pp. 252–9.
62 Patent 3,207, 5 November 1869.
63 Uhland, *Corliss Engines*, p. 213.
64 Harrisburg Foundry & Machine Works, *Catalogue* (Harrisburg, Pennsylvania 1900) pp. 52–3.
65 Allis-Chalmers Manufacturing Co. (Milwaukee, Wisconsin) *Bulletin 1529-B*, April 1925.
66 Farey, *Steam Engine*, Vol. II, p. 14.
67 Dickinson, *Short History*, p. 110.
68 J. Evans, *The Endless Web. John Dickinson & Co. Ltd, 1804–1954* (London 1955) p. 96.
69 A. E. Seaton, *A Manual of Marine Engineering* (London, 7th edn 1888), p. 9.
70 Patent 466, 25 February 1854.
71 E. C. Smith, *A Short History of Naval and Marine Engineering* (Cambridge 1937) pp. 146–7.
72 G. Watkins, *The Textile Mill Engine* (Newton Abbot 1970) Vol. I, p. 48 and Williamson, 'Rochdale', p. 14.
73 Haeder, *Handbook*, p. 18.
74 Watkins, *Mill Engine*, Vol. I, p. 32.

Chapter 11

1 U.S. Patent 20,894, 13 July 1858.
2 O. Mayr, 'Yankee Practice and Engineering Theory; Charles T. Porter and the Dynamics of the High-Speed Engine', *Technology and Culture*, Vol. 16, No. 4, October 1975, p. 576.
3 U.S. Patent 14,991, 27 May 1856.
4 U.S. Patent 35,068, 29 April 1862.
5 U.S. Patent 35,069, 29 April 1862.
6 Mayr, 'Yankee Practice', pp. 572–3.

7 *Ibid.*, pp. 570–2 and C. T. Porter, *Engineering Reminiscences* (New York 1908, reprint Bradley 1985) p. 5.
8 H. W. Dickinson, *A Short History of the Steam Engine* (Cambridge 1938) p. 142 and Porter, *Reminiscences*, pp. 14ff.
9 O. Mayr, *Feedback Mechanisms in the Historical Collections of the National Museum of History and Technology*, Smithsonian Institution Press, No. 12, 1971, p. 14.
10 See G. Watkins, *The Stationary Steam Engine* (Newton Abbot 1968) pictures 2, 3 & 47, G. Watkins, *The Steam Engine in Industry* (Ashborne 1978) Vol. I, p. 88 and G. Watkins, *The Textile Mill Engine* (Newton Abbot 1970) Vol. I, pp. 14, 18, 21 & 22.
11 U.S. Patent 20,894, 13 July 1858 and British Patent 1,422, 23 June 1858.
12 C. T. Porter, 'Improved Centrifugal Governor', Leaflet, 1859.
13 D. K. Clark, *The Steam Engine. A Treatise on Steam Engines and Boilers* (London 1890) Vol. II, p. 71 and A. Rigg, *A Practical Treatise on the Steam Engine* (London 1888) p. 207.
14 *Scientific American*, 9 October 1858, p. 36.
15 Porter, 'Governor'.
16 Mayr, 'Yankee Practice', p. 574.
17 G. Zeuner, *Treatise on Valve-Gears with Special Consideration of the Link-Motions of Locomotive Engines* (London 1884) pp. 152 & 165 and J. Rose, *Modern Steam Engines* (Philadelphia, 3rd edn 1893) pp. 28 & 145.
18 Zeuner, *Treatise*, p. 167, note.
19 D. K. Clark, *The Exhibited Maichinery of 1862* (London 1864) p. 321.
20 Rigg, *Practical Treatise*, pp. 99–100.
21 See Patent 2,199, 25 August 1866.
22 C. T. Porter, 'The Allen Engine', *Franklin Institution Journal*, Vol. 9, No. 6, December 1870, p. 385.
23 Z. Colburn, *Locomotive Engineering and the Mechanism of Railways* (London 1871) p. 51.
24 Porter, 'Allen Engine', p. 324.
25 *Ibid.*, p. 385.
26 Porter, *Reminiscences*, p. 160.
27 Patent 1,669, 21 June 1865.
28 Mayr, 'Yankee Practice', p. 585.
29 *Ibid.*, p. 577.
30 *Cassier's Magazine*, February 1903, p. 538.
31 *Proceedings of the Institution of Mechanical Engineers*, 1868 and Rigg, *Practical Treatise*, Plate 95.
32 Southwark Foundry Machine Co., *Catalogue* (Philadelphia *c.* 1900) p. 44.
33 Porter, *Reminiscences*, p. 221.
34 W. H. Uhland, *Corliss-Engines and Allied Steam-Motors* . . . (London 1879) Vol. II, p. 76.
35 Porter, 'Allen Engine', p. 249.
36 Mayr, 'Yankee Practice', pp. 585–601 for a full discussion of Porter's theory.
37 Porter, 'Allen Engine', pp. 251–2.
38 *Ibid.*, p. 252.
39 Porter, *Reminiscences*, p. 204.
40 Southwark Foundry, *Catalogue*, p. 48.
41 Rigg, *Practical Treatise*, p. 280.
42 *Ibid.*, p. 278.
43 Clark, *Steam Engine*, Vol. I, pp. 500–4 and Dickinson, *Short History*, p. 142.
44 Southwark Foundry, *Catalogue*, p. 52.
45 H. Haeder, *A Handbook of the Steam Engine with Especial Reference to Small and Medium-Sized Engines* (London, 2nd edn 1896) see pp. 133ff. for illustrations of various types of governors.

46 Sir J. A. Ewing, *The Steam-Engine and Other Heat Engines* (Cambridge, 4th edn 1926) p. 428.
47 Haeder, *Handbook*, p. 141.
48 W. H. Fowler, *Stationary Steam Engines* (Manchester *c.* 1908) p. 200 and G. W. Sutcliffe, *Steam Power and Mill Work* (London 1895) p. 563.
49 Haeder, *Handbook*, p. 142.
50 M. McLean, *Modern Electric Practice* (London, no date) p. 83.
51 T. W. Croft, *Steam-Engines Principles and Practice* (New York 1922) p. 228.
52 *Ibid.*, p. 209.
53 Patent 20,892, 1894.
54 Fowler, *Steam Engines*, pp. 191–2.
55 Sutcliffe, *Steam Power*, p. 551.
56 Uhland, *Corliss Engines*, p. 146.
57 Ewing, *Steam-Engine*, p. 443.
58 Patent 14,431, 9 November 1886 and Patent 13,871, 13 October 1887.
59 Patent 21,847, 3 December 1900.
60 L. Lumb, 'Governing and Regulating of Steam Engines', *Bolton Managers' and Overlookers' Association*, Lecture, 10 October 1908, pp. 4 & 17.
61 Sutcliffe, *Steam Power*, p. 552.
62 Patent 947, 21 January 1888.
63 Ewing, *Steam-Engine*, p. 444.
64 See L. C. Hunter, *A History of Industrial Power in the United States, 1780–1930* (Charlottesville 1979) Vol. I, Waterpower, p. 466.
65 W. Fairbairn, *Treatise on Mills and Millwork* (London 1861) Vol. II, p. 2 and Sutcliffe, *Steam Power*, p. 854.
66 J. & W. Wood, Order 106, 1877.
67 Galloway Drawing 3660, 4 August 1876, Order 1540.
68 *Ibid.*, Drawing 3895, 16 June 1877, Order 1640.
69 *Ibid.*, Drawing 5073, 10 March 1881, Order 2090.
70 W. Kenyon & Sons Ltd, *Ropes and Rope Driving* (Manchester 1924) pp. 8 & 9.
71 Sutcliffe, *Steam Power*, pp. 708 & 710.
72 Galloway Drawing 3932, 26 July 1877, Order 1655.
73 J. & W. McNaught, Order 709, 1881, Melland & Coward for the first rope drive and Order 1012, 1901, A. Brierley & Sons for the last spur drive.
74 Patent 1,741, 6 April 1883.
75 J. Musgrave & Sons Ltd, *Illustrated Catalogue of Vertical Quadruple Expansion Engines, Horizontal & Vertical Triple Expansion Engines, Compound Engines, etc.* (Bolton, no date) pp. 194–5.
76 Patent 3,694, 24 August 1881 and Patent 48, 3 January 1883.
77 Patent 3,867, 9 August 1833, Patent 2,926, 8 February 1884 and Patent 6,702, 23 April 1884.
78 Patent 4,387, 20 March 1890.
79 W. English, *The Textile Industry* (London 1969) p. 166.
80 E. Leigh, *The Science of Modern Cotton Spinning* (Manchester, 3rd edn 1875) Vol. II, p. 223.
81 H. Catling, *The Spinning Mule* (Newton Abbot 1970) p. 187.

Chapter 12

1 A. A. Jackson, *Volk's Railway Brighton, 1883–1968* (London 1968) p. 4.
2 R. H. Parsons, *The Early Days of the Power Station Industry* (Cambridge 1939) p. 12.
3 *Ibid.*, p. 53.
4 J. G. White, 'The Evolution of the Modern Power Station', *Street Railway Journal*, Vol. 24, No. 15, 8 October 1904, p. 556.

5 Allis-Chalmers Company, *Bulletin* No. 1503, 'Allis-Chalmers Reliance-Corliss Engines Direct Connected Type', May 1906, p. 3, and M. McLean, *Modern Electric Practice* (London, no date) p. 74.
6 White, 'Evolution', p. 556 and L. Bell, 'The Development of Direct-Connected Generators', *Street Railway Journal*, Vol. 24, No. 15, 8 October 1904, p. 588.
7 Smithsonian Steam Engine Archives, Washington, D.C., typescript by T. H. Fehring, 'Technological Contributions of Milwaukee's Menomonee Valley Industries', p. 7.
8 Patent 974, 19 March 1874.
9 Patent 1,572, 16 April 1880.
10 Patent 4,901, 14 October 1882.
11 Parsons, *Early Days*, p. 45.
12 Patent 13,769, 17 October 1884.
13 Parsons, *Early Days*, p. 90.
14 *Ibid.*, pp. 72, 107 & 103.
15 *Ibid.*, pp. 163–5.
16 Willans & Robinson Ltd, *The Willans Engine for Rope & Belt Driving Purposes* (Rugby, no date) p. 30.
17 *Ibid.*, p. 8.
18 Patent 7,397, 1890 and Patent 11,432, 1892. See H. W. Dickinson, *A Short History of the Steam Engine* (Cambridge 1938) p. 149, Sir A. J. Ewing, *The Steam-Engine and Other Heat-Engines* (Cambridge 1926) p. 558 and W. H. Fowler, *Stationary Steam Engines* (Manchester *c.* 1908) p. 267.
19 Patent 13,498, 17 June 1898 and Ashworth & Parker, *The Parker Engine* (Bury, no date) p. 2.
20 Ashworth, *Parker Engine*, pp. 19–20.
21 A. Ridding, *S. Z. de Ferranti, Pioneer of Electric Power* (London 1964) p. 5.
22 Parsons, *Early Days*, pp. 22–5.
23 J. E. Smart, *The Deptford Letter-books. An Insight on S. Z. de Ferranti's Deptford Power Station* (London 1976) p. 2.
24 *The Electrician*, 26 October 1888.
25 A. E. Seaton, *A Manual of Marine Engineering* (London, 7th edn 1888) pp. 457 & 9.
26 R. W. M. Clouston, 'The Development of the Babcock Boiler in Britain up to 1939', paper read to the Newcomen Society, 11 February 1987, p. 2 and British Patent 1,154, 27 March 1873.
27 Clouston, 'Babcock Boiler', p. 5.
28 Parsons, *Early Days*, p. 131.
29 *Ibid.*, p. 29.
30 See Smart, *Letter-books* for comments on these engines.
31 Greater Manchester Museum of Science and Industry Archives, J. & W. McNaught Collection, Order 1047.
32 Ferranti Archives, B21/1.7.
33 H. Scholey, 'Sebastian Ziani de Ferranti', *Cassier's Magazine*, February 1897, reprint p. 15.
34 'Deptford. A short history to commemorate the first high voltage central generating station in the world, conceived, designed and built by Sebastian Ziani de Ferranti', reprinted in 1969 from *Notes and Records of the Royal Society of London*, Vol. 19, No. 1, June 1964 and Ferranti Archives, B5/2.4.
35 Patent 1,902, 2 February 1889.
36 Ferranti Archives, Letter by S. Z. de Ferranti in December 1929 quoted in *The Engineer*, 16 June 1944.
37 C. Day, 'Engines for Driving Large Dynamos', *Manchester Association of Engineers*, paper read on 12 October 1901, pp. 2–3.
38 Ferranti Archives, B17/4.4.
39 Day, 'Engines', pp. 7 & 16.

40 Ferranti Archives, B5/2.4, p. 3.
41 Smithsonian Archives, note on reverse of photograph of Wakefield engine.
42 Allis-Chalmers Co., *Bulletin*, No. 1503, 'Allis-Chalmers Reliance-Corliss Engines, Direct-Connected Type', May 1906, p. 4.
43 Patent 2,374, 29 January 1898 and Patent 3,536, 22 February 1900.
44 Day, 'Engines', p. 7.
45 *Ibid.*, pp. 8–11.
46 Ferranti Archives, B5/2.4, p. 4.
47 Day, 'Engines', pp. 12–15.
48 Ferranti Archives, B5/2.4.
49 Parsons, *Early Days*, p. 117.
50 Ferranti Archives, B8/1.8.
51 Ferranti Archives, B5/2.3.
52 Ferranti Archives, B5/2.4.
53 Ferranti Archives, B5/2.3.
54 Scholey, 'Ferranti', pp. 14 & 15.
55 Parsons, *Early Days*, p. 127.
56 *Ibid.*, p. 167.
57 *Ibid.*, p. 116.
58 Smithsonian Archives, Allis-Chalmers, *Bulletin*, No. 1500, 1906.
59 Smithsonian Archives, Sulzer Co., *Catalogue*, 1908, pp. 7–8.
60 *Power*, Vol. 19, No. 4, April 1901.
61 *Scientific American*, 11 January 1902, p. 21.
62 *Street Railway Journal*, Vol. 19, No. 5, February 1902.
63 *American Electrician*, October 1904.
64 *Interborough Rapid Transit. The New York Subway, Its Construction and Equipment* (New York 1904) p. 95.
65 *Street Railway Journal*, 8 October 1904, p. 615.
66 Parsons, *Early Days*, p. 168.
67 Ferranti Archives, L7, Letter from J. Johnston McGreggor to W. E. W. Millington, 22 October 1952.
68 Ferranti Archives, *ibid.*, 29 October 1952.
69 *Power*, 20 April 1915, p. 547.
70 See Willans & Robinson Ltd, *Catalogue of Turbines* (Rugby *c.* 1906).

Chapter 13

1 D. Gurr & J. Hunt, *The Cotton Mills of Oldham* (Oldham 1985) p. 8.
2 G. Watkins, *The Textile Mill Engine* (Newton Abbot 1907) Vol. I, p. 48.
3 E. Leigh, *The Science of Modern Cotton Spinning* (Manchester, 3rd edn 1875) Vol. II, pp. 196–7.
4 Watkins, *Mill Engine*, Vol. I, pp. 32, 88 & 100.
5 Gurr & Hunt, *Cotton Mills*, these figures are taken from various pages.
6 E. Alban, *The High-Pressure Steam Engine. An Exposition of Its Comparative Merits and an Essay towards an Improved System of Combustion* (German edn 1843, translated London 1848) p. 221.
7 W. J. M. Rankine, *A Manual of the Steam Engine and Other Prime Movers* (London 1861) p. 395.
8 A. Rees, *The Cyclopaedia; or Universal Dictionary of Arts, Sciences and Literature* (London 1819, reprint Newton Abbot 1972) Vol. V, p. 105.
9 *Ibid.*, p. 395.
10 Alban, *High Pressure*, p. 221.
11 G. W. Sutcliffe, *Steam Power and Mill Work, Principles and Modern Practice* (London 1895) p. 304 and C. J. Galloway & J. H. Beckwith, Patent 16,917, 24 December 1884.
12 Alban, *High Pressure*, p. 221 and Rankine, *Manual*, p. 396.

13 C. H. Peabody, *Thermodynamics of the Steam-Engine and Other Heat-Engines* (London 1889) p. 304.
14 J. T. Henthorn, *The Corliss Engine* (New York 1897) p. 5.
15 W. H. Fowler, *Stationary Steam Engines* (Manchester *c.* 1908) p. 21.
16 R. Appleyard, *Charles Parsons. His Life and Work* (London 1933) p. 22 and Sir A. J. Ewing, *The Steam-Engine and Other Heat-Engines* (Cambridge 1926) p. 216.
17 Ewing, *Steam-Engine*, p. 29, W. H. Uhland, *Corliss Engines and Allied Steam-motors*... (London 1879) p. 91, D. Adamson, Patent 52, 9 January 1861, J. Elder, Patent 1,214, 25 April 1862 and W. Hargreaves & W. Inglis, Patent 11,307, 4 August 1888.
18 Appleyard, *Parsons*, pp. 109ff.
19 J. & E. Wood Archives, *Engine Book*.
20 F. Colyer, *A Treatise on Modern Steam Engines and Boilers* (London 1886) pp. 44–5 and Fowler, *Steam Engines*, p. ii.
21 Fowler, *Steam Engines*, p. 29.
22 H. Haeder, *A Handbook on the Steam Engine, with Especial Reference to Small and Medium-sized Engines* (London, 2nd edn 1896) p. 4.
23 J. & W. McNaught Drawings, Orders 892, 898 and 955.
24 B. Goodfellow Archives, details from Order Book.
25 J. Musgrave & Sons, *Reference Book*.
26 Uhland, *Corliss Engines*, p. 91 and D. Adamson, Patent 1,214, 25 April 1862.
27 See various patents of W. Y. Flemming and P. Ferguson, Patent 2,605, 19 February 1887, Patent 13,843, 3 September 1889, Patent 20,137, 10 December 1890 and Patent 3,165, 21 February 1891 and J. Musgrave & G. Dixon, Patent 15,395, 12 August 1893.
28 Rankine, *Manual*, p. 242.
29 Patent 5,344, 14 March 1828. See H. H. P. Powles, *Steam Boilers, Their History and Development* (London 1905) pp. 84–5.
30 Patent 8,583, 1 August 1840.
31 Patent 10,755, 21 July 1845.
32 Rankine, *Manual*, p. 429. For an illustration of marine use, see N. P. Burgh, *Practical Illustrations of Modern Land and Marine Engines*... (London 1864) Plate 17 for a marine boiler with superheating tubes.
33 Rankine, *Manual*, p. 435.
34 Ewing, *Steam Engine*, p. 210.
35 D. Adamson's Patent 14,259, 12 August 1852 and C. Normand's Patent 635, 17 March 1856.
36 R. H. Thurston, *A Treatise on Friction and Lost Work in Machinery and Mill Work* (New York, 1st edn 1885, 3rd edn 1889) pp. 118 & 141.
37 A. Rigg, *A Practical Treatise on the Steam Engine* (London 1888) p. 45.
38 D. K. Clark, *The Exhibited Machinery of 1862* (London 1864) p. 358.
39 Thurston, *Friction*, p. 133.
40 Sutcliffe, *Steam Power*, p. 794.
41 Ewing, *Steam-Engine*, p. 31.
42 R. H. Parsons, *The Early Days of the Power Station Industry* (Cambridge 1939) p. 175.
43 Electricity Council, *Electricity Supply in Great Britain. A Chronology from the Beginning of the Industry to 31 December 1976* (London 1977) p. 9 and Parsons, *Early Days*, p. 107.
44 *Engineering Record*, Vol. 51, No. 8, p. 209.
45 Babcock & Wilcox Co., *Steam, Its Generation and Use* (New York 1899) p. 87 and Patent 6,221, 1895.
46 Fowler, *Steam Engine*, p. ii.
47 T. W. Croft, ed., *Steam Engine Principles and Practice* (New York 1922) p. 417.
48 M. McLean, *Modern Electric Practice* (London, no date) p. 50.

49 See Chapter 12, p. 231.
50 J. Stumpf, *The Una-Flow Steam Engine* (London 1912) p. 7.
51 Croft, *Steam Engine*, p. 423.
52 Rankine, *Manual*, p. 465.
53 *Engineering*, 4 January 1867, 29 May 1868 and R. L. Hills & D. Patrick, *Beyer, Peacock, Locomotive Builders to the World* (Glossop 1982) p. 56.
54 D. K. Clark, *The Steam Engine: A Treatise on Steam Engines and Boilers* (London 1890) Vol. I, p. 689.
55 R. W. M. Clouston, 'The Development of the Babcock Boiler in Britain up to 1939', Newcomen Society Lecture, 11 February 1987, typescript, p. 5.
56 Colyer, *Treatise*, p. 46 and J. Musgrave & Sons, *Illustrated Catalogue of Vertical Quadruple Expansion Engines, Horizontal & Vertical Triple Expansion Engines, Compound Engines, Simple Engines, etc.* (Bolton, no date) p. 285.
57 Haeder, *Handbook*, p. 46
58 Harrisburg Foundry & Machine Works, *Catalogue* (Pennsylvania 1900) p. 57.
59 G. C. V. Holmes, *The Steam Engine* (London, 2nd edn 1888) p. 212.
60 Hills, *Beyer, Peacock*, p. 80.
61 A. Richardson, *The Evolution of the Parsons Steam Turbine* (London 1911) p. 83.
62 T. Allen, *Uniflow, Back-Pressure and Steam Extraction Engines. A Complete Treatise for Designers, Works Engineers and Students* (London 1931) p. 48.
63 Fowler, *Steam Engines*, pp. 266ff.
64 Patent 16,700, 17 August 1899.
65 Hills, *Beyer, Peacock*, p. 9.
66 Appleyard, *Parsons*, p. 22.
67 Patent 2,878, 27 July 1877.
68 Patent 5,175, 13 March 1894.
69 Fowler, *Steam Engines*, p. 280 and McLean, *Electric Practice*, p. 195.

Chapter 14

1 Skinner Engine Co., *Catalogue of Vertical Engines* (Erie, Pennsylvania, 1943) p. 3.
2 Patent 7,301, 1885.
3 J. Stumpf, *The Una-Flow Steam-Engine* (London 1912) p. iii.
4 Patent 2,132, 13 February 1886.
5 Patent 6,666, 6 May 1887.
6 *Ibid.*
7 Patent 7,301, 1885.
8 *Ibid.*
9 Patent 2,132, 13 February 1886.
10 Stumpf, *Una-Flow*, p. iv.
11 *Ibid.*, p. iv and H.-J. Braun, 'The National Association of German-American Technologists and Technology Transfer between Germany and the United States, 1884–1930', *History of Technology*, Vol. 8, 1983, p. 21.
12 A. Throp, 'Some Notes on the History of the Uniflow Steam Engine', *Transactions of the Newcomen Society*, Vol. 43, 1970–1, p. 20.
13 Throp, 'Notes', p. 22 and H. Pilling, 'The Uniflow Steam Engine', *The Engineer*, Vol. 129, 30 April 1920, p. 456.
14 T. Allen, *Uniflow Back-Pressure and Steam Extraction Engines. A Complete Treatise for Designers, Works Engineers and Students* (London 1931) p. 183.
15 Stumpf, *Una-Flow*, p. 15.
16 Patent 8,371, 15 April 1908.
17 Patent 26,020, 2 December 1908.
18 Patent 8,372, 15 April 1908 and Patent 21,174, 15 April 1908.
19 Patent 8,380, 15 April 1908.

20 Patent 26,020, 2 December 1908.
21 Pilling, 'Uniflow', p. 456.
22 *Ibid.*, p. 456 and Throp, 'Notes', p. 24.
23 *Ibid.*, p. 26 and 'British Empire Exhibition, Engineering Exhibits', *The Engineer*, Vol. 138, 4 July 1924, p. 228.
24 Robey & Co. Ltd, *Instructions for Erecting and Working the 'Robey' Patent Uniflow Engine* (Lincoln, no date) p. 21 and Skinner Engine Co., *The Universal Unaflow Steam Engine* (Erie, Pennsylvania 1927) p. 20.
25 Stumpf, *Una-flow*, p. iv and Throp, 'Notes', p. 19.
26 Stumpf, *Una-flow*, p. v.
27 F. B. Perry, 'The Uniflow Steam-Engine', *Proceedings of the Institution of Mechanical Engineers*, 1922, p. 733.
28 T. Allen, *Uniflow*, p. 176.
29 J. Musgrave & Sons Ltd, 'Reference Book'.
30 Hick Hargreaves, 'List of Central Exhaust Engines made by Hick Hargreaves & Co. Ltd'.
31 Throp, 'Notes', p. 38.
32 Galloway Archives, Order 7860, Messrs Barratt & Co., Wood Green, London, 25 September 1914.
33 Throp, 'Notes', p. 32.
34 Allen, *Uniflow*, pp. 412–16 and Robey, *Instructions*, p. 20.
35 Galloways Ltd, *Uniflow Steam Engines and 'Heat Extraction' Engines for Combined Power and Heating* (Manchester, no date) p. 19, *The Engineer*, Vol. 132, 26 August 1921, p. 223 and '800 H.P. Heat-Extraction Uniflow Compound Engine, Galloways', *Engineering*, Vol. 125, No. 3244, 16 March 1928, p. 318.
36 Skinner Engine Co., *Universal*, pp. 6 & 8.
37 Braun, 'Technology Transfer', p. 22.
38 Stumpf, *Una-flow*, p. 84.
39 Skinner Engine Co., *Bulletin*, No. 109 (Erie, Pennsylvania 1923) p. 2.
40 Skinner Engine Co., *Universal*, pp. 16–17.
41 Perry, 'Uniflow', pp. 745–6.
42 Allen, *Uniflow*, p. 454.
43 Pilling, 'Uniflow', p. 481.
44 Stumpf, *Una-flow*, p. 19.
45 *Ibid.*, p. 20.
46 Sir J. A. Ewing, *The Steam-Engine and Other Heat-Engines* (Cambridge, 4th edn 1926) p. 480.
47 J. Musgrave & Sons Ltd, *The 'Unaflow' Patent Steam Engine* (Bolton 1910) p. 13.
48 Galloways Ltd, *Uniflow Steam Engines* (Manchester 1920) p. 16.
49 Galloways, *Uniflow and Heat Extraction*, p. 27.
50 Pilling, 'Uniflow', p. 456.
51 Perry, 'Uniflow', p. 734.
52 *Ibid.*, p. 750.
53 Patent 152,863, October 1919, by Galloways Ltd and H. Pilling.
54 *The Engineer*, Vol. 132, 26 August 1921, p. 223.
55 Stumpf, *Una-flow*, p. 88.
56 Patent 2,132, 13 February 1886.
57 *Ibid.*
58 T. W. Croft, ed., *Steam Engine Principles and Practice* (New York 1922) p. 286 and *Engineering*, 10 November 1922, reprinted by Galloways Ltd.
59 Allen, *Uniflow*, p. 548 and *Engineering*, Vol. 125, No. 3244, 16 March 1928, p. 318.
60 Allen, *Uniflow*, p. 45.
61 *The Engineer*, Vol. 129, 7 May 1920, p. 481.
62 Stumpf, *Una-flow*, pp. 173 & 203.

63 Skinner Engine Co., *Marine Unaflow Steam Engines* (Erie, Pennsylvania 1947) p. 11.
64 *Ibid.*, pp. 9–11.
65 Skinner Engine Co., *The 'Skinner' Compound Unaflow Marine Steam Engine* (Erie, Pennsylvania, no date) p. 3.
66 *Ibid.*, pp. 4–5.
67 *Ibid.*, p. 5.
68 *The Times*, Wednesday 23 October 1985.

Chapter 15

1 G. Watkins, *The Textile Mill Engine* (Newton Abbot 1971) Vol. II, p. 94.
2 Patent 6,734, 23 April 1884. See R. Appleyard, *Charles Parsons, His Life and Work* (London 1933) pp. 69f. For the history of earlier attempts, see R.M. Neilson, *The Steam Turbine* (London 1902) pp. 7–37.
3 N. C. Parsons, 'The Origins of the Steam Turbine and the Importance of 1884', *Transactions of the Newcomen Society*, Vol. 56, 1984–5, pp. 25–6.
4 *Ibid.*, p. 26.
5 Appleyard, *Parsons*, pp. 42 & 40.
6 *Ibid.*, p. 34 and Patent 7,143, 1889.
7 Sir J. A. Ewing, *The Steam-Engine and Other Heat-Engines* (Cambridge, 4th edn 1926) p. 239 and Neilson, *Turbine*, pp. 90f.
8 Ewing, *Steam-Engine*, p. 240.
9 A. Richardson, *The Evolution of the Parsons Steam Turbine* (London 1911) pp. 20 & 30.
10 *Ibid.*, pp. 51 & 57 and Patent 1,120, 1890.
11 Appleyard, *Evolution*, p. 85.
12 R. H. Parsons, *The Early Days of the Power Station Industry* (Cambridge 1939) p. 175.
13 Richardson, *Evolution*, p. 58 and Patent 840, 1902.
14 Richardson, *Evolution*, pp. 7–8.
15 Parsons, *Early Days*, p. 175.
16 Ewing, *Steam-Engine*, p. 211 and Neilson, *Turbine*, p. 88.
17 Richardson, *Evolution*, p. 7.
18 Appleyard, *Parsons*, p. 50, Neilson, *Turbine*, p. 130 and Richardson, *Evolution*, p. 63.
19 Richardson, *Evolution*, pp. 198–9.
20 Appleyard, *Parsons*, p. 104, see also F. E. C. Jarrett, 'Application of the Steam Turbine to Marine Propulsion', *T.N.S.*, Vol. 56, 1984–5, pp. 38–44.
21 Neilson, *Turbine*, p. 142 and Richardson, *Evolution*, p. 100.
22 Ewing, *Steam-Engine*, p. 32.
23 Appleyard, *Parsons*, p. 198.
24 R. L. Hills & K. A. Barlow, *The Changing Face of Engineering* (Milton Keynes 1984) for a fuller discussion of the development of machine tools.
25 Richardson, *Evolution*, p. 177.
26 R. H. Parsons, *The Development of the Parsons Steam Turbine* (London 1936) p. 354.
27 Scott & Hodgson, see drawing 9948, Order 140, 23 September 1914.
28 Watkins, *Textile Mill Engine*, Vol. I, p. 12.
29 Appleyard, *Parsons*, p. 57.

Bibliography

Books

Aikin, J., *A Description of the Country from Thirty to Forty Miles Round Manchester*, J. Stockdale, Manchester 1795, reprint David & Charles, Newton Abbot 1968.
Alban, E., *The High Pressure Steam Engine. An Exposition of Its Comparative Merits and an Essay towards an Improved System of Construction*, trans. W. Pole, J. Weale, London 1848.
Alderson, M. A., *An Essay on the Nature and Application of Steam, with an Historical Note of the Rise and Progressive Improvement of the Steam Engine*, Sherwood, Gilbert & Piper, London 1834.
Allen, T., *Uniflow, Back-Pressure and Steam Extraction Engines. A Complete Treatise for Designers, Works Engineers and Students*, Pitman, London 1931.
Appleyard, R., *Charles Parsons. His Life and Work*, Constable, London 1933.
Arago, M., *Historical Eloge of James Watt*, J. Murray, London 1839.
Armstrong, R., *A Practical Essay on Steam Engine Boilers erected at the East London Waterworks*, J. & J. Thomson, Manchester 1838.
A Rudimentary Treatise on Steam Boilers: their Construction and Practical Management, J. Weale, London, 3rd edn 1857.
Baines, E., *History of the Cotton Manufacture in Great Britain*, H. Fisher, R. Fisher & P. Jackson, London 1835.
Banks, J., *A Treatise on Mills in Four Parts*, W. Richardson, London 1795.
On the Power of Machines, W. Pennington, Kendal 1803.
Barton, D. B., *The Cornish Beam Engine*, D. Bradford Barton, Truro 1965, new edn 1969.
Bauer, G. & Robertson, L. S., *Marine Engines and Boilers, Their Design and Construction*, Crosby Lockwood, London 1905.
Birkbeck, G. & Adcock, H. & J., *The Steam-Engine Theoretically and Practically Displayed*, J. Murray, London 1827.
Blakey, W., *A Short Historical Account of the Invention, Theory and Practice of Fire Machinery; or Introduction to the Art of Making Machines, Vulgarly Called Steam-Engines*, London 1793.
Bourne, J., *A Catechism of the Steam Engine*, Longmans, Green, London, new edn 1876.
Recent Improvements in the Steam Engine, Longmans, Green, London, new edn 1880.
A Treatise on the Steam Engine in Its Application to Mines, Mills, Steam Navigation and Railways, Longmans, Green, London 1846.
Browne, C. L., *The Fitting and Erecting of Engines*, Emmett, Manchester, 2nd edn 1918.
Brunel, I., *I. K. Brunel*, 1870, David & Charles, Newton Abbot, reprint 1971.
Buchanan, R., *Practical Essays on Millwork and Other Machinery*, J. Weale, 3rd edn, revised by G. Rennie, London 1841.
Buchanan, R. A. & Watkins, G., *The Industrial Archaeology of the Stationary Steam Engine*, Allen Lane, London 1976.
Burgh, N. P., *Practical Illustrations of Modern Land and Marine engines, showing in detail the improvements in high and low pressure, ordinary and surface condensing; together with Cornish Land and Superheating Marine Boilers*, E. & F. N. Spon, London 1864.

A Practical Treatise on Boilers and Boiler-Making, London, revised edn 1881.
Practical Treatise on the Condensation of Steam, E. & F. N. Spon, London 1871.
Burn, R. S., *The Steam-Engine. Its History and Mechanism*, H. Ingram, London 1854.
Cardwell, D. S. L., *From Watt to Clausius. The Rise of Thermodynamics in the Early Industrial Age*, Heinemann, London 1971.
Technology and Change, 1750–1914, Open University Technological Essay, Milton Keynes 1985.
Carnot, S., *Reflections on the Motive Power of Fire*, Paris 1824, E. Mendoza, ed., New York 1960.
Catling, H., *The Spinning Mule*, David & Charles, Newton Abbot 1970.
Chaloner, W. H., *National Boiler, 1864–1964: A Century of Progress in Industrial Safety*, National Boiler, Manchester 1964.
Clark, D. K., *Railway Locomotives, and Their Progress, Mechanical Construction and Performance*, Blackie, Glasgow 1860.
The Exhibited Machinery of 1862, Day & Son, London 1864.
The Steam Engine. A Treatise on Steam Engines and Boilers, Blackie, London 1890.
Clark, S., *Chorlton Mills and Their Neighbourhood. A Background Study to Manchester Industrial Archaeology*, private publication, Manchester 1979.
Colburn, Z., *Locomotive Engineering and the Mechanism of Railways*, W. Collins, London 1871.
Colyer, F., *A Treatise on Modern Steam Engines and Boilers*, E. & F. N. Spon, London 1886.
Corlett, E., *The Iron Ship. The History and Significance of Brunel's "Great Britain"*, Moonraker, Bradford-on-Avon 1974.
Cornwell, E. L., ed., *The Pictorial Story of Railways*, Hamlyn, London 1974.
Croft, T. W. ed., *Steam Engine Principles and Practice*, McGraw-Hill Book Co., New York 1922.
Crowley, T. E., *The Beam Engine. A Massive Chapter in the History of Steam*, Senecio Pub., Oxford 1982.
Beam Engines, Shire Pub., no date.
Curr, J., *The Coal Viewer, and Engine Builder's Practical Companion*, J. Northall, Sheffield 1797.
Day, C., *Indicator Diagrams and Engine and Boiler Testing*, Technical Pub. Co., Manchester 1895.
Derry, T. K. & Williams, T. I., *A Short History of Technology from the Earliest Times to A.D. 1900*, Clarendon Press, Oxford 1960.
Desaguliers, J. T., *A Course of Experimental Philosophy*, London, 3rd edn 1743.
Dickinson, H. W., *Matthew Boulton*, Cambridge University Press, Cambridge 1936.
Sir Samuel Morland, Diplomat and Inventor, 1625–1695, Heffer, Cambridge 1970.
Thomas Newcomen, Engineer, 1663–1729, Newcomen Association, Dartmouth.
A Short History of the Steam Engine, Cambridge University Press, Cambridge 1938.
James Watt, Craftsman and Engineer, Cambridge University Press, Cambridge 1935.
Dickinson, H. W. & Jenkins, R., *James Watt and the Steam Engine*, 1st edn 1927, Moorland, Ashbourne, reprint 1981.
Dickinson, H. W. & Titley, A., *Richard Trevithick. The Engineer and the Man*, Cambridge University Press, Cambridge 1934.
Dunsheath, P., *A History of Electrical Engineering*, Faber & Faber, London 1962.
Electricity Council, *Electricity Supply in Great Britain. A Chronology from the Beginning of the Industry to 31 December 1976*, The Electricity Council, London 1977.
Evans, J., *The Endless Web. John Dickinson & Co., 1804–1954*, J. Cape, London 1955.
Evans, O., *The Young Mill-Wright and Miller's Guide*, Lea & Blanchard, Philadelphia, 13th edn 1850.
Evers, H., *Steam and the Steam Engine, Land, Marine and Locomotive*, W. Collins, London, 4th edn 1880.
Ewing, Sir J. A., *The Steam Engine and Other Heat-Engines*, Cambridge University Press, Cambridge, 4th edn 1926.

Fairbairn, W., *Report on the Consumption of Fuel and the Prevention of Smoke*, British Association for the Advancement of Science, R. & J. E. Taylor, London 1845.
Treatise on Mills and Millwork, Longman, Green, Longman & Roberts, London 1861.
Useful Information for Engineers, Longman, London, 4th edn 1864.
Farey, J. A., *A Treatise on the Steam Engine, Historical, Practical and Descriptive*, Vol. I, 1827, Vols. I and II, David & Charles, Newton Abbot 1971.
Fitton, R. S. & Wadsworth, A. P., *The Strutts and the Arkwrights, 1758–1830. A Study of the Early Factory System*, Manchester University Press, Manchester 1958.
Fowler, W. H., *Stationary Steam Engines*, Scientific Pub. Co., Manchester 1908.
Galloway, E., *History and Progress of the Steam Engine*, T. Kelly, London 1830.
Goodeve, T. M. *Text-Book on the Steam Engine*, Crosby Lockwood, London, 6th edn 1884.
Gurr, D. & Hunt, J., *The Cotton Mills of Oldham*, Oldham Leisure Services, Oldham 1985.
Haeder, H., *A Handbook for the Steam Engine with Especial Reference to Small and Medium-Sized Engines*, trans. H. H. P. Powles, Crosby Lockwood, London 1893.
Hair, T. H., *Sketches of the Coal Mines in Northumberland and Durham*, London 1844.
Harris, T. R., *Arthur Woolf, The Cornish Engineer, 1766–1837*, D. Bradford Barton, Truro 1966.
Hayes, G., *A Guide to Stationary Steam Engines*, Moorland, Ashbourne 1981.
Stationary Steam Engines, Shire Pub., Aylesbury 1979.
Henthorn, J. T., *The Corliss Engine*, Spon & Chamberlain, New York 1897.
Hiller, E. G., *Steam Boiler Construction, Rules of the National Boiler and General Insurance Co. Ltd*, Chorlton & Knowles, Manchester, 3rd edn 1931.
Working of Steam Boilers, Bethell, Manchester, 6th edn 1923.
Hills, R. L., *Machines, Mills and Uncountable Costly Necessities. A Short History of the Drainage of the Fens*, Goose, Norwich, 1967.
Power in the Industrial Revolution, Manchester University Press, Manchester 1970.
Hills, R. L. & Barlow, K. A., *The Changing Face of Engineering*, Open University Technological Essay, Milton Keynes 1984.
Hills, R. L. & Patrick, D., *Beyer, Peacock, Locomotive Builders to the World*, Transport Pub., Glossop 1982.
H.M.S.O., Ministry of Fuel and Power, *The Stoker's Manual*, H.M.S.O., London 1945.
Hoblyn, R. D., *A Manual of the Steam Engine*, Scott, Webster & Geary, London 1842.
Hodge, P. R., *Analytical Principles and Practical Application of the Expansive Steam Engine*, J. Williams, London 1849.
Hollowood, B., *Cornish Engineers, Holman Bros. Ltd, Camborne*, Holman Bros., Camborne 1951.
Holmes, G. C. V., *The Steam Engine*, Longmans, Green, London, 2nd edn 1888.
Interborough Rapid Transit, The New York Subway, Its Construction and Equipment, Interborough Rapid Transit Co., New York 1909.
Jamieson, A., *Elementary Manual on Steam and the Steam Engine*, Griffin, London, 6th edn 1898.
Jenkins, R., *The Collected Papers of . . .*, Cambridge University Press, Cambridge 1936.
Jeremy, D. J., *Transatlantic Industrial Revolution. The Diffusion of Textile Technologies between Britain and America, 1790–1830s*, Blackwell, London 1981.
Jevons, W. S., *The Coal Question: an Inquiry concerning the Progress of the Nation, and the Probable Exhaustion of our Coal-Mines*, Macmillan, London 1865.
Johnson, W., *The Imperial Cyclopaedia of Machinery, being a series of plans, sections and elevations of stationary, marine and locomotive engines, spinning machinery, grinding mills, tools. etc. . . .* Mackenzie, Glasgow 1854.
Kolin, I., *The Evolution of the Heat Engine*, Longmans, London 1972.
Lardner, D., *Lectures on the Steam Engine*, J. Taylor, London, 4th edn 1832.
A Rudimentary Treatise on the Steam Engine for the Use of Beginners, J. Weale, London 1848.
The Steam Engine, Taylor & Walton, London, 6th edn 1836.

Law, R. J., *James Watt and the Separate Condenser*, H.M.S.O., London 1969.
The Steam Engine, H.M.S.O., London 1965.
Lean, T., *On the Steam Engines in Cornwall*, Simpkin, Marshall, London 1839, reprint Barton, Truro 1969.
Leigh, E., *The Science of Modern Cotton Spinning*, Palmer & Howe, Manchester, 3rd edn 1875.
Lindqvist, S., *Technology on Trial. The Introduction of Steam Power Technology into Sweden, 1715–1736*, Almqvist, Sweden 1984.
McGarvie, M., *Bowlingreen Mill. A Centenary History, 1878–1978*, Avalon Leather-board Co. Ltd, Street 1979.
von Mackensen, L., *The Introduction of English Steam Engine and Metallurgical Technology into Germany during the Industrial Revolution prior to 1850*, Colloques Internationaux C.N.R.S. No. 538, L'acquisition des techniques par les pays non-initiateurs, 1973.
McKenzie, P., *W. G. Armstrong. The Life and Times of Sir William George Armstrong, Baron Armstrong of Cragside*, Longhirst Press, Newcastle-upon-Tyne 1983.
McLean, M., *Modern Electric Practice*, Gresham Pub., London, no date.
Marsden, R., *Cotton Spinning, Its Development, Principles and Practice, with an Appendix on Steam Engines and Boilers*, G. Bell, London, 4th edn 1891.
Mayr, O., *The Origins of Feedback Control*, M.I.T. Press, Cambridge, Mass. 1970.
Feedback Mechanisms in the Historical Collections of the National Museum of History and Technology, Smithsonian Institution Press, Washington 1971.
Meyer, H. C., *Steam Power Plants, their Design and Construction*, McGraw Pub., New York, 2nd edn 1905.
Montgomery, J., *The Theory and Practice of Cotton Spinning, or the Carding and Spinning Master's Assistant*, Glasgow, 3rd edn 1836.
Mott, R. A., ed. by P. Singer, *Henry Cort, The Great Finer*, The Metals Society, London 1983.
Muirhead, J. P., *The Origins and Progress of the Mechanical Inventions of James Watt*, J. Murray, London 1854.
Musson, A. E. & Robinson, E., *Science and Technology in the Industrial Revolution*, Manchester University Press, Manchester, 1969.
Nasmith, J., *Recent Cotton Mill Construction and Engineering*, J. Heywood, Manchester, 2nd edn 1900.
The Students Cotton Spinning, J. Nasmith, Manchester 1892.
Neilson, R. M., *The Steam Turbine*, Longmans, Green, London 1902.
Nicholson, J., *The Operative Mechanic and British Machinist*, Knight & Lacey, London 1825.
Pacey, A. J., *The Maze of Ingenuity, Ideas and Idealism in the Development of Technology*, Allen Lane, London 1974.
de Pambour, G., *A New Theory of the Steam Engine*, London 1839.
Parsons, Sir C. A., *The Steam Turbine, as a Study in Applied Physics*, An address on the occasion of the centenary celebration of the founding of the Franklin Institute, Philadelphia, 17–19 September 1924.
Parsons, R. H., *The Development of the Parsons Steam Turbine*, Constable, London 1936.
The Early Days of the Power Station Industry, Cambridge University Press, Cambridge 1939.
Partington, C. F., *An Historical and Descriptive Account of the Steam Engine*, J. Taylor, London 1822.
The Century of Inventions of the Marquis of Worcester from the Original Manuscripts with Historical and Explanatory Notes and a Biographical Memoir, J. Murray, London 1824.
Peabody, C. H., *Thermodynamics of the Steam-Engine and Other Heat-Engines*, Macmillan, London 1889.
Pickworth, C. N., *The Indicator Handbook*, Emmott, Manchester, 4th edn 1910.

Pole, W., ed., *The Life of Sir William Fairbairn, Bart*, Longmans, Green, London 1877, David & Charles, reprint 1970.
Porter, C. T., *Description of Richard's Improved Steam Engine Indicator, with Directions for Its Use*, Longmans, Green, Reader & Dyor, London 1868.
Engineering Reminiscences, J. Wiley, New York 1908, Lindsay Pub., Bradley reprint 1985.
A Treatise on the Richards Steam-Engine Indicator and the Development and Application of Force in the Steam-Engine, E. & F. N. Spon, London, 5th edn 1894.
Powles, H. H. P., *Steam Boilers, Their History and Development*, Constable, London 1905.
Rankine, W. J. M., *A Manual of the Steam Engine and Other Prime Movers*, Griffin, Bohn, London 1861.
Rees, A., ed., *The Cyclopaedia; or Universal Dictionary of Arts, Sciences and Literature*, London 1819, David & Charles reprint 1972.
Rennie, G., *Illustrations of Mill Work and Other Machinery together with Tools of Modern Invention; Atlas to the New Edition of Buchanan's Work*, J. Weale, London 1841.
Reynolds, M., *Stationary Engine Driving, a Practical Manual for Engineers in Charge of Stationary Engines*, Crosby Lockwood, London 1885.
Reynolds, O., 'The Transmission of Energy', Society for the Encouragement of Arts, Manufctures and Commerce, Cantor Lecture 1883.
Richardson, A., *The Evolution of the Parsons Steam Turbine*, Engineering, London 1911.
Ridding, A., *S. Z. de Ferranti, Pioneer of Electric Power*, H.M.S.O., London 1964, 2nd edn 1975.
Rigg, A., *A Practical Treatise on the Steam Engine*, E. & F. N. Spon, London 1888.
Ripper, W., *Heat Engines*, Longmans Green, London, new edn 1909.
Robinson, E. H. & Musson, A. E., *James Watt and the Steam Revolution (documents)*, A. M. Kelly, New York 1969.
Rolt, L. T. C., *Isambard Kingdom Brunel. A Biography*, Longmans, Green, London 1957.
James Watt, Batsford, London 1962.
Thomas Newcomen. The Prehistory of the Steam Engine, David & Charles, Dawlish 1963.
Rolt, L. T. C. & Allen, J. S. A., *The Steam Engine of Thomas Newcomen*, Moorland, Hartington 1977.
Rose, J., *Modern Steam Engines*, H. C. Baird, Philadelphia, 3rd edn 1893.
Savery, T., *The Miners Friend, or An Engine to Raise Water by Fire*, S. Crouch, London 1702.
Scott, D., *The Engineer and Machinist's Assistant, being a series of plans . . . of steam engines, spinning machines . . .* , Blackie, Glasgow 1849.
Scott, E. K., *Matthew Murray, Pioneer Engineer, Records from 1765 to 1826*, E. Jowett, Leeds 1928.
Seaton, A. E., *A Manual of Marine Engineering*, C. Griffin, London 1883, 7th edn 1888.
Skempton, A. W., ed., *John Smeaton, F.R.S.*, T. Telford, London 1981.
Smart, J. E., *The Deptford Letter-Books, an insight on S. Z. de Ferranti's Deptford Power Station*, Science Museum, London 1976.
Smiles, S., *Lives of the Engineers: Harbours, Lighthouses, Bridges: Smeaton and Rennie*, J. Murray, London 1874.
Lives of the Engineers: The Steam Engine: Boulton and Watt, J. Murray, London 1878.
Men of Invention and Industry . . . William Murdock, J. Murray, London, new edn 1890.
Smith, C. C., *A Short History of Naval and Marine Engineering*, Cambridge University Press, Cambridge 1937.

Spratt, H. P., *The Birth of the Steamboat*, London 1958.
Stuart R., *A Descriptive History of the Steam Engine*, Knight & Lacey, London, 2nd edn 1824.
Historical and Descriptive Anecdotes of Steam-Engines and of their Inventors and Improvers, Wightman & Cramp, London 1829.
Stumpf, J., *The Una-Flow Steam Engine*, R. Oldenbourg, Munich 1912.
Sutcliffe, G. W., *Steam Power and Mill Work, Principles and Modern Practice*, Whittacker, London 1895.
Tann, J., *The Development of the Factory*, Cornmarket, London 1970.
Templeton, W., *The Millwright and Engineer's Pocket Companion*, Simpkin Marshall, London, 9th edn 1856.
Thurston, R. H., *A History of the Growth of the Steam-Engine*, D. Appleton, New York 1878.
A Treatise on Friction and Lost Work in Machinery and Mill Work, J. Wiley, New York 1885, 3rd edn 1889.
A Manual of Steam Boilers: Their Design, Construction and Operation, J. Wiley, New York 1888.
Tredgold, T., *The Steam Engine, Its Invention, and an Investigation of Its Principles for Navigation, Manufactures and Railways*, J. Weale, London 1838.
Trevithick, F., *Life of Richard Trevithick, with an Account of his Inventions*, E. & F. N. Spon, London 1872.
Triewald, M., *Short Description of the Atmospheric Engine*, Stockholm 1734, English Translation by the Newcomen Society, Cambridge 1928.
von Tunzelmann, G. N., *Steam Power and British Industrialization to 1860*, Clarendon Press, Oxford 1978.
Uhland, W. H., *Corliss Engines and Allied Steam-motors Working With and With-out Automatic Variable Expansion-gear Including the Most Approved Designs of All Countries with Special Reference to the Steam-Engines of the Paris International Exhibition of 1878*, trans. A. Tolhausen, E. & F. N. Spon, London 1879.
Vale, E., *The Harveys of Hale, Engine-builders, Shipwrights and Merchants of Cornwall*, D. Bradford Barton, Truro, 1966.
Walker, J., *Report to the Directors on the Comparative Merits of Loco-motive and Fixed Engines, as a Moving Power*, J. & A. Arch, London 1829.
Warren, J. G. H., *A Century of Locomotive Building by Robert Stephenson & Co., 1823–1923*, A. Reid 1923, David & Charles, Newton Abbot reprint 1970.
Watkins, G., *The Stationary Steam Engine*, David & Charles, Newton Abbot 1968.
The Steam Engine in Industry, Moorland Pub., Ashbourne 1978.
The Textile Mill Engine, David & Charles, Newton Abbot, 1970.
White, J. H., *A History of the American Locomotive*, Dover, New York 1968, reprint 1979.
Wicksteed, T., *Cornish and Boulton & Watt Engines erected at the East London Water-works*, J. Weale, London 1842.
Williams, C. W., *On the Combustion of Coal and Generation of Smoke, being an abstract from the lectures, delivered at the Victoria Gallery, Conversazione, Manchester 1841*.
The Combustion of Coal and the Prevention of Smoke Chemically and Practically Considered, Simpkin Marshall, London, 3rd edn 1841.
Winton, J. G., *Modern Steam Practice and Engineering*, Blackie, London 1885.
Zeuner, G., *Treatise on Valve-Gears, with Special Consideration of the Link-Motions of Locomotive Engines*, trans. J. F. Klein, E. & F. N. Spon, London 1884.

Patents

Appropriate American and English patents have been consulted but are too many to list individually.

Archives, Catalogues, etc.

Allis-Chalmers Manufacturing Co., Milwaukee, Wisconsin,
Allis-Chalmers Reliance-Corliss Engines, Bulletin 1503, 1906.
Heavy Duty Corliss Engines, Bulletin 1529-B, reprint April 1925.
Ames Iron Works, Oswego, New York,
The Ames Center Crank Engine, Bulletin 73-A.
The Ames Four-cylinder Vertical Unaflow Engine. Bulletin 1925-V.
Ashworth & Parker Ltd., Bury,
The Parker Engine.
Babcock & Wilcox Co., New York,
Steam, Its Generation and Use, 1899.
Beyer, C. F.,
Notebooks Nos. 1 & 2, 1850–70 (G.M.M.S.I.).
Notebook, 1841, lent by L. T. C. Davenport.
Boulton & Watt,
Archives etc. (Birmingham Library).
Patents.
Buckley & Taylor, Oldham,
Archives (G.M.M.S.I.).
Patents.
Ferranti Ltd., Hollinwood,
Archives.
Patents.
Galloway Ltd., Manchester,
Archives (G.M.M.S.I.).
Boiler Catalogue *c*. 1900.
Patents.
Uniflow Steam Engines, catalogue *c*. 1920.
Goodfellow, B., Hyde,
Archives (G.M.M.S.I.).
Patents.
Hardie-Tynes Manufacturing Co., Birmingham, Alabama,
Balanced Valve Engines, Bulletin 103, 1921.
Harrisburg Foundry & Machine Works, Harrisburg. Pennsylvania,
A System of Engines and Products of the Harrisburg Foundry and Machine Works, 1900.
Hewes & Phillips Iron Works, Newark, New Jersey,
Twentieth Century Improved Double Port Corliss Steam Engines.
Hick Hargreaves & Co., Bolton,
Archives (Bolton Library).
First Order Book, 1833.
Patents.
Kenyon, W. & Sons Ltd, Duckinfield,
Ropes and Rope Driving, 1924 edn.
Kingsford Foundry & Machine Works, Oswego, New York,
Unaflow Engines, Bulletin 22.
Mattoon Engine Works, Mattoon, Illinois
Chuse Six Cylinder Vertical Uniflow Engines, Bulletin 109.
Musgrave, J. & Sons Ltd, Bolton,
Archives (Bolton Library).
Illustrated Catalogue of Vertical Quadruple Expansion Engines, Horizontal & Vertical Triple Expansion Engines, Compound Engines, Simple Engines, Barring Engines, Lancashire Boilers, Globe Safety Boilers, Sectional Boilers, Mill Gearing, Indicators etc., no date.
Patents
The "Unaflow" Patent Steam Engine, 1910.

McNaught, J. & W., Rochdale,
Archives (G.M.M.S.I.).
Patents.
Nagle Engine & Boiler Works, Erie, Pennsylvania,
Catalogue, 1915.
Petrie, J. & Co., Rochdale,
Archives (G.M.M.S.I.).
Patents.
Porter, C. T.,
Patents.
Porter's Improved Centrifugal Governor, Leaflet *c.* 1859.
Providence Engineering Works,
Rice & Sargent Corliss Type Engine.
Robey & Co. Ltd, Lincoln,
Instructions for Erecting and Working the "Robey" Patent Uniflow Engine.
Scott & Hodgson, Guide Bridge,
Archives (G.M.M.S.I.).
Skinner Engine Co., Erie, Pennsylvania,
Exhaust Valves on Unaflow Engines, Their Proper Location, Bulletin 109, 1923.
Lubrication of the "Universal Unaflow" Steam Engine.
The Skinner Compound Unaflow Marine Steam Engine.
Skinner "Universal Unaflow" Engines, Erecting and Maintenance Instructions.
The Universal Unaflow Steam Engine, 1927.
Universal Unaflow vs. High-speed Four-valve Economy, Bulletin 102.
Southwark Foundry & Machine Co.,
Catalogue, *c.* 1900.
Porter-Allen Automatic Engine.
Troy Engine & Machine Co., Troy, Pennsylvania,
Troy Vertical Engine, Type SH, Bulletin H, 1923.
Willans & Robinson Ltd, Rugby,
Description of the Willans Compound Engine.
The Economy of the Willans Engine, *c.* 1904.
Instructions for Working the Willans Compound Launch Engine, 1886.
Patents.
The Willans Engine for Rope and Belt Driving Purposes, *c.* 1904.
The Willans & Robinson (Willans-Parsons) Steam Turbine.
Wood, J. & E. Ltd, Bolton,
Archives (Bolton Library).
Patents.
Types of Horizontal Condensing Corliss Engines made by J. & E. Wood, Victoria Foundry, Bolton.

Journals, Periodicals, etc.

American Electrician
The American Historical Review
Bolton Managers' and Overlookers' Association
British Journal for the History of Science
Cassier's Magazine
The Electrical Engineer
The Electrician
Emporium of the Arts and Sciences
Engine Builders Association of the United States
The Engineer
Engineering
The Engineering Review
The Flywheel, Journal of the Northern Mill Engine Society

Franklin Institution Journal
History of Technology
Industrial Archaeology
Industrial Archaeology Review
Institution of Mechanical Engineers, Proceedings
International Congress on the Conservation of Industrial Monuments, Transactions
Manchester Association of Engineers, Proceedings and Transactions
Manufacturer and Builder
Newcomen Society, Transactions
Philosophical Transactions
Power
Quarterly Journal of Science
The Railroad Advocate
Relf's Philadelphia Gazette and Daily Advertiser
Rochdale Literary and Scientific Society, Transactions
Royal Society of London, Notes and Records
Scientific American
Street Railway Journal
Technology and Culture

Unpublished articles, theses, etc.

Bailey, M. R., 'Robert Stephenson & Company, 1823–36', M.Sc. Thesis, Newcastle-upon-Tyne 1984.
Bathe, G., 'An Engineering Miscellany', typescript at Smithsonian Institution.
Boucher, C. T. G., 'Barney's References Examines', typescript.
Broadhurst, W. N., 'A Consideration of Steam Engine Thermodynamics and Valve-Gears, as applied to Petrie-McNaught Engines between the Years 1892 and 1910', 3rd Year Mechanical Engineering Project, U.M.I.S.T., 1967.
Chapman, S. D., 'The Midlands Cotton and Worsted Industry, 1769–1800', Ph.D. Thesis, London 1966.
Cookson, K., 'An Experimental Determination of the Power and Efficiency of a Model of an Early Newcomen Engine', 3rd Year Mechanical Engineering Project, U.M.I.S.T., 1971.
Eddison, C., 'The Uniflow Engine', 3rd Year Mechanical Engineering Project, U.M.I.S.T., 1983.
Elton, A. & Drummond, B., 'The Life and Times of William Murdock, Frederic Winsor and Philippe Lebon', typescript.
Fehring, T. H., 'Technological Contributions of Milwaukee's Menomonee Valley Industries', typescript in Smithsonian Institution.
Hearn, D. A., 'Steam Power and Cotton', monograph sponsored by Courtaulds Ltd, 1968.
Kanefsky, J., 'The Diffusion of Power Technology in British Industry, 1760–1870', Ph.D. Thesis, Exeter 1979.
Minchinton, W., 'The Energy Basis of the British Industrial Revolution', typescript of lecture, November 1985.
Pacey, A. J., 'Some Early Heat Engine Concepts and the Conservation of Heat', typescript.
'Vis Viva and Power. A History of the Concept of Mechanical Energy, 1600–1800', typescript.
Storr, F., 'The Development of the Marine Compound Steam Engine, Ph.D. Thesis, Newcastle-upon-Tyne 1982.
von Tunzelmann, G. N., 'Steam Power and Textiles in Britain to 1856', typescript of lecture presented to the VI Economic History Congress, Copenhagen, 1974.
Turner, T., 'History of Fenton, Murray & Wood', M.Sc. Thesis, U.M.I.S.T., 1966.

Index

Numbers in square brackets indicate an illustration.

Printed in the United States
By Bookmasters